Praise j

The C.R. Pa

(Second edition)

"A masterful and compelling deep dive into indelible chapters of untouched and never explored American history, stunning your sensibilities of what you thought you knew.

In over a decade of filmmaking, **I've never seen more replete research for a Hollywood project!**

Christopher Nelson takes you on a winding journey through the past with elements still entirely relatable to the present. A must-read for generations to come as these pioneers are finally given a well-deserved place in the annals of history."

~ Madisun Leigh - Scenario Entertainment & Pergola Pictures

Support from first-line family descendants of the Patterson Company

"I am Richard P. Patterson, the Great Grandson of C.R. Patterson and the Grandson of Frederick Douglass Patterson. Christopher Nelson **captured the essence of "the Patterson and Sons Company"** in his first book and expands it in the second edition. The book is historically correct as **Christopher has painstakingly verified the facts about the family and the business**. In the second edition, Christopher includes personal family stories shared by my sister Betty P Gardin and myself. This book would be a great addition to anyone's library and will enhance our understanding of American history."

~ Richard Postell Patterson – Grandson of Frederick D. Patterson

---

9.4/10 Review Snippet

Something we really respect and admire about this book is that it is incredibly objective. We know that might sound strange, but it's true! This book doesn't attempt to make some big, huge, profound socio-political point. Almost every book is trying to do that these days. No, instead **this book is written in the way that history SHOULD be written: with an objective lens—detailing people and events just the way they are and presenting whatever relevant information is available.**

If it wasn't for authors/historians like Christopher Nelson and books like this, we might not have a history.

~ Outstanding Creator Awards 2/28/2023

"Author Christopher Nelson offers an exciting narrative about America's first Black Auto Manufacturers in his book 'The C. R. Patterson and Sons Company'"

"This is a masterful book that delves deep into the history of a specific family, into pioneers who have not been given credit so far. **The story has been researched in-depth and updated, featuring a great deal of top-quality work from author Christopher Nelson.** It is easy to see the parallels and learn from the past for the present."

"Nelson offers **a sound, exciting narrative about the Patterson family and makes a great case as to why they need more recognition**, contrasting their business with a white-owned one and exploring the power behind their work and contributions."

"The company was a pioneer in the business with Frederick Patterson, the son of founder C.R., who became the first and only Black manufacturer to have built an automobile. This success was linked to a strong development in other areas, like politics and promoting business for Black people during a time when it was very difficult."

"Christopher Nelson is an award-winning author whose book (originated) as a master's thesis. Since then, **it has been revised and reworked to create a meticulously crafted and researched book that comes with plans for the big screen**."

~ paxjones.com (Inspiring Stories from Around the World) blog

---

5-Stars "An exhilarating survey for an auto enthusiast"

Award-winning author, Christopher Nelson, writes with precision and candor, as he offers a clear-eyed chronicle of this company's

transition from making carriages, automobiles, buses, and trucks in a bid to keep pace with the whirlwind of change that was happening in the motoring field.

The book does not tally in its progress but progresses with firm details and information on the geographical and family background of the company owners. **With a stirring preface and a spirited introduction, Nelson engages readers from the first sentence with his laity-friendly language to the last sentence capturing all the awe and fascination of a truly lionized period in automotive history.**

The author further provides visual history through numerous vintage photographs that shine authenticity as they accentuate the topic herein. Throughout, he adds essential details not found anywhere else, including pointing out some of the challenges that the Patterson Company struggled with, including ostracism, fierce competition, the changing demands of the economy, and lack of capital.

"The C.R. Patterson and Sons Company: Black Pioneers in the Vehicle Building Industry, 1865- 1939" is a comprehensive and magisterial study that makes an irrefutable addition to Black History. **Nelson's thorough research is evident during the reading and manages to give this five-star work a towering grace.** Long-term car enthusiasts and new car owners alike will find an enlightening enchiridion in this award-winning feat.

*~ Lily Andrews for **Reader Views** (2022)*

The C.R. Patterson and Sons Company: Black Pioneers in the Vehicle Building Industry, 1865 - 1939 is an award-winning biographical work by author Christopher Nelson unearthing the impact the Patterson family had on America's automobile industry. After collecting reams of historical data, Nelson shows how the Pattersons–across three generations–were instrumental in automotive manufacturing–staying productive, competitive, and innovative as the industry transitioned from horse carriages to horseless carriages down to modern cars and buses. **This book is not only a biographical work about the contribution of the Patterson family to the automotive industry, but also an important recovery of a certain piece of Black history that otherwise was in danger of being forever forgotten.**

*~ **Indie Reader** 4/7/2023*

Selected review snippets from the 2010 First edition:

"The Story Black History Missed! This definitive work on Patterson's company and the Patterson family deserves far more recognition by Black historians than what has been given."

~ Larry Chapman (Retired History teacher from Greenfield)

"Black Enterprise" The C. R. Patterson and Sons Company book was a good period-of-time story spanning the end of the horse drawn carriage into the horseless carriage era. Christopher Nelson did a particularly good comparison and contrast of a Black owned business and a White owned business during this era. The force and contributions of Black owned businesses have too often been overlooked in American history. This book should be recommended reading for the youth of today."

~ RZ Texas

AWARDS AND RECOGNITION

2022 American Writing Awards

WINNER

- History – General

FINALIST

- Best New Nonfiction
- History – United States
- Biography

2023 Indie Reader Discovery Awards

WINNER

1st Place Overall Non-Fiction 2023

Book Excellence Awards

WINNER

- Biography

Winter 2023 Outstanding Creator Awards

WINNER

- History
- Educational & Reference
- Best Research - Special Award
- 2nd Place Best Non-Fiction Book of Winter 2023

Reader Views 2022

WINNER

History – Classics (Gold Medal Winner)

The Bookfest 2023

WINNER

- History (Silver Medal)
- Biography (Silver Medal)

(Highest score in both categories – no gold medal winners)

2022 Best Book Awards
Sponsored by American Book Fest

FINALIST

History – General
History – United States

The C.R. Patterson and Sons Company

Black Pioneers in the Vehicle Building Industry, 1865-1939

(Second Edition)

Christopher Nelson

Lebanon, Tennessee, USA

Hurricane Creek Publishing
Lebanon, Tennessee, United States of America

LIBRARY OF CONGRESS CONTROL NUMBER:
2022916261

ISBN: 9798848332070

First Printed 2010, Second Edition Published 2022

Printed in the United States of America
Bulk purchases, book signing events, or speaking engagements, please contact: hurricanecreekpublishing@yahoo.com

Cover design: www.BookCoverDesign.us
Author photo: Vance Imaging, Lebanon, TN

I dedicate this work to the three most important men in my life.

First of all, I dedicate this to Harold F. Nelson, my father: the man who taught me to never give up.

Secondly, I dedicate this to Noah Eli Thomas Nelson and Benjamin Patrick Nelson, my sons: the "little men" who make me never want to give up.

Preface

Living in Clinton County most of my life, which neighbors the Highland County setting of this history, I occasionally heard mentions of the story of the Patterson family. For a few years during the early parts of this century, I lived in Greenfield, Ohio where this family's story took place. The story became a popular topic of discussion once again in the community in 2003 while the old Patterson factory was removed to make way for a modern structure. It was during this time that I became much more familiar with the story of the family and company and found it quite intriguing.

As I continued my education and was in the process of acquiring a Master of Science degree in the Industrial Archaeology program at Michigan Technological University in Houghton, Michigan, I began to look at the Patterson story once again. As I began to research the topic, I found that the Patterson story dropped from the public eye for around 40 years from the time they closed the factory in 1939. Except for a brief 1965 article and another in 1976, the story began circulating once again more widely with a series of brief articles during the early 1980s. These articles became the basis for the Patterson story that has been told for the last few decades in articles and one small book written in 2006 all telling this now generally accepted story about the Patterson family.

Being an Industrial Archaeologist, I decided to build upon the traditional story and include as many details as possible about the technical specifications of the Patterson carriages, automobiles, buses, and trucks as well as the factory and its setup. As I delved into the research looking for any new details on these topics that had not already been told in the previous works, I began to discover an entirely different story than the traditional version that has circulated the last few decades. Although many portions of the traditional story and the current work mirror each other, there are many details that have been missed or were erroneous in previous versions. Although some of the details are minor, enough of them together made several significant changes to the story. These versions of the story have also been spread widely across social media platforms in recent years. As some say, "the devil is in the details." This prompted a change in my research as I then not only focused on the technical details, but I attempted to collect enough information to tell the most accurate history of this family and company as possible.

My research strategy focused on gathering as many Patterson specific sources from 1939 or before as possible. This method allowed me to utilize sources from the time period when the Patterson family and company were actively in business and aided in providing information that was quite different than that provided in articles from later dates. Some of these changes in the story have been pointed out in the notes for each chapter at the end of each chapter and the reader is encouraged to browse the notes occasionally for additional information that was removed from the main text. The Pattersons received good coverage in the local newspapers and occasionally some widespread publications. This helped to piece together a better idea of what was happening with the family and business during this time. The Patterson Company frequently used advertising for their products. Over 1,200 advertisements from the company were collected during this research and are now housed at the Greenfield Historical Society in Greenfield, Ohio. Much information was gleaned from these advertisements and the many early newspaper and journal articles about the company. Other information came from Records of the National Negro Business League where correspondence was located between Frederick and NBL officials as well as transcripts of several speeches that Frederick made at the annual meetings. In these speeches, Frederick discussed financial details and processes used at the factory. Several sources such as these were located, sometimes in obscure places that most would not think to look, such as medical journals and farm magazines.

This research took me to several archives, libraries, and associations across several states. Endless hours of scrolling through microfilm reels of newspapers and other records led to the story that is presented here. Although it is recognized that there may be some occasional holes in the story, this has occurred due to a lack of sources on those details and the intentional refraining from presenting speculation as fact in this history. The attempt has been made to provide the most accurate history possible about this family who managed to survive for 74 years through making transitions to the current industry trends and their willingness to adapt and try to seek their own niche within the business world.

In 2021, resulting from relationships developed as part of the induction of two members of the Patterson family into the Automotive Hall of Fame, the idea of a possible documentary and feature film was presented to me. A screenwriter asked me about characteristics and personality traits of the people involved in the Patterson story. I really had to think about it and realized that I didn't know as much about those aspects of the people involved in the story as previously thought. The original research for this story was for a master's thesis in Industrial Archaeology and the focus of the research was more on technical details of the factory and the different types of vehicles that the Patterson Company built. Some personal details of the people were thrown in as they were discovered, but there was not enough for having a full understanding of the family personalities and involvement outside of the company.

This led to a deep dive into additional research over the past year, which uncovered many new details. I must admit that it was a lot of fun to go back through the original research while looking at it with a new focus. Many details were missed during the original research because they weren't part of the focus at that time. Additionally, many new sources were discovered during the current research as the internet has increased the amount of readily available materials over the past 13-14 years since the original research was conducted. Additionally, members of the Patterson family were able to provide several details and documents that helped to piece more of the story together. The new details discovered during the current research, as well as those provided by Patterson descendants, were numerous. The collection of this much additional information resulted in the revision and addition of many of these details into this second edition of the original book from 2010.

Note to the reader on terminology: The term "Black" is used throughout this book rather than African American. African American was not a term used during the period that this history takes place and was therefore deemed not an appropriate use in this document. The terms "colored" or "Negro" are only used in this book when within a quote or as part of a proper title, such as the National Negro Business League.

Legal note: Images from the Patterson Company catalogs, as well as various family photographs were used with express written permission from Richard Patterson, grandson of Frederick Patterson. Richard graciously provided the author with the documents and gave permission to use the images for the purpose of telling and representing the history of the C.R. Patterson and Sons Company, as appropriate, to illustrate the most accurate history possible about the company and family from approximately 1833 to 1939. As privately owned documents not readily available in public repositories, the privately owned images in this book may not be reproduced without the written consent of Richard Patterson, Betty Patterson Gardin, or Eric Patterson. These images are all annotated within this book as having a source of R.P.

Acknowledgements

The basic story and history of this company has been told in several articles that have appeared over the last few decades. These have been very brief and preliminary, at best, and have lacked the factual details that are necessary to illustrate the rise of this Black owned company and how they overcame adversity and entered the business world just after the Civil War. Scholarly research to accurately detail the 74-year history of this company is almost non-existent, so this project has been an enormous task to say the least. By starting anew at the beginning and working through the entire history, the accomplishment of this research project would not have been possible without the aid and support of many people and institutions.

Since the original version of this manuscript was created as my master's thesis, I only feel that it is proper to recognize those members of my committee that aided me through this research. First, I would like to express my appreciation and gratitude to my advisor, Dr. Larry Lankton, who provided expert guidance throughout this project, answered tons of questions, and read through multiple drafts. I would also like to thank the other members of my thesis committee, Dr. Susan Martin, Dr. Louise Dyble, and Erik Nordberg. I appreciate the work and effort that each member has made as well as the encouragement, criticism, and support that they have given during this project. Other individuals I would like to recognize include Dr. Thomas Kinney of Bluefield College in Virginia, for his advice and information regarding the carriage trade; Katherine Magruder from the Carriage Museum of America in Lexington, Kentucky, also for information on period carriages and their builders. I'd like to express my gratitude to Dana Chandler of the Tuskegee University Archives for his assistance in locating related Negro Business

League documents while I was conducting research at that institution. Harold Schmidt of the Greenfield Historical Society assisted on several occasions through telephone and email conversations as well as during my several visits to the Society.

There are several institutions that provided information and support during this project. There are too many to list, but some of the major contributors include the Ohio Historical Society, which houses the State Archives of Ohio; the State Library of Ohio; the archives of the Antique Automobile Club of America in Hershey, Pennsylvania; the National Automobile History Collection in Detroit, Michigan; the Benson Ford Research Center in Dearborn, Michigan; the Icabod Flewellen Collection at the East Cleveland Public Library in Cleveland, Ohio; the archives at the Ohio State University and at Central Michigan University; and the offices of the Highland County Courthouse in Hillsboro, Ohio.

I'd like to express a special thanks to Ed Welburn, Madisun Leigh, and Nadia Fugazza for their encouragement to dig deeper into the Patterson story. Their discussions and resulting questions have caused me to follow leads about specific topics of the story, which has helped me to discover many new details. I'd also like to thank automotive historian extraordinaire Brian Baker of the National Corvette Museum in Bowling Green, Kentucky. His encouragement and knowledge have helped keep me moving forward on this quest to discover new aspects to the Patterson story. I also wish to thank the Automotive Hall of Fame in Detroit for acknowledging the Patterson Company's special place in transportation history. The induction of C.R. and Frederick Patterson into the Automotive Hall of Fame was a highlight of my career, and I am so glad that my master's research and first edition of this book was able to play a part in that moment.

I'd also like to express a very special thanks to my colleagues at Bright Morning Star Films in Nashville who have partnered with me to bring this amazing Patterson story to the big screen as a feature film and documentary, currently under the working title "Driven." Specifically, I'd like to thank my fellow producers Jack Hager, Michael Giancana, Joan Uselman, Weiss Night, and Johnny Reeves, but I'd be remiss to not extend my thanks to all of the crew at the studio that are doing their part to make this happen.

Many of the new details provided in this story would not have been possible without the assistance of several descendants of the Patterson family. My sincerest gratitude goes out to Richard Patterson, grandson of Frederick Patterson, who provided several documents to assist in piecing more of the Patterson story together. Thank you for the documents, photographs, as well as the many phone and email conversations. Additionally, I'd like to thank Richard's sister, Betty Patterson Gardin, granddaughter of Frederick Patterson for also providing information, especially some very revealing details that I would have never dreamed of and wouldn't be in available records. I'd also like to thank additional Patterson descendants, Margaret Patterson Carpenter and Lynda Patterson Baker, for their contributions and assistance with piecing together more of the story.

Last, but not least, of course, I wish to thank my entire family. They have provided support in many ways throughout the research project to include emotional support, editing, criticism of proposed hypotheses, topic ideas, and even financial support at times throughout the months of research in Ohio for the original research. A special thanks goes to my wife, Stephanie, for her patience with my endless hours of research and writing and talking about some family of strangers from Greenfield, Ohio. At least I kept my promise to her not to name our first son "Patterson" when he was born during the original research for the master's thesis. My family has been especially supportive over the past year during the additional research for the second edition of the book.

Contents

C. R.
PATTERSON
& SONS
GREENFIELD, OHIO.

INTRODUCTION

On a September day in 1915, the big wooden doors of the local Greenfield, Ohio, carriage shop opened wide. The C.R. Patterson and Sons Carriage Company had opened these doors many times before to let the thousands of finished carriages they had built over the years roll out into the street ready for sale. The vehicle that rolled out of the doors on this particular day was different; it was an automobile. Just as many carriage makers did during this time, they made the transition from building carriages to producing automobiles. It had become apparent that the auto age was here to stay, and they had to keep up with the times.

This automobile, dubbed the Patterson-Greenfield, was similar to most cars of the time as it had no significant difference in body style, components, or performance from most of the others, yet at the same time, it was entirely unique. Frederick D. Patterson did something that no other Black man had done; he became the first and only Black auto manufacturer on the continent. The C.R. Patterson and Sons Company had been in business for 50 years before this occurred and all the experiences in life and business for the Patterson family had led up to this day. Although car production didn't last – the company only built automobiles for about three years – it was an important step toward the company's later endeavors. But the story of the C.R. Patterson and Sons Company began a long time before the first cars ever rolled out of their shop. This company went through several transitions and manufactured many types of vehicles throughout its 74 years in business.

Charles Richard (C.R.) Patterson was born in 1833 in the State of Virginia. While traditional versions of this story have misled many to erroneously believe that he was born a slave and had escaped to freedom in the North, he was born a free person of color, just as his father and grandfather were born free. There is ample documentation that shows that C.R. and the two generations before him were free, but it is still uncertain whether earlier generations prior to that had always been free or if slavery was ever part of their lineage. As free Blacks living in the South, there were struggles for Charles and Nancy Patterson (C.R.'s parents). While not treated as inhumanely as their enslaved brethren, free Blacks (particularly in the South) were still discriminated against and in many cases treated poorly. Many obstacles stood in their way of success, such as limited, if any, access to money from banks and general discriminatory practices that held them back from finding success in life and business. According to noted historian Ira Berlin, he described free people of color living in the South as "slaves without masters." According to Berlin, free Blacks living in the South faced struggles "for community, liberty, economic independence, and education within an oppressive society."[1] According to author Sherri Burr, as time went on following the American Revolution and the number of free Blacks grew in the South, States such as Virginia passed laws that made it more difficult for them to prosper in the South. Once such example was the restriction of movement. In Virginia, for example, laws were passed that stated that if free Blacks were to leave the State for any reason, including to receive an education, they could never return to Virginia.[2]

There were several ways that Blacks could become free prior to the Civil War. Burr mentions five categories that led to freedom. The first included children that were born to free parents. Second was mulatto children born of free Black mothers. Third included mulatto children born of White servant or free women. Fourth were those children of free Black and Indian mixed parentage. The fifth and final way included manumitted slaves, defined as those that had been officially freed by their masters.[3]

One may question why free Blacks became such a concern in States like Virginia that their legislature troubled themselves with passing laws oppressive to that portion of the population. Following the American Revolution, the number of free Blacks grew exponentially. Originally, many free Blacks came to the United States on their own and worked as indentured servants until they paid off their debt and became truly free. In 1790 Virginia, for example, free Blacks made up made up only 4.2% of the Black population. However, by 1820, the population of free Blacks had nearly doubled to 7.98%. The free Black population continued to grow and in 1840 they made up 9.99% of the Black population. This was a significant number, meaning that one out of every 10 Blacks in 1840 Virginia was free. These numbers stayed steady in 1860 with 10.6% of the Black population being free. These numbers tell a very different story than most are aware of as the typical narrative is that all Blacks in the South were slaves, and now we know that this was not the case. The Patterson family was among the free Black population of Virginia until they made the decision to move to Ohio.

A decision was made in late 1841 or early 1842 to move north to Ohio and the so-called "free" land, likely for the promise of better opportunities, both personally and professionally. This timeframe for the move was narrowed down based on the birth of Charles and Nancy's son Henry in 1841 in Virginia, but they were in Ohio by the time that son John Edward was born in 1842. After reaching Ohio, they made their way to the small town of Greenfield in the southwest portion of the state, which luckily for them, had been a town long known for holding strong abolitionist views.[4] After reaching adulthood, C.R. eventually gained employment as a blacksmith at a local carriage maker. Within a few years he held the position of foreman with several Whites working under him.

C.R. became very well respected in the community and a prominent citizen of the town. In 1873, he entered a partnership with a White man named James P. Lowe and they opened their own carriage making business. In 1893, Lowe sold his share of the business to Patterson. His business became very successful, being known for its quality work and fine craftsmanship. He became one of the wealthiest men in the town and he raised his family there. His sons became partners in the business and the company passed through several generations of the Patterson family.

The Patterson carriages became quite popular in the local area and their business extended to reach customers across several states. This became possible by the company's heavy use of advertising in both local newspapers and publications that reached both statewide and nationwide audiences. Some of these advertisements targeted a specific audience, such as physicians or the Black community, while others were meant for a general audience. The family involvement with certain organizations also aided the increase in sales for the company. Other times, sales were based entirely on the merit of their products, such as a special winter buggy that the company developed and patented that was seen as being superior to many other styles available at the time. A horse-drawn school wagon became another popular vehicle that the company built. As school districts expanded through consolidation, proper transportation was needed for the students. The company recognized this need and provided a vehicle for transporting students.

The elder Patterson died in 1910 at the age of 77, and his son Frederick took over the company. At this time the carriage business was rapidly dwindling owing to the rise of the automobile. Frederick began a company transition into the automobile era by incorporating automobile repair into the services provided by the company, while at the same time, still building carriages. In 1915, he started building these "horseless carriages" which had become so popular across the United States and had practically replaced the horse and carriage by this time. The automobiles were high quality, just as his carriages had been, but when business had declined by 1918, he realized that he could not compete with the large automakers in Detroit and discontinued the Patterson-Greenfield automobile.

After ending automobile production, the Pattersons shifted their focus to building special purpose vehicles such as buses and trucks, which were in high demand. Learning from the short-lived attempt at manufacturing automobiles, they did not completely depend on this new focus of the company alone for income and still built carriages, as well as keeping and expanding the automobile repair business. They built carriages into the 1920s until the bus business finally took hold and became a success and, by the mid-1920s, they even discontinued automobile repair services to concentrate on the rapidly growing bus and truck business.

The Patterson Company, according to some, built the first buses used in some of the larger Ohio cities such as Cincinnati and Cleveland.[5] The Patterson buses were even sold internationally; some were built for the government of Haiti. The C.R. Patterson and Sons Company, also known in the 1920s and after as the Greenfield Bus Body Company, was successful well into the 1930s, when the Great Depression and other factors caused the company to fall on hard times. The Pattersons did what they could to keep ahead of the crumbling economy, including a return to automobile repair and they also incorporated body and fender work into their business. Finally, in 1939, the company closed its doors after being in the transportation business for over 70 years. The C.R. Patterson and Sons Company remained in the motorized vehicle field much longer than many of the other small pioneering auto builders by changing its focus toward producing specialty vehicles such as buses and trucks and, through it all, the Pattersons never lowered their standards or the quality of their vehicles.

All of this occurred at a time that it was difficult for Blacks to enter business, let alone become successful. Many obstacles stood in the way of Black business owners, whether they were in the north or the south. Many Blacks learned trades from working for White establishments and observing the White businessmen in how to direct the business. After several years of observation, many of these men felt that they were prepared enough to start their own establishment. The problem with this was usually a lack of capital to get started in the business. Limited credit and buying power hindered Black-owned businesses and they were unable to offer a competitive price for their products. The business location was an important factor to success as well. "The unprecedented industrial development of the North and West during this period (after 1865), moreover, tended to leave the small Negro businessman behind," resulting in the black business having less opportunity to compete in the business world in these areas.[6]

According to Harmon, another problem during this time was that Blacks had been programmed from the days of slavery to believe that the Whites were superior, and many felt that it would be difficult to compete against the White businessmen. Many had doubts like this before they even began their business, which many times doomed their venture from the beginning. They felt that Whites would not "abandon their former connections to patronize novices in businesses; and the freedmen themselves could not easily abandon the thought that the White businessman was more reliable and could give the most in return for one's money."[7] It was difficult for Blacks to dismiss the thoughts that they could not compete with the Whites and that they could succeed and compete in business.

According to a 1903 article by Carl Kelsey, one advantage that Blacks had was the fact that they had learned to work at many trades or tasks. Although this came about in a very negative way through the bonds of slavery, it could now be looked upon as a positive skill that the Black man held. Early African American leaders such as Booker T. Washington admitted that the institution of slavery had its benefits. He stated, "American slavery was a great curse to both races, and I would be the last to apologize for it; but, in the presence of God, I believe that slavery laid the foundation for the solution of the problem that is now before us...During slavery, the Negro was taught every trade, every industry, that constitutes the foundation for making a living." Although many of the tasks of the slave were menial and dull, many slaves had learned the artisans' trades such as blacksmithing, carpentry, and masonry. Kelsey stated that in many parts of the south, these slaves held a "practical monopoly of the trades. In technical knowledge; they, of course, soon outstripped their masters and became, as compared with other slaves, independent and self-reliant. The significance of this training appeared in the generation after freedom was declared." This gave many former slaves the capability to start a business and successfully compete with White businesses. As discussed in detail in Chapter 2, many artisan skills, such as blacksmithing, carpentry, and shoemaking, became concentrated almost exclusively in the hands of Blacks, whether slave or free. This appears to have been the case with the Patterson family, starting with Charles, C.R.'s father, who passed the blacksmith trade down to several of his sons. This tradition continued through four generations of this family in Greenfield. According to author Frank Mather in 1915, the C.R. Patterson and Sons Company became the "largest plant owned by colored people in the United States."[8] This all began with a Southern-born blacksmith who had a dream and pursued it.

The full story of this unique company has never been told. Several newspaper accounts and one small book written in 2006 tell a very general history of the company. Many interesting elements add to this story, such as Frederick's entrance into politics and his relationships with African American greats such as Booker T. Washington in helping to form the National Negro Business League. Frederick also broke color barriers in sports, as he became the first African American to play football for the Ohio State University. Many times, Frederick's public involvement had direct ties with the success of the company at one point or another.

A wealth of information can be learned from this company and its rise as a Black-owned and operated enterprise. The company was able to keep afloat during its several transitions by never completely abandoning its previous focus for the new one. A constant crossover and several additions to the services offered by this company allowed it to remain in business for 74 years. During a period when technology had rapidly changed the transportation landscape of America, the C.R. Patterson and Sons Company adapted to this change to stay active in the vehicle manufacturing business. This study explores these details and outlines how this Black-owned company survived for over seven decades through the transitions that it made and how its products evolved to meet the changing demands of the economy.

INTRODUCTION NOTES

---

[1] Ira Berlin, *Slaves Without Masters: The Free Negro in the Antebellum South* (New York, 1974).
[2] Sherri L. Burr, "The Free Blacks of Virginia: A Personal Narrative, A Legal Construct," *The Journal of Gender, Race & Justice* 1, 2016.
[3] Ibid.
[4] Charlotte Pack, *Time Travels: 200 Years of Highland County History Told Through Diaries, Letters, Stories, and Photos* (Fayetteville, Ohio: Chatfield Publishing Company, 2007), 183-185.
[5] Leland J. Pennington, "The Patterson-Greenfield Automobile," *The Highland County Magazine,* Vol. 1, No. 3 (1984): 33; Steve Konicki, "150 Built: Better than Model T," *Dayton Daily News,* March 21, 1976; Pack, *Time Travels,* 232; Juliet E. K. Walker, *Encyclopedia of African American Business History* (Westport, Connecticut, 1999), 50; Bill Naugton, "Former Slave, Descendants Were Early Vehicle Pioneers in Ohio," *Tri-State Trader,* July 5, 1980: 12.
[6] J. H. Harmon, "The Negro as a Local Business Man," *The Journal of Negro History,* Vol. 14, No. 2 (1929): 128, 152-153, 125, 119.
[7] Ibid., 122.
[8] Carl Kelsey, "The Evolution of Negro Labor," *Annals of the American Academy of Political and Social Science,* Vol. 21 (1903): 56-57; Frank Lincoln Mather, *Who's Who of the Colored Race: A General Biographical Dictionary of Men and Women of African Descent* (Chicago, 1915), 212.

1

GEOGRAPHICAL SETTING AND FAMILY BACKGROUND

The Patterson Family

Charles Richard Patterson (C.R.) founded the C.R. Patterson & Sons Company in Greenfield, Ohio. Although he did not become the sole owner of the company until 1893, it is generally accepted that the company's founding occurred, or at least Patterson service began, in 1865. That is the year claimed in early Patterson advertisement, as well as other references made by C.R. or his son, Frederick, during various presentations. While Patterson service started in 1865, the story begins at an earlier time.[1]

According to U. S. Census records, C.R. was born in Virginia in 1833. His father and mother, Charles and Nancy Patterson, were free persons of color, or "free Negroes" as the terminology from most references of the period would indicate. As discussed in the Introduction, at least two generations prior to C.R. were free persons of color. It is possible that earlier generations of the family were also free, but the records are scarce and sufficient information is not currently known to indicate for certain. C.R.'s grandparents were David Dover Patterson (born 1774) and Fanny Dover Patterson (born 1790), both born in Virginia. David and Fanny are both shown as "Free Inhabitants" in the 1850 Census and living in Greensburg, Kentucky, by that time. Both died in Kentucky during the 1860's. Charles was born in Bedford County, Virginia, in 1806. His wife, Nancy, was born in Virginia in 1822. It is suspected that the 1822 birth date may be incorrect as it is inconsistently shown in records as late as 1827 or 1835. Charles and Nancy first married in Lynchburg, Virginia, in 1832. They married again in 1845 at Ross County, Ohio. It is possible that they married twice because of differing State laws regarding free Blacks at the time. However, if the 1822 birth date for Nancy is correct, her age at marriage, 10 years old, may have also been a factor that required a second marriage after arriving in Ohio.

In late 1841 or early 1842, Charles and Nancy Patterson made it to Greenfield, Ohio, as a family. According to the 1850 Census, C.R.'s parents and their eight children were living in Greenfield by that time. C.R.'s mother and father (Charles and Nancy), he, and three other siblings were all born in Virginia, while the other four siblings were born after arriving in Ohio. The 1860 Census shows that five more siblings were born in that decade. C.R. had eight brothers (William David, Henry, John Edward, James, Thompson Alexander, George, Franklin, and Allen) and four sisters (Mary, Sarah, Marie, and Martha) by 1860. In 1865, the youngest of C.R.'s siblings, Robert Sterling Patterson, was born. In total, Charles and Nancy Patterson had 14 children. Three of C.R.'s brothers enlisted on the Union side for the Civil War. These included John Edward, Henry L., and William David. Two of the brothers, Henry L. and John Edward, died during the war, but the date or place of death is unclear. William David returned from the war and continued his career as a blacksmith, following in his father's footsteps, just as brother John had done and served as a blacksmith during the war until the time of his death.[2]

C.R. married Josephine Outz of Greenfield on July 4, 1864. After the Civil War, the family of C.R. Patterson and Josephine began to grow.[3] It is unclear whether C.R. took part in the Civil War, but it seems unlikely since he was in Greenfield in 1862 when he purchased his first property there and in 1864 at the time of his marriage to Josephine. Little is known about Josephine beyond her being a Virginia-born light-skinned mulatto woman partially of German descent, hence her maiden name of Outz. Some recent accounts have erroneously reported that Josephine was a White woman, which enraged the citizens of the town, but this was not the case. Born in 1847, she had five sisters, but it is unknown if she had any brothers or who her parents were. It is also unclear regarding her arrival in Greenfield or when that occurred, but we know she was there during the Civil War when she married C.R. in 1864. At least one of her sisters also lived in Greenfield.

In 1866, C.R. and Josephine's first daughter, Mary (sometimes referred to as Maimie), was born. Sometime around 1882, she married Clifford A. Napper and they had a son, Charles (Charlie) W. Napper, who was born on June 6, 1883. Maimie died on September 29, 1886. Her obituary appeared in the *News Herald* (Hillsboro) on October 6, 1886. The obituary indicated that she and her husband "were doing well in life, just getting a good start, and looking forward to a pleasant and happy future, when death came and blasted their hopes. She was highly respected here, by the White as well as citizens of her own color. She was a Christian lady whose death will be mourned by all who knew her." It continues to state that her funeral was "largely attended by both White and colored friends." This obituary provides the first documented glimpse into how the Patterson family was perceived in Greenfield and the surrounding area at that time. C.R. became the legal guardian of his grandson, Charles W. Napper, after her death and it is unclear as to why he didn't continue to live with his father, Clifford. Charlie Napper became an integral part of the Patterson Company in later years, serving as the sales manager and becoming a partner.

Charles Richard (C.R.)
Patterson
(1833-1910) (R.P.)

Josephine Outz Patterson
(1847-1938) (R.P.)

THIS CERTIFIES THAT
The RITE of
Holy Matrimony
WAS CELEBRATED BETWEEN
Witness
Witness

Marriage certificate of Charles R. Patterson and Josephine Outz.
Dated July 4, 1864.

Photograph of a young Frederick Douglass Patterson (1871-1932). (R.P.)

Photograph of a young Charles (Charlie) W. Napper (1883-1924).
(public domain)

A second daughter, Nellie Beatrice, was born in 1868 but died four years later. Their first son, Frederick Douglass Patterson, was born in 1871 and later became a driving force behind the Patterson Company. Details of his life will be provided throughout the remaining chapters. A daughter, Dollie, was born in 1872. She married John Rudd on June 29, 1908. She died in 1954.

In 1875, a second son, Samuel Claude, was born, but died at the age of 24 on May 3, 1899, under potentially mysterious circumstances. The traditional story is told that Samuel fell seriously ill around 1896 or 1897. This illness prompted older brother Frederick to return to Greenfield from his teaching position in Kentucky to help his father and brother with the carriage factory. It is unclear as to whether Samuel had an illness that lasted into May 1899 or not, but for certain Samuel died on May

3, 1899. According to a May 11, 1899, article in the *News Herald* (Hillsboro) headlined "Suicide at Greenfield," there were mysterious circumstances surrounding his death. Samuel was found dead in Paint Creek at the foot of 2nd Street on the eastern edge of Greenfield. He had a noose around his neck but was floating in the water and the rope was not attached to anything. It was ruled as a suicide, yet it was found that he had several personal belongings on his person at the time, which is not typical of a person committing suicide. The noose also seems an odd element to this story as it was ruled a clear case of suicide by drowning.

The Mayor of Greenfield at the time, Albert M. Mackerley, was a lawyer by profession and had no medical background, yet he was the acting Coroner at the time and made the ruling. Interestingly, the Mackerley family was known to be from a long line of competing carriage manufacturers in Highland County. According to a May 18, 1899, article in the *News Herald* (Hillsboro), the inventory of the personal belongings found on Samuel's body were being filed with the Probate Court by Mayor Mackerley. However, the Highland County Courthouse has no records on receiving this inventory or concerning Samuel's death -- not even a death certificate. Interestingly, the two articles concerning Samuel's death were in the Hillsboro newspaper. Since Samuel was local to Greenfield, it was assumed that Greenfield newspapers would provide a fuller account of his death. Following a review of newspapers for three months following Samuel's death, there was not a single mention of his death in any of the Greenfield newspapers -- not even an obituary, which seems unusual considering that Samuel was from a somewhat prominent family in Greenfield. The original May 11, 1899, article indicated that there was no accounting for Samuel to commit suicide as he was a well-respected colored citizen of the town. Perhaps Samuel really did have a long illness that lasted until this time, and he felt that suicide would end his suffering, but some of the details just don't seem to reflect the whole story. Interviews with living Patterson descendants has revealed that none of them were aware of details surrounding Samuel's death, including any mention of suicide or anything nefarious.

In 1877, daughter Fannie was born, but died in 1897. The final child of C.R. and Josephine was a daughter, Kathleen (Kate), born in 1878. Kate married author Greene B. Buster on December 30, 1913. In 1954, Greene wrote a book, *Brighter Sun*, which is an account of his grandfather's struggles to free himself and his family from the bonds of slavery. Kate died in California in 1971.[4] According to Richard Patterson, sometime around 1965, Kate penned a short 4-page history titled "Origin of the C.R. Patterson Carriage Business," a handwritten document that the author has a copy of in his possession that was passed down through Patterson family descendants. While the short document does not add a lot to the body of knowledge on the history of the Patterson Company, it does assist in confirming some of the findings of the author that stand apart from the traditional accounts shared about the Patterson Company and family. Information from this document will be presented in the appropriate portions of this work where it can be used to reinforce statements made about covered topics.

Photograph of Kathleen (Kate) Patterson Buster (1878-1971), youngest daughter of C.R. and Josephine. Pictured here with her daughter, Josephine, ca. 1915. (public domain)

Photographs of the company leadership in 1914. The company still paid homage to their founder, who had died 4 years earlier. (*JNMA* Vol. 6, No. 2 [1914])

Frederick D. Patterson married Bettie Estelline Postell of Hopkinsville, Kentucky on September 11, 1901. Bettie Estelline was born to Peter Postell and Pauline Buckner Postell on July 20, 1873, in Hopkinsville, Kentucky. Bettie Estelline's father, Peter Postell, was born into slavery in South Carolina in 1841 and would eventually become the wealthiest Black man in Kentucky by the turn of the century. Born a mulatto slave, resulting from Captain Elijah Coachman Postell and slave Selina Kirkpatrick having a child together. It is unclear whether this was a consensual union, but was unlikely to be consensual, as Selina was married to another slave named Edward Kirkpatrick. Peter was sold in 1858 to a Christian County, Kentucky, farmer named

Posey J. Glass. During the Civil War, he escaped and enlisted in the Union Army under the name of Peter Glass. At the end of the war, Peter was traveling through Hart County, Kentucky, and passed by the farm of Confederate General Simon Bolivar Buckner. Having thirst from his long march, Peter observed a young girl at a well and asked for water. The girl, named Pauline, provided the water and a lifelong relationship began.

Upon reaching an area where Peter could settle, he immediately started working in a barber shop and he was able to raise the money to purchase Pauline Buckner's freedom. Under Kentucky law, slaves were not freed by the 13th Amendment, so he was forced to purchase her freedom. Later, the 14th Amendment was passed, which freed slaves in all States. Peter purchased Pauline's freedom in late 1868 and brought her home to be with him. Peter and Pauline were married on February 24, 1872. Peter and Pauline had 10 children together, several of which died prior to Pauline. Peter formed a partnership with a White man, and they started a grocery business together in 1868. A year later, the White partner retired, and Peter assumed full ownership. He grew a prosperous grocery store and was able to amass a large amount of wealth and real estate holdings between 1871 and 1898.

Peter died of dropsy at his home on May 22, 1901. His extensive obituary was titled "Richest Colored Man in Kentucky, Passes Away." At the time of his death, his estate passed to his wife, Pauline Buckner Postell. His will had the provision that his entire estate was to be left to Pauline if she does not remarry. Regarding the thought of Pauline remarrying, he stated in the will "I cannot conceive that she would. My estate shall be equally divided among my children. I close this will expressing the hope that all my children will unite in helping their mother to earn a support, as I know she will help them; that no contention or strife or ill feeling shall exist by reason of money or property, and my earnest prayer is that they may be useful and honorable in the community in which they live." News of his death reached as far west as Washington State, where a headline read "Rich

Negro Dead" in the May 23, 1901, edition of the *Tacoma Daily Ledger*. The article indicated that "Peter Postell, probably the richest negro in the South, died suddenly of heart disease today. After the war Postell opened a grocery, in which he accumulated a fortune estimated at $500,000." News also reached as far east as New York when the *New York Times* ran an article with the headline "Death of a Wealthy Negro" on May 23, 1901. Just a few years later, in 1904, Pauline and the younger children moved away from Kentucky to live with her daughter, Bettie Estelline, and son-in-law, Frederick Douglass Patterson, at their home in Greenfield, Ohio.[5]

Photograph of Peter Postell, father of Estelline Postell and wealthiest Black man in Kentucky at the time of his death in 1901. (public domain)

Frederick and Bettie Estelline lived in Greenfield and had two sons together. The first was Frederick Postell Patterson, born on July 22, 1903, and the other, Postell Patterson, born in 1906. As the sons got older, they both married and became involved in the Patterson Company along with their wives. Fred P. married Bernice Coleman of nearby Lyndon in 1928. Bernice and Fred knew each other as children growing up in the Greenfield area. Bernice's family owned a farm just outside of Greenfield in the nearby town of Lyndon. After high school, they both attended college at the same time, but at different colleges. While Fred was at Ohio State, he would travel to Wilberforce University where Bernice attended. After receiving an A.A. degree from Wilberforce, Bernice took a position teaching home economics at the Colored Normal, Industrial, and Mechanical College of South Carolina at Orangeburg, which would later become South Carolina State University.[6] While Bernice was teaching in South Carolina, Fred had returned to Greenfield and was working at the Patterson shop. Eventually, in 1928, Fred and Bernice came together in Philadelphia and got married. They both returned to Greenfield following their wedding.

Fred and Bernice had three children together. The first child was a boy, Frederick (Freddie) Coleman Patterson, born in 1935. Freddie died in 1998 due to complications of Lou Gehrig's Disease after spending many years as a college professor. The second child was a girl, Betty Kathleen, born on February 3, 1938. The third child, Richard (Butch) P. Patterson, was born in 1942 just a few years after the Patterson Company closed its doors. Betty Kathleen and Richard P. are still living and are the closest remaining ties to the Patterson Company as their father had a direct role in running the company with his brother Postell. Freddie and Betty were very young during the last few years of the company. However, they would see their Uncle Postell driving a bus home after work, which he did frequently, and the children started referring to him as "Uncle Bus." Once Richard was born, a few years after the company closed, he began calling his uncle by that name as he learned it from his older siblings.

This nickname stuck with Postell until he died in 1991. Other Patterson descendants remain but are not as closely tied (within one generation) to the company lineage.

Postell married Kathleen Wilson, a childhood friend from Greenfield, in 1935. Postell was married twice. First was to a woman named Marguerite from Jacksonville, Florida. Little is known about their marriage, but it is known that they were married by 1928 as Marguerite shows up as a Patterson on mortgage documents starting that year. The last mention of Marguerite is in 1931 so it is likely that they remained separated for a few years. After at least 6-7 years of marriage, Postell divorced Marguerite on February 1, 1935, four days before his marriage to Kathleen Wilson on February 5, 1935, in New York City. According to Kathleen Wilson Patterson's autobiography, she was not aware of his marriage to Marguerite, or at least that he was still married to her so soon before their marriage. Postell had no children from either of his marriages.

Kathleen was born in Greenfield but had moved to New York City as a teenager. As younger children, Postell and Kathleen used to play together in Greenfield. She worked in various positions in New York and became independent. This independence, especially with money, would become an issue for her when she returned to Greenfield as Postell's wife and had an allowance given to her from her mother-in-law, Bettie Estelline. Kathleen worked in the Patterson factory and was the last living Patterson that had worked at the factory as she lived until 2003. She was possibly the last living person to have been employed at the company altogether.

Kathleen wrote an autobiography at age 92 in 2001. As she married into the family in 1935, during the last five years of the company's existence, her autobiography provides a nice window of details into those last years as the company was failing to thrive. When comparing some portions of Kathleen's narrative with actual Patterson documents from earlier in the company's existence, it was very clear in her autobiography that she was not provided the entire story from the family that she married into.

In her book and in later articles/interviews, Kathleen repeatedly indicates that the company partially failed because banks would not give money to Blacks at that time. In actuality, for decades the Pattersons had always had an excellent credit rating and obtained loans fairly easily from multiple sources, especially since they owned so many properties across town in which they could use as collateral to obtain mortgages. However, during the late 1920s, at the onset of the Great Depression, the Pattersons misused their credit (many small and large businesses did at that time due to the Great Depression) and ruined their creditworthiness. Of course, when Kathleen married into the family several years later in 1935, it appears that they kept those skeletons in the closet and provided her with the excuse that they couldn't get money from the banks because they were Black. This is an understandable "white lie" as nobody wants to look bad in front of their new spouse. By the time she came into the picture, it was true that the banks wouldn't give them loans, but not because they were Black. Due to the truths that were hidden from Kathleen, her autobiography has some inaccuracies regarding that portion of her life. They were very real to her since that is the story she was told and generally experienced during those last five years of the company's existence, so this is through no fault of her own and her autobiography is a valuable resource to the Patterson story.[7]

One determination of economic status was land ownership. This is often used as an index to economic stability and even a value of citizenship. In the years just after the Civil War, and even into the early 20th century, land ownership was not common for Blacks. Those who did own property were typically in a higher class within the Black community. C.R. Patterson owned property from the 1860s on. Appendix C lists the properties that were purchased by the Patterson family and company. C.R.'s first property documented in the courthouse records was purchased in 1862. During that same decade, he also bought a property in 1865 and two more parcels in 1869. This indicates that C.R. was somewhat wealthy during this time. In support of

Photograph of Frederick and Bettie Estelline with their two young sons, Frederick Postell and Postell, ca. 1908. (public domain)

Photograph of C.R. with his grandsons Frederick Postell and Postell, ca. 1907. (R.P.)

Frederick Postell Patterson
(1903-1973) (R.P.)

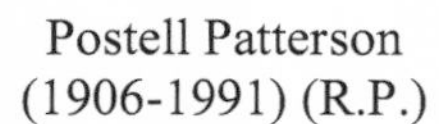

Postell Patterson
(1906-1991) (R.P.)

Kathleen Wilson Patterson
(1909-2003)
Wedding day photo.
(from Kathleen Patterson
autobiography, 2003)

Bernice Coleman Patterson
(1902-1981) (R.P.)

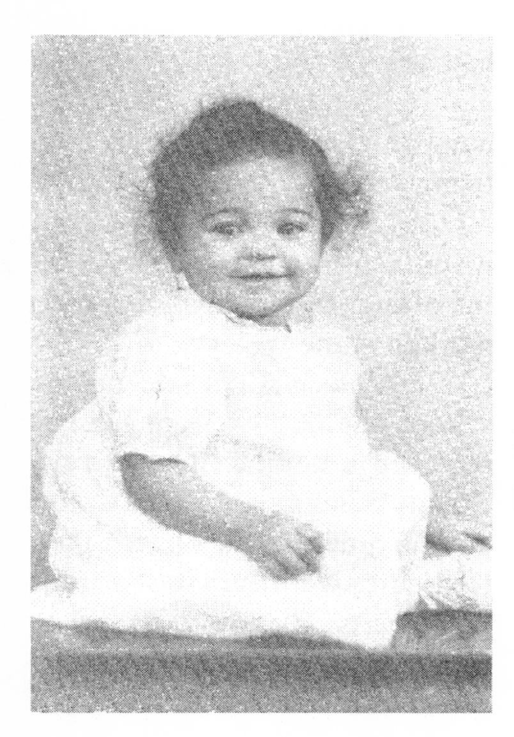

Betty Kathleen Patterson, daughter of Frederick P. and Bernice. (1938-living) (R.P.)

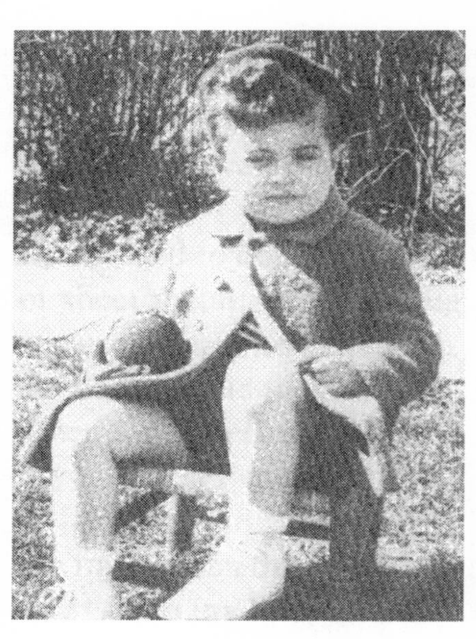

Richard Postell Patterson, son of Frederick P. and Bernice. (1942-living) (R.P.)

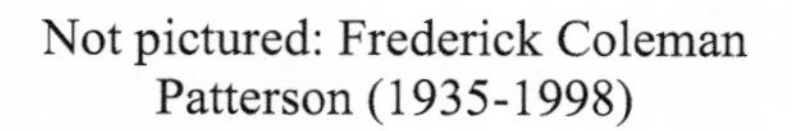

Not pictured: Frederick Coleman Patterson (1935-1998)

this, the 1870 U. S. Census showed values of real estate owned and the value of personal property. For C.R., his real estate value was listed at $3,500 and his personal property was listed at $1,450. His real estate value, when compared to the entire population of Greenfield during that census year, was higher than the average and placed C.R.'s real estate value in the top 22.6% of Greenfield property owners.[8] His personal property value was slightly lower than the average and placed C.R. in the top 22.8%. These numbers suggest that, even as early as 1870, C.R. held a high economic status among all residents, White or Black, in Greenfield.[9]

These numbers were then compared to only the Black population of Greenfield. C.R.'s real estate value was $3,500, which placed him first among Black property owners with the next closest value being $2,000.[10] C.R.'s personal property value of $1,450 placed him at the top of this list as well, with the next closest value being $800.[11] These numbers indicate that C.R. had already become the wealthiest Black citizen of Greenfield by 1870 yet was still in the top 22.8% of all citizens, White or Black. One last group compared to C.R. were those identified as being blacksmiths, carriage makers, or others related to the carriage trade. C.R.'s real estate value of $3,500 topped this list as well.[12] C.R.'s personal property value was $1,450, well above the average, and placed him in the top 15% of this group. Even within his occupational group, C.R. fared very well in relation to economic status at this early date.[13]

Courthouse records were examined to track property purchases throughout the company history. Initially, the research was expected to produce only three properties: the family home on Jefferson Street, the company's east building, and the west building. According to the records, the Pattersons purchased 30 properties in Greenfield.[14] Of course, ownership of these properties did not occur at the same time, but the total reflects all the properties owned between 1862 and 1938. An illustration on the following pages shows a chart that demonstrates the pattern of property purchases by decade. It clearly shows a pattern of economic stability and strong purchasing power from 1890 to 1910, but a sharp decrease in purchases after this time seems to coincide with an economic decline in the company. These properties were used strategically in company planning as a method of obtaining capital. It seems that when business was good, the Pattersons purchased properties. Whenever money was needed, the properties were either sold or mortgaged to obtain the funding the company needed at that time.

Although a major obstacle for a Black business owner was obtaining capital through credit to conduct the business, this does not seem to have been the case with the Patterson Company.[15] The fact that the family owned so many properties speaks a lot for the credit worthiness of the family, at least in the first several decades of the company's existence. These properties were mortgaged, as necessary, to obtain the money needed for business operations. Several sources had claimed that the Patterson Company failed owing to a lack of capital. Highland County mortgage records indicate that this was not the case; the Pattersons had received a total of 67 mortgages on their properties through the years amounting to $184,904 (see Appendix D for a list of Patterson mortgages). An illustration on the following pages shows the pattern of mortgage amounts by decade, and for the most part, it coincides with the property purchase pattern as it plummets rapidly after the 1920s. This indicates that during the 1920s, the economic stability of the company had reached its peak.

Until the mid-1920s, the Patterson mortgage amounts were low and could easily be managed by the company. Several mortgages prior to this time were paid off early, sometimes in less than a year, although the mortgage amounted to several thousand dollars and was on a multi-year payment plan. These coincided with the alternating busy and slow seasons. The company sometimes needed funds for the upcoming busy season and would obtain a mortgage, but then immediately repaid it once the busy season hit and the funding was available. This protected the Patterson Company's credit and allowed them to obtain

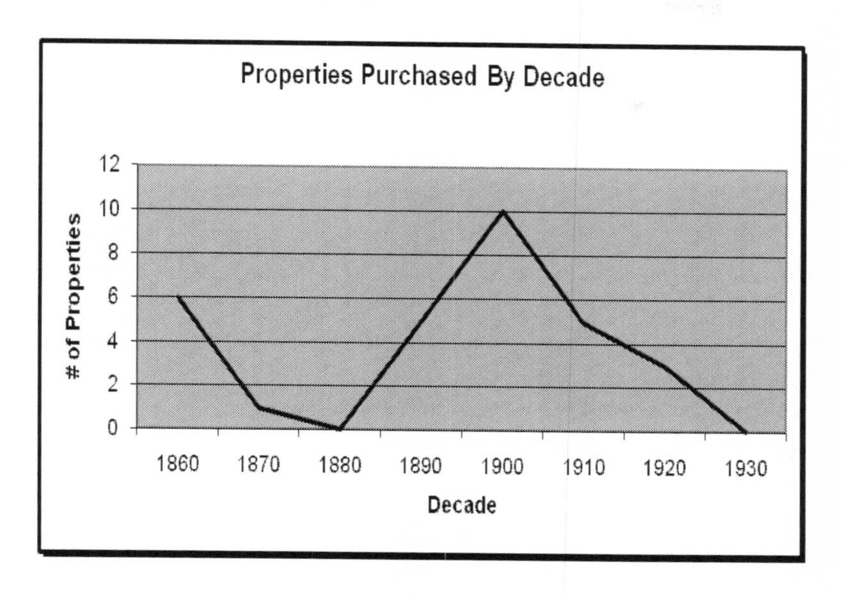

Patterson properties chart showing purchases by decade.

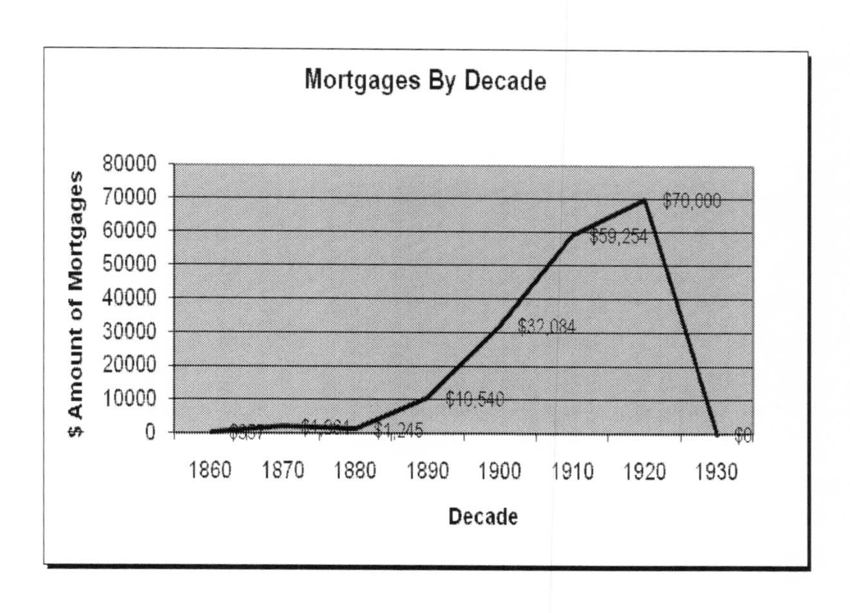

Patterson mortgage chart showing mortgages by decade.

capital when needed. This eventually hurt the company in the 1920s and 1930s when it borrowed more than it could afford considering the business profits had decreased by this time.

Author Juanita Johnson has provided an excellent take on the Patterson Company financial condition in her 2021 book, *The Audacity Code: Coloring in Black Outside the Lines.* In the book, she expanded upon information provided in the first edition of this book and provides a more in depth look at how the Pattersons broke down barriers typically standing in the way of Black businesses during the period.

C.R. Patterson died in 1910 at the age of 77 years. His wife, Josephine, lived until 1938. Their son Frederick D. Patterson died in 1932 at the age of 61 as the result of what seems to have been a stroke. Frederick P. Patterson lived until 1973 and his brother Postell passed in 1991. Fred P.'s wife, Bernice, survived until 1981, and Postell's wife, Kathleen, outlived every other family member that worked at the company as she reached the age of 94 and passed in 2003.

While there were several sons, daughters, and grandchildren in the Patterson family in Greenfield, only a select few were major players in the Patterson Company. The company passed through three generations of the family and each person had a hand in the company in one way or another throughout its span of operations. Some, such as C.R.'s daughters Maimie, Dollie, and Kate, were given stock in the company, but that seemed to be the extent of their involvement. One exception is Kate, who indicated in her handwritten history of the company that she started working in the office after she finished high school in 1897 and worked there until her marriage in 1913. Another person that Kate pointed out that worked in the office was Ocie Pettiford, who eventually married Charlie Napper. Kate described her as "one of our hometown girls who, at the time, was one of the efficient employees in our office." C.R.'s sons Frederick and Samuel became partners with their father in the 1890's, but Samuel died in 1899, which left Frederick as the second generation to run the company from 1910, when C.R. died, until his own death in 1932. Frederick's two sons, Frederick Postell Patterson and Postell Patterson became the third generation to run the company after their father's death and were in control until the company ceased operations in 1939.

Geographical Setting

Greenfield, Ohio, is in Highland County, which is one of the northernmost counties of Appalachian Ohio. Greenfield is situated in Madison Township and is the primary town in that township, so the histories of the two are essentially the same. The town was incorporated in 1841 and grew from approximately 447 residents in 1840 to 2,104 by 1880. Greenfield has always been a small town yet very productive. Throughout the years several manufacturing companies have been located there and several townspeople have produced useful patents through their inventiveness. Some major industries in Greenfield during the period under study were The American Pad and Textile Company (TAPATCO), Waddel Manufacturing, and J. A. Harps. Some of these industries, such as E. L. McClain's TAPATCO made their owners very wealthy. Greenfield made products including detachable horse collar pads, furniture, oil cans, and a range of other products. Two rail lines provided easy access to Greenfield, one running north and south, the Baltimore and Ohio, and the other east and west, the Detroit, Toledo, and Ironton, so goods were easily transported in and out of town.[16]

Perhaps one of the most important aspects of Greenfield that pertains to this study was its history of abolitionism. Reverend Samuel Crothers, who is considered "one of the fathers of the abolitionist movement," called Greenfield his home. Rev. Crothers published many letters, books, and pamphlets and was an active member of the Ohio Anti-Slavery Society. His writings were influential in the national movement to end slavery, and he became "one of the first writers west of the Alleghenies to take a firm stand against slavery." Rev. Crothers founded the

First Presbyterian Church in Greenfield in 1820, and it is still in existence today. With such an influential anti-slavery person in the town's midst, it is no surprise that Greenfield was a documented stop for the Underground Railroad and helped guide many freedom seekers northward away from the bonds of slavery. Frederick Douglass, the famous ex-slave and orator who openly spoke out against slavery in the nation, had been attracted to Greenfield at one point to give a speech.[17]

Greenfield's abolitionists attracted many Blacks to the area. Madison Township had 13 Blacks in 1820, but this number quickly grew to 124 by 1860. After the Civil War, the Black population of Greenfield jumped to 373 by 1870. The Black population for the entire area of Highland County topped out at 1,763 in 1880, which was 5.8% of the total population of the county. The 1850 U. S. Census showed that 33% of Blacks in Highland County had been born in Virginia and another 11% had come from four other slave holding states. Those that had been born in Ohio totaled 55% of the Black population and 1% came from other northern states.[18] The Pattersons were among those from Virginia.

Family Education

C.R. Patterson became known as a very competent blacksmith, carriage maker, and all-around businessman. He was successful as a craftsman, but it is unlikely that he attended high school as there was no high school for Black students in Greenfield during his teenage years. However, it was discovered during an interview with his great-granddaughter, Betty, that C.R. did attend college just prior to the Civil War but did not finish with a degree. Betty was under the impression that C.R. had attended Wilberforce University just east of Xenia, Ohio. Wilberforce University has a storied history and is the country's oldest private historically Black university, with origins dating to before the Civil War. The land on which the university was located was originally a health resort for wealthy patrons. The resort was founded by Elias Drake, a lawyer and former speaker of the Ohio General Assembly. Drake started the resort, called Tawawa Springs, in 1850. Southern slaveholders traveled to the resort for relaxation, and they oftentimes brought their slaves. Local Northerners didn't approve of this occurrence and business soon declined. The resort closed in 1855.[19]

The property was immediately purchased by the Methodist Episcopal Church, which was a church known for its ties to the Underground Railroad. In 1856, the hotel and cottages were transformed and made into a school where they could "provide an intellectual Mecca and refuge from slavery's first rule: ignorance," according to the university's historical narrative. The school was named in honor of William Wilberforce who was a well-known abolitionist during the eighteenth-century. Within two years, more than 200 students from around the country had enrolled in the school. However, shortly after the Civil War began, the university was forced to close when enrollment declined in 1862. The following year, Bishop Daniel A. Payne negotiated the purchase of the property for the African Methodist Episcopal Church. Payne reopened the school in July of 1863 and he became the second president of Wilberforce University and the first Black educator to lead a university.[20]

Attempts to verify that C.R. attended Wilberforce University during its early years were fruitless. An unfortunate event occurred on April 21, 1865, where an arsonist had set fire to several of the buildings and destroyed much of the university. Ironically, this was the exact date that Abraham Lincoln was assassinated at Ford Theater in Washington D.C. Few records survived the fire, and Wilberforce University archivists were unable to locate records of the earliest students at the university. However, the timeline of events does support a strong plausibility that C.R. attended during the first period of the university. During a review of archival documents, C.R. was not found in the 1860 U.S. Census. A search of census records across the United States did not result in identifying C.R. One explanation is that he may have been attending university at this time. It is unlikely that census takers of the time would visit the Wilberforce campus given the makeup of its student body. Therefore, C.R., and other campus students, would not have appeared in the 1860 census records. This likely places C.R. at Wilberforce starting in 1859, since we know that he didn't complete college. He likely would have attended through the closure of the original university in early 1862, which provided him only three years of college and, therefore, didn't finish with a degree. Additional support of this theory is that the very first property that C.R. purchased in Greenfield was in early 1862, perhaps when he returned to Greenfield from the defunct Wilberforce University.

C.R. wanted to provide opportunities for his sons to receive a good education. He became so concerned with this that, at times, he was willing to fight to provide these opportunities for his sons. When his oldest son, Frederick Douglass Patterson, graduated from Greenfield's "colored" middle school, he received a promotion card stating that he had been "promoted to the High School Grade." Since no high school existed for the Black children, he arrived at the White high school on the first day of classes in 1886. Superintendent William G. Moler refused Frederick admittance to the school that day and each day afterward for about a month. C.R. Patterson filed a suit against the Board of Education on September 30, 1886.[21]

This suit went back and forth throughout most of the school year. The Pattersons made arguments that were then refuted by the school board. The most obvious argument dealt with the promotion card that stated that Frederick had been promoted to high school. Since there was no other high school in town, the school board must provide Frederick the opportunity to attend the White high school. Another argument in favor of Patterson was that owing to his light complexion, his "Negro blood was scarcely visible" and that he should be "entitled to the same privileges as other White students."[22]

The Board of Education's best argument was that according to the "Black Laws" of Ohio, which were part of Ohio's 1803

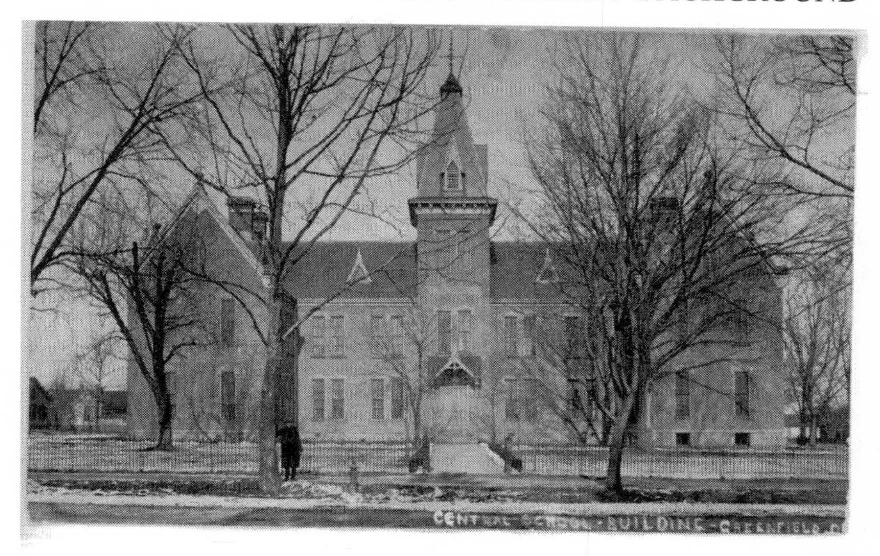

Greenfield High School that Frederick fought to attend. Replaced in 1915.
(courtesy of Greenfield Historical Society)

Constitution (limiting the rights of Blacks in Ohio), "C.R. Patterson had no legal right to sue and that there were no grounds for the complaint." The arguments continued until April of 1887. A court ruling required "the Board of Education and Superintendent immediately admit Frederick under the same regulations they admit White pupils. He further ordered the defendants to pay costs." Frederick began his high school career as soon as he could and became the first Black to graduate from the high school in Greenfield in 1888. According to his sister Kate in a 1965 interview, "Fred entered school and 'made good,' graduating as Valedictorian of his class."[23] This small victory set in motion a long line of successes for Frederick Patterson.

Wanting to break away from immediately going into his father's business and to find his own way, Frederick moved to Columbus and worked as train porter for a year. While in Columbus, he made the decision to enroll at Ohio State University starting in the fall of 1889. Frederick entered the Ohio State University on September 11, 1889. While some sources claim that he became the first African American to enter the university,

Sherman Hamlin Guss and William Clark became the first two Black students when they entered a year earlier in 1888. While at the university, Frederick did achieve a major first for Black students at the university. He became the first to play on the Ohio State University Football team, breaking the color barrier for that sport. He played in 1890 and 1891. According to a 1992 article about African American sports firsts in Columbus, "Patterson [was] so light-skinned that he doesn't stand out among his White teammates in an 1891 photo."[24] This can be clearly seen in the 1891 team photo. Frederick did not start the 1890 season with the team and is, therefore, not shown on the 1890 team photo that was taken earlier in the season. The early OSU football team was struggling and needed more students to form a practice squad. A call for students to form a practice squad was answered by Frederick and several other students. Frederick made enough of an impression during the practice sessions that he was eventually added to the regular team roster as a substitute.[25]

During the scrimmages, Frederick was credited with having "several fine runs". Although a substitute, there were occasions where he played in games as a starter. Records indicate that Frederick played in several of the games during the 1890 and 1891 football seasons. Frederick saw varsity playing time at both end positions, halfback, and even quarterback on occasion. The first game on record with Frederick in the varsity starting lineup was the very first Thanksgiving Day game in 1890 when OSU hosted Kenyon College in Columbus. Frederick started at end for that game, both on the offensive and defensive side of the ball. It was a snowy day, and the conditions were slippery. While on defense, Frederick scooped up a fumble from a Kenyon player on a punt return and ran the ball in for an OSU touchdown. This would only be the sixth touchdown ever scored in a game by the OSU team and would be the first defensive touchdown for the team. It was also the first OSU touchdown by a Black player in the history of OSU football. After Frederick scored the touchdown, a Kenyon player gave him an angry shot to the shoulder, an action that did not prompt a response from Frederick.[26]

In a November 1891 game, OSU was playing Adelbert in Columbus. Adelbert was a school with considerable experience in playing football, in stark contrast to the fledgling OSU team at the time. By halftime, Adelbert was winning by a score of 32-0 in the one-sided contest. Just before the end of the first half, frustrations became too much for the OSU quarterback, Frank Haas, which prompted him to throw a punch and get disqualified from participating in the remainder of the game. Frederick was sent in to play quarterback for the rest of the game. As described by the local newspapers, the OSU team played through the second half as if they "didn't know when they were beaten." The team showed grit and kept pushing forward to the delight of the OSU fans in attendance. Near the end of the game, OSU forced an Adelbert fumble. On offense, OSU began driving the ball down the field. When still 40 yards from the goal line, the center snapped the ball to Frederick who in turn tossed it over to Hobart Beatty who dashed down the field for a touchdown. The Columbus Dispatch newspaper called that last scoring drive, in the face of a clearly superior opponent, "one of the grandest fights ever seen on a football field." The final score showed OSU losing 50-6, but that last drive inspired changes in the football program that made OSU football move toward improvements and in the direction that OSU football is known for today. During the Spring of 1892, Frederick also played on the OSU baseball team, or at least as part of the 2nd group of nine, which generally served as a practice squad.[27]

The second Black athlete to play varsity on the OSU football team was Julius B. Tyler in 1896. Julius was the brother of Ralph W. Tyler, the journalist who would later play a factor in Frederick's political aspirations. The next Black athlete to play varsity on the OSU football team after Tyler was William "Big Bill" Bell who played in 1929, 38 years after Frederick played. Big Bill didn't get quite the reception that Frederick had received as he reportedly had to endure being spiked with metal cleats by both opponents and his own teammates unhappy about having to share an integrated playing field. Unlike the OSU of today, like many other major universities, OSU was segregated during the first half of the twentieth century. OSU finally ended their segregation in 1946 and considers it a dark stain on their past.[28]

Frederick engaged in many other activities at the university as well. He was an active member of the Horton Literary Society for at least the 1891 and 1892 school years. In 1891, he served as the Corresponding Secretary for the society, while in 1892, he served as the Vice President of the Horton Literary Society. In 1892, he served as the Assistant Business Manager of *The Lantern*, which was the school's student newspaper. The previous year, he had served as an editor for the Personal and Local sections of that same newspaper. Perhaps the biggest position that he served in while a student at Ohio State was that of Class President for the class of 1893, which speaks a lot about how Frederick Patterson was viewed among his peers.[29] These opportunities also allowed him to gain some managerial experience, along with his education, which helped him to become a successful businessman in later years when he took control of the carriage business.

Frederick lived on the OSU campus in the North Dorm, which was the main dormitory on campus at the time. He lived in the North Dorm all three years that he attended OSU. See photo on following pages of Frederick and Sherman Hamlin Guss with their 60 or so White dorm mates. William Clark also normally lived in the North Dorm but was absent from the photo as he was spending the semester in Los Angeles at that time. Previously assuming that Frederick was forced to live off-campus, it was a pleasant surprise to discover that he lived in the main dorm while he attended the school. This likely partially explains how he ended up being voted as President of the class. When you live with someone, you naturally get to know them better than just having some classes with them. This forced exposure to each other likely allowed the White dorm mates the opportunity to get

Photograph of the 1891 Ohio State University Football Team. Frederick D. Patterson is in the back row, third to the right. His light skin color allows him to practically blend in with all the other players on the team. (Photograph courtesy of the Ohio State University Archives, location 157-189)

to know Frederick as a person and a mutual camaraderie formed. Considering that the Olympic great Jesse Owens, who won four gold medals at the 1936 Olympics in Berlin and stared down Hitler himself, was forced to live off-campus during his time at Ohio State when he returned to the school from Berlin, it was a bit surprising to discover the many documented successes of Frederick at the university four decades earlier.

In 1890, Frederick was a member of the OSU Military Science class. He is listed as a Corporal in that year's Makio yearbook. In fact, of the four companies that made up the Battal-

Photograph of the North Dorm at The Ohio State University.
(Photograph courtesy of the Ohio State University Archives, location 100-026)

ion on campus, one company (Company A) was specially formed for a military drill competition that took place on July 4,1890, in Portsmouth, Ohio. The temporary Company A was formed from the best of the battalion, which included Frederick. Company A won the competition and Frederick was shown in the photograph of the winning team. Military Science was a requirement for all males attending land grant universities, so it is not a surprise that he was in the class, however, Frederick apparently excelled at the drill techniques since he was selected as part of the competition team.[30]

As part of the literary society, Frederick made many speeches and participated in debates across campus. Several of his orations and debates were highlighted in school newspaper, *The*

Photograph of the students living in the North Dorm at The Ohio State University ca. 1890. Frederick is sitting on the rail at the top right of the photo. Sherman Hamlin Guss is standing at the top of the stairs just left of center. (Photograph courtesy of the Ohio State University Archives)

Lantern, and the *Makio* yearbook. Frederick received accolades in The Lantern following his first oration as a new member of the Horton Literary Society. The article stated that it "was a fine opening performance by a new member." Later that same year, after another of Frederick's orations where he read "Rienzi's Address to the Romans," apparently a popular selection for the time, Frederick was described as having a "pleasing and forcible oration style which characterizes all Mr. Patterson's performances." One debate that was highlighted was one that involved arguing that cities are a menace to American civilization. While

Photograph of the 1892 Lantern Staff at The Ohio State University. Frederick is in the back row to the far right. (Photograph courtesy of the Ohio State University Archives, location 100-018)

Frederick made the best argument that cities are a menace, his opponent aptly argued that we cannot get on without them, so it appears to have been a bit of a draw.[31]

It appears that Frederick was a well-liked student on campus. On occasion that further reflects his status on campus took place in the Spring of 1891. The OSU baseball team had a stellar year. By June 3rd, they had a 5-0 record. The team then went on a 3-game road trip, playing Kenyon, Buchtel, and Adelbert on consecutive days. A few minutes after the Adelbert game, the team sent a telegraph back to Columbus indicating that they had swept all three games. Fred and Charlie Powell were waiting in the telegraph office and, upon receiving the news, the two grabbed a pair of drums and began marching across campus spreading the news of the team wins. Obviously, Fred did not have issues with making a bit of a spectacle on campus – in other words, he was

Frederick as member of the OSU, Company A military drill team that won the 1890 competition at Portsmouth, Ohio, on July 4th of that year. (Photograph courtesy of the Ohio State University Archives, location 99-1890s)

quite comfortable and confident on campus.[32]

There was an interesting quote about Frederick that made it into the 1891 *Makio* OSU yearbook. The quote revolves around not judging outward appearances against the inner man. This could be a very powerful quote and could provide a glimpse of how his classmates viewed him. The quote stated: "Now the saw is, 'you can't judge a horse by its harness,' nor our class by its looks, for in it genius comes in clusters, and its greatness is not alone in Patterson's head; looks betray not the inner man."[33]

After Frederick's junior year at Ohio State, he took a teaching job in Kentucky. Because of this, he decided to not return to Columbus for his final year. Frederick wrote to the Editor of *The*

Lantern, Walter J. Sears, explaining his reasons for not returning:

> I am down here in this land of good whiskey, fast horses, and fair women, and to tell you honestly, W. J., I am making hay in a way that is pleasant to me. My eyes are being opened and I am getting a glimpse of the field which I will have to tread in a year or two and when I return to O. S. U. I think it will be a matter of ease to direct my effort in a line that will most benefit me hereafter. I would have liked to [have] returned, Walt, and finished with my class, but you can appreciate the fact that the O. S. U. is not the world and that my battles must be fought upon ground different from that occupied by your fellows. For that reason, I am trying my hand at experience, and let me tell you, it is the best of teachers.[34]

Since he left the university, Frederick had to relinquish his position as Class President. The story is unclear as to exactly what prompted Frederick to leave OSU, but he made it very clear in the letter above that he must fight his battles upon a different ground than at a university. The letter is very telling that he feels that, for him, experience is going to serve him better than continuing his education, although he did mention an intent to return to OSU. With the worst racial violence since the Civil War occurring in that year, some have suggested that the "lynching outbreak of 1892" was a possible reason for Fred to leave the university. While Ohio wasn't known as one of the many States that experienced that widespread issue that year, it may have influenced Fred and compelled him to move to Kentucky to educate Black students.

It is also possible that Frederick was recruited to teach in Kentucky. In writing a story about Frederick's contemporary at Ohio State, Sherman Hamlin Guss, retired Ohio State Professor Kenneth Goings indicated that very few Black students attended college at all at the time, let alone earn a degree. He stated, "Any formal education, particularly for Black students, was such a leg up that they were often able to secure very good jobs with just a partial college education and they didn't need the college degree then. In fact, the college degree was really very, very rare at the time." Goings went on to say, "There was such a desperate need for teachers at the time and such a thirst for education on the part of the Black community…that anyone with even a partial education was recruited into the teaching profession."[35]

There was a brief announcement included in *The Lantern* newspaper in September of 1892 indicating that Frederick would not be back at the school that year. The article indicated "Mr. Patterson has made many friends in the University both among students and faculty and he will be greatly missed by all." A few years later, in January of 1895, *The Lantern* newspaper provided a brief alumni update indicating that "F.D. Patterson is one of several O.S.U. alumni who are enjoying prosperity in Louisville, Ky."[36]

Frederick remained in Kentucky as a professor of History at the Louisville Central High School for four years. While in Kentucky he met Bettie Estelline Postell of Hopkinsville, Kentucky, who earned her teaching degree from Fisk University in Nashville in 1894. She will hereafter be referred to as Estelline as that appears to have been her preferred name in a review of historical documents. She taught at Hopkinsville for a short time and then abroad at a mission school for four years for the American Missionary Association, which was responsible for the founding of Fisk University.[37] When Estelline's father died, his obituary mentioned that one of his daughters was "a teacher of unusual talents and culture." While this didn't specify which daughter by name, it is quite possible that this statement was about Estelline in reflection of her Kentucky and missionary school teaching work. Estelline and Frederick married on September 11, 1901.

Frederick and Estelline's two sons, Frederick Postell and Postell, both followed in the footsteps of their father and attended Ohio State. Fred P. studied mechanical engineering, which gave him the skills that he used in the 1920s and 1930s designing Patterson commercial vehicles. Postell was one of twelve founders of IOTA PSI in 1926, a Black fraternity at Ohio State. Its "main emphasis has been helping to foster a comforting environment for young Black men at one of the largest predominantly White universities in the United States." Postell also involved himself with boxing while at OSU. A December 11, 1925, article in The Lantern named the winners of a boxing context. Postell was listed as the winner for the 145-pound class.[38]

Charles W. Napper, C.R.'s grandson who lived with him, also attended Ohio State. Like his uncle before him, he spent at least two years in the OSU Military Science class and served as a Private in Company B of The Battalion military cadets. He graduated in the class of 1906 with a bachelor's degree. The class was the largest graduating class to date at the university with 225 degrees conferred. Charles was only the 8th Black student to graduate from the university. Charles studied geology and was active as a member of the Ohio Academy of Science. He presented papers at their meetings and still published papers through Ohio State years after he graduated. Charles and Frederick both remained involved with the university long after their departure. They made a gift of two fine rugs to the Ohio State University Association for their offices in 1913. They both remained members of the Association for several years, with Charles serving as a director of the local Highland County chapter of the Association for several years until 1915.[39]

Involvement in the Community

Patterson family members were involved in the community in many ways. Frederick was one of the most active, or at least his actions were more often publicized in the local newspaper. The Patterson family became known for their involvement with the Greenfield African Methodist Episcopal Church. C.R. Patterson was listed as one of the church's Trustees in 1880. A 1908 *Greenfield Republican* article, published two years before C.R.'s death, noted that he served as a Sunday School teacher at the church and that he was "the oldest, but most active" teacher.[40]

The family also supported the traveling evangelistic revivals that came to town. In 1915, when the Stephens-Storrs Evangelistic Party was set up in Greenfield, a newspaper announcement noted that the C.R. Patterson and Sons Company were one of the businesses that had agreed to close their shops promptly at 7:00, so that workers could attend the revival meetings during the evenings. In 1917, Estelline became the President of a group called the "Willing Workers," which supported repairs and beautification of the A. M. E. church building, both inside and out. This group organized dinners and fundraisers to assist in the maintenance of the building. In 1926, the A. M. E. District Conference was held in Greenfield. Along with the pastor of the local A. M. E. church, Frederick was a guest speaker at the conference, and they were said to have "delivered fine and timely addresses of welcome."[41]

Frederick enlisted in the 3rd Ohio Volunteer Infantry to serve in the Spanish-American War. He enlisted as a Private in Company F, which formed in Hillsboro, the county seat of Highland County. The 3rd Ohio Volunteer Infantry was organized in Columbus on April 26, 1898, four days after the U.S. declares war on Spain, and was mustered into federal service on May 10, 1898. The regiment originally consisted of 43 officers and 913 enlisted men under the command of Colonel Charles Anthony of Springfield, Ohio. Eight days after being mustered in, the regiment departed for Tampa, Florida, arriving on May 23. In Florida, the regiment was assigned to the First Brigade, First Division of the Seventh Army Corps, under the command of Maj. Gen. Fitzhugh Lee. On June 11, the regiment was transferred to the Second Brigade, Third Division of the Fourth Army Corps. The regiment was later transferred to Fernandina, Florida, arriving on July 25. On August 29, the regiment was again transferred, this time to Huntsville, Alabama, where it arrived on

August 31. On September 14, the regiment departed for home, arriving at Columbus on September 15. After being furloughed for one month, the regiment was mustered out on October 26, 1898. At the time of muster out, the regiment consisted of fifty-three officers and 1,297 enlisted men. During its term of service, the regiment had two officers and eight enlisted men die from disease, and six enlisted men discharged on disability. The 3rd Ohio Volunteer Infantry never served outside of the United States.[42]

Frederick was a member of the Greenfield Businessmen's League. In February of 1915, he and many others joined the push for Greenfield to pave its streets. He said, "I am in favor of street paving and consider the movement the best that can be made for Greenfield's good." It is likely that Frederick had already made up his mind to start building automobiles, which would be introduced later this year. With this in mind, he saw the benefits to having paved streets in town, not only for the good of the town, but also for his new venture. Frederick also belonged to the Greenfield Commercial Club in 1917. During World War I, he tried to help his city in other ways. He was a member of an organizing committee to form a company of 50 to 60 "colored" men that would function in a similar manner as the "Home Guards" who were an organized group that stood ready to assist local or state officials during homeland emergencies that came because of the war. Their goal was to put themselves in a "position where they could render a public service should they be called upon" to support the war effort. The Patterson Company also supported the war effort by sponsoring an advertisement that encouraged citizens to purchase Liberty Bonds that would make sure "America's brave soldiers will never lack anything that they need to help them to win the war and preserve Liberty."[43]

In October of 1908, Frederick served as umpire for a game between the Cincinnati Reds and the local Greenfield team for a game played in Greenfield. Another Greenfield Republican article mentioned Fred and his reputation as an umpire, which is apparently something he did quite often, stating that Fred "maintained his reputation for fairness as an umpire and rendered his decisions with such accurateness that no possible chance was given for a kick to be registered."[44]

Frederick became quite a speechmaker through the years. He commonly involved himself in public meetings and took his turn at the lectern. Frederick delivered a speech in 1921 to a "colored alumni group" from nearby Wilmington College. The Patterson family also participated in the annual "Emancipation Day" celebrations in the local area. These tri-county celebrations were held in different communities on a rotating basis on September 22nd of each year. The purpose of the celebration was to "make a program befitting the importance of the occasion to the colored race, one that will be up-lifting in character and which will not only appeal to the people of their race and create a desire to lift themselves up to a higher plane, but will also appeal to the members of the White race and give them a better idea of what the members of the colored race can accomplish if they take advantage of the opportunities afforded them." Estelline participated in the 1917 celebration by directing a "juvenile operetta" and Frederick presided over the program. Both in 1919 and 1920, Frederick was a featured speaker at the event. The 1919 event had a focus on "colored" soldiers that had fought in the recent World War I, as well as the earlier Spanish-American War, and Civil War. This Emancipation Day celebration was planned to be the largest ever held in Greenfield. There were parades, a baseball game, bands, demonstrations, and a dance. Additionally, there were several speeches from dignitaries across the State and one from Georgia. Frederick was one of the principal speakers, as well as Ralph W. Tyler, of Cleveland, who would later lead the push for Frederick to be one of Harding's Delegates-at-Large, which will be discussed later.[45]

C.R. Patterson and his son Frederick were both active members of the Cedar Grove Lodge No. 17, Free and Accepted Masons. At this time, there were two separate chapters of the Masons, one for Whites and a separate lodge for "colored" citizens. Greenfield's Cedar Grove Lodge No. 17 had been chartered in 1871. C.R. Patterson involved himself in the activities of the lodge from its inception. He served as Worshipful Master, Senior Warden, Junior Warden, Treasurer, and Secretary, each for multiple terms between 1871 and 1899.[46]

Frederick also served as the Worshipful Master of the Cedar Grove Lodge in Greenfield. Because of his speechmaking ability, Frederick served as the Grand Orator at several lodge functions. In a 1982 book, *Great Black Men of Masonry, 1723-1982*, author Joseph Mason Andrew Cox paid tribute to Patterson. Cox dedicated his book to seven "Great Black Brothers who Illuminated Freemasonry," and Cox lists Frederick as one of the seven. Others were Thurgood Marshall and Prince Hall (the first Black man to be accepted in Freemasonry).[47]

National Negro Business League

At the turn of the century, Frederick joined a fledgling organization known as the National Negro Business League. Professor Booker T. Washington started this organization at Tuskegee Institute in Alabama. A 1902 history of Highland County mentions that Frederick "takes great interest in the grand work of Booker T. Washington and is associated with that famous educator in his efforts to establish the National Negro Business League." Booker T. Washington was born into slavery and received no formal education, but in the 1870s, he developed the Tuskegee Institute in Alabama as a place of higher learning for Black Americans. Through his good business sense and ability to raise substantial capital, he took Tuskegee from a one-room school to a multimillion-dollar educational institution. Washington became a person who "by his personal influence…was a tremendous factor in the Black south by 1900." W. E. B. Dubois stated, "the most striking thing in the history of the American Negro since 1876 is the ascendancy of Booker T. Washington."[48] The following image shows Frederick with

PRESIDENTS OF NATIONAL NEGRO BUSINESS LEAGUE.
(1) BOOKER T. WASHINGTON, PRESIDENT. (2) CHAS. BANKS, MOUND BAYOU, MISS., 1ST VICE-PRESIDENT. (3) FRED D. PATTERSON, GREENFIELD, O., 2D VICE-PRESIDENT. (4) S. G. ELBERT, M. D., WILMINGTON, DEL., 3D VICE-PRESIDENT. (5) HARRY T. PRATT, BALTIMORE, MD., 4TH VICE-PRESIDENT. (6) J. A. LANKFORD, WASHINGTON, D. C., 5TH VICE-PRESIDENT.

Photograph of the Presidents of the National Negro Business League in 1907. The caption identifies all the men in the photograph to include Frederick D. Patterson and Booker T. Washington. (Booker T. Washington, *The Negro in Business* (New York: Boston, Hertel, & Jenkins, 1907)).

Booker T. Washington and other early leaders of the National Negro Business League.

The main purpose of the League, as described at the top of their letterhead, was "to promote the commercial and financial development of the Negro," through measures such as patronizing each other's businesses to keep revenue within the Black community. A speech made by Frederick at the 6th Annual Meeting in 1905 showed that he was in favor of expanding this idea. He stated, "it won't do to depend on your own people to *push* your business, but you must conduct it upon such business principles as will enable you to draw trade from the general public and, competing with any other man in a similar line of business, resting your claim for patronage upon the merit of the goods to be sold or the services to be rendered."[49] Although he favored marketing his products to all buyers, Black or White, Frederick apparently had no problem in taking the opportunity to sell to his "own people," as there are several instances of him hustling business at the meetings of the League.

Frederick also favored making an organizational effort to increase interstate trade between the Black owned businesses. He wanted to form a group within the League to encourage "the development of interstate traffic and commerce; in other words, to see if we cannot be of mutual help in advertising and extending each other's business throughout the various states." Frederick also commended the League in 1907 for how the annual meetings had evolved since the initial meetings. He stated, "it seems clear that we are all getting the correct idea that these League meetings should be largely educative in character and not merely an opportunity to boast of what we have done…men and women come to this convention from all over the country to learn something and to be benefited by the discussion of their various subjects."[50]

While serving as 4th Vice President of the League, Frederick spoke at the 1904 meeting held in Indianapolis. During the speech he covered a topic regarding making and selling carriag-

Tomlison Hall in Indianapolis where Frederick gave speech at the 1904 meeting of the Negro Business League. (public domain)

President Roosevelt giving speech at the 1910 conference of the Negro Business League. (public domain)

Frederick shown with Booker T. Washington and other officers of the Negro Business League in 1905. (public domain)

(next page) Large group photo of the attendees of the 1909 Negro Business League conference in Louisville, Kentucky. Frederick is several rows up behind Booker T. Washington. (public domain)

es. "His remarks elicited a great deal of interest and a lively discussion ensued."[51]

In introducing Frederick as a speaker at the 6th Annual Meeting in New York in 1905, Booker T. Washington jokingly added "Mr. Patterson has not only a reputation for building first-class carriages, buggies, and all sorts of vehicles, but he has an equally fine reputation for making short and good speeches (laughter from the audience)." Frederick started his speech by stating, "In response to the gentle reminder of our President to be brief, I will say to the President that I am sure I shall not forget my warning (laughter from the audience)."[52] Also for this same conference, President Roosevelt had penned a letter in support of the mission of the Negro Business League, which was read at the 1905 conference in New York. Roosevelt would later show enough support that he personally spoke at the 1910 conference.[53]

During this speech, Frederick stated that "our achievements along all lines in the past forty years shall be an [indicator] of still greater progress in the future." League members believed that even the smallest business venture was a step in the right direction for Blacks. In a 1902 speech, Booker T. Washington stated, "we want to learn the lesson of small things and small beginnings. We must not feel ourselves above the most humble occupation or the simple, humble beginning. If our vision is clear, our will strong, we will use the very obstacles that often seem to beset us as stepping-stones to a higher and more useful life."[54]

During Frederick's years with the National Negro Business League, he held several prominent positions for the organization. From 1904 to 1907, he served as one of the Vice-Presidents of the League. Frederick was named as the 4th VP of the NBL at the time of its incorporation through the State of New York on December 19, 1905. While the NBL was organized earlier in 1900, it was not officially incorporated until 1905, an action in which Frederick took part and the records of this transaction were found in the archives at Tuskegee University. From 1908 to

1909, he served as a member of the Executive Committee that helped to organize the meetings and handled other logistical tasks performed by the League. From 1910 to 1919, Frederick was not present at any of the meetings of the League, nor did he serve in any official capacity. The only position that he held through this time included serving as the President of the local chapter of the Negro Business League in Greenfield. Frederick was listed as a lifetime member of the National Negro Business League.[55]

With Frederick's strong involvement and high positions with the Negro Business League during the first decade of their existence, by today's standards it is clear that he would be considered as part of the "Tuskegee Machine," a term first coined by W.E.B. Du Bois. The Tuskegee Machine referred to "the financial control used over Black education, particularly, over Black newspapers and periodicals by Booker T. Washington. By 1904, Washington had successfully surrounded himself with what was called the 'Tuskegee Machine.' It enabled him to be influential in many political decisions and he became viewed as the key national advisor for the African American community. As a result of this media control, readers of Black newspapers and magazines rarely encountered materials unfavorable to Washington's philosophy of accommodation."[56]

W.E.B. Du Bois described the Tuskegee Machine in the following terms: "Tuskegee became the capital of the Negro nation. Negro newspapers were influenced and finally the oldest and largest was bought by White friends of Tuskegee. Most of the other papers found it to their advantage certainly not to oppose Mr. Washington, even if they did not wholly agree with him. Negroes who sought high positions groveled for his favor."[57] Many sources indicate that the Negro Business League, and therefore the Tuskegee Machine, favored Black businesses in the South. Those in the North were typically viewed as being part of the industrial complex rather than the agrarian based South. Frederick's strong involvement seems a bit out of place, however, he may have been viewed by Washington as an ally that could help to tie the South and North Black businesses together. Whatever Frederick's original reasons were for his involvement with the Negro Business League, his ties abruptly halted in 1910 except for some local community involvement at the Greenfield chapter of the league. This likely is related to the death of C.R. and Frederick's fully taking control of the company dominating his time, however, there may have been disagreements between Frederick and the direction that the league was heading that caused him to cut his ties. There is no evidence of Frederick communicating with Washington after 1909.

Frederick's Involvement in Political Affairs

It is unclear exactly how Frederick became involved in politics. There may have been some influence from his father, C.R., who was involved with local county politics as early as 1872. In that year, C.R. was named as one of the delegates to represent Madison Township at the Republican County Convention. At that convention, C.R. was named to the Republican Central Committee of Highland County. This was repeated in 1879.[58]

While at Ohio State University, Frederick stood up at an 1891 meeting of the Horton Literary Society and supported McKinley to become the next President of the United States. He gave a lengthy speech about Free Coinage and the Tariff, which were big issues at the time, and explained how McKinley would deal with these situations. Frederick was one of 5 students selected by the McKinley Club to represent OSU at the Intercollegiate Convention of Republican Clubs in May of 1892. Issues that arose regarding travel prevented the group from attending. A 1902 history of Highland County mentions that Frederick was a delegate for the Republican State convention in 1902 that was held in Springfield. The historian had this to say about Frederick: "He is looked upon as a rising young man, whose popularity aided by his marked ability promises for him high honors in the ranks of his party."[59]

Also in 1902, Frederick spoke at a meeting of the Foraker Club of Greenfield that supported Ohio's Republican Senator Foraker. The newspaper article that announced that he would be speaking mentioned, "It is hardly necessary to ask that a large crowd be present to hear Mr. Patterson, who has on previous occasions delighted the citizens of Greenfield and other towns with his eloquence. The citizens of Greenfield know what command of language he has, the strong grasp he has on subjects of widespread interest, the clear, incisive manner of delivery, the felicity of expression – all these combined making Mr. Patterson a popular speaker."[60]

The newspaper edition that followed his speech printed its full text. The topic of the speech covered "the relation of the Republican Party to the Negro." Frederick said that the party has a lot of relevance to the Negro. "The Republican Party claims to have sprung from a positive declaration of principles, and one of these principles was an inborn hatred of slavery and it never stopped until this principle was put into practice and the shackles fell from the forms of ten million slaves." Patterson also said, "The Republican Party gave the Negro liberty because it thought that liberty was the right of every man Black, Yellow, or White and the Republican Party is ready to defend this principle at all hazards." "As Senator Foraker says," according to Frederick, "there has come a time when there is no Black, no White, no rich, no poor, as long as he is a man. Let any Negro show ability and he will be respected."[61]

In 1902 and 1903, Frederick served as a City Councilman for Greenfield. During his very first month as a Councilman, Frederick was already making arguments on behalf of the people, particularly fighting for those improvements that would benefit both families and local businesses. A request was made from E.L. McClain Manufacturing to extend Fourth Street so that 25 new houses could immediately be built for 25 new families coming to town to work at the McClain factory. Regarding Mr. McClain's request, Frederick took to the floor and said: "We have credit, and to my mind, it behooves us to borrow for the purpose of making Improvements which will draw people to Greenfield. Twenty-five extra families will benefit Mr. Head and Mr. Konnecker from whom they will get coal, Mr. Styerwalt from whose mill they will get flour. They will benefit all of us. Why then should we feel chary about borrowing a little money to insure the settlement of twenty-five families who will bring immediate benefit to our merchants. It is true their properties will not go on the tax list until next year, but they will begin to spend their money at once for the necessaries of life."[62] This attitude of working for the good of the town and its people was what got Frederick elected in the first place.

In September of 1896, Frederick was the Chairman of a meeting involving Republican nominee for Prosecuting Attorney for Highland County. In 1901, Frederick was named one of two special inspectors that would audit the Highland County Treasury account books. This showed that the Pattersons were viewed for their trustworthiness, even in public matters. Frederick was paid $30 for his services in performing this task. These instances, along with being a member of the Foraker Club, seems to have been the extent of his local political affiliations.[63]

He did have larger political ties as well. One source claimed that Frederick was at one time offered the Presidency of the country of Liberia by U. S. President Warren G. Harding.[64] Although the U. S. Government was highly involved in the affairs of the Liberian Government during this time, the Presidency of Liberia was decided by free elections. The United States sent presidential advisors to Liberia and this position had at one time been offered to Booker T. Washington. If there is any truth to this claim about Frederick at all, it is likely that Booker T. Washington may have recommended him to President Harding for the advisor position.

Frederick had ties to Warren G. Harding in other ways as well. Leading up to the 1920 Presidential election, the Black contingent in Ohio pushed for Frederick Patterson to be one of Harding's Delegates-at-large at the Republican National Convention. The *Greenfield Republican* stated that only one Black man had been selected to represent the entire State of Ohio. This improved the chances of that one man being chosen, owing to the full support of Ohio's Blacks going toward that single representative. A Cleveland newspaperman by the name of Ralph W. Tyler had pushed hard for Frederick to be that Black representative. Tyler had earned considerable respect among journalists as he was the only Black war correspondent that was allowed to join the forces in Europe during World War I and his focus was to report on Black contributions to the war effort – something largely overlooked by the mainstream correspondents. After the war, Tyler became editor of the Cleveland Advocate. He was also involved with the National Negro Business League during Frederick's time with the Negro Business League. Tyler was the older brother of Julius Tyler, who had become the second Black student to play football at OSU six years after Frederick. Despite the efforts of Tyler and other Black leaders in Ohio to get Harding to accept Frederick, Harding refused and stated that he would not give in to pressure if he felt that there was a better person for the job. After this, Tyler turned his attention toward another Republican candidate for the Presidency, General Leonard Wood. Tyler sent out letters to Ohio Blacks describing "how the Harding managers flatly refused to recognize the race's plea for representation, but the Wood forces did, and left a place vacant on their ticket for Patterson."[65] A Wood-Patterson Club was formed among Ohio Blacks in efforts to "fulfill the long-time hope of the race for a voting representation in a national convention." Although Wood did not receive the nomination, which went to Harding, Frederick did have his chance once again in 1924, when he was an alternate delegate-at-large for the Ohio Republican Party at the 1924 Republican National Convention.[66]

Frederick was involved in the 50th anniversary of President Lincoln's Emancipation Proclamation, a matter of national scale. A bill introduced and passed by the 63rd Congress of the United States in 1914 required a national celebration to be planned to "illustrate the history, progress, and present condition of the Negro race, and to celebrate the fiftieth anniversary of the proclamation of emancipation by President Lincoln." A large organizing committee was appointed to develop the program and details of the celebration and Frederick received a request to be on the advisory committee, which were chosen of "the line of the leading men of this country, including both the White and colored races, to make this a national exposition and that it should not only be an exposition for statistical information, but be in every particular a practical demonstration of the Negroes ability and achievement in every phase of human endeavor."[67] It was an honor for Frederick to be chosen as part of this group of advisors. It shows that he had made quite a name for himself even before he had built his first automobile the following year.

Much of the focus has been on Frederick's story through the years since more is known about him. This stems from the extensive media coverage of his public involvement and his business exploits, such as building an automobile. It has been a natural tendency for writers to focus on Frederick. This story began much earlier with his father, C.R. Patterson. The Patterson Company operated for 45 years in the carriage industry under his father before Frederick took over. This story starts at the beginning with C.R. Patterson entering the carriage business in Greenfield.

CHAPTER 1 NOTES

[1] The traditional story has been told over the last few decades of how C.R. was an escaped slave that made it to Ohio before the Civil War. Others have claimed that he arrived just after the war in 1865. With the increased amount of information on the internet and social media, this false narrative has been shared repeatedly in recent years. See Reginald Larrie, "From Slave to Auto Manufacturer: The Black Family Who Built Cars," *Ward's Auto World* Vol. 17, No. 4 (1981): 60; Bill Naugton, "Former Slave, Descendents Were Early Vehicle Pioneers in Ohio," *Tri-State Trader* July 5, 1980. We now know that C.R. was born as a free person of color.

[2] 1850 United States Federal Census; 1860 United States Federal Census; 1870 United States Federal Census. Other documents that support that the entire Patterson family was present in Greenfield include: Charles B. Galbreath, *History of Ohio* (Chicago: The American Historical Society, 1925), 242; *Journal of the National Medical Association* (hereafter shown as *JNMA*), Vol. 6, No. 4 (1914): 268; J. W. Klise, *The County of Highland: A History of Highland County, Ohio, From the Earliest Days, with Special Chapters on the Bench and Bar, Medical Profession, Educational Development, Industry and Agriculture, and Biographical Sketches* (Madison, Wisconsin: Northwestern Historical Association, 1902), 428; Wayne L. Snider, *All in the Same Spaceship: Portions of American Negro History Illustrated in Highland County, Ohio, U.S.A.* (New York: Vantage Press, 1974), 29.

[3] An additional argument that the Patterson family was free, rather than escaping from slavery, is the inclusion of their names in the U. S. Census. The Fugitive Slave Act of 1850 required the return of any escaped slaves to their owners. It seems likely that any escaped slave would avoid being in the public eye, such as including their name (and whereabouts for that matter) in a census document. Also in support of this is the fact that the cen-

sus was for "free inhabitants" as these words are included at the top of the 1850 and 1860 United States Census. In the southern states, a separate type of record was kept for slaves since they were considered property. Also, when John, C.R.'s brother, enlisted in the Army for service during the Civil War, he enlisted as a "free Negro" according to records from Highland County. In the larger scheme of things, it doesn't really matter whether or not the family were free or had come from slavery. As indicated in the Introduction, free people of color, especially those living in the South, were in no way considered equals to White citizens. All of their successes that followed their arrival in Greenfield are still a reason for celebrating this family and their achievements. Most people that came from a similar background did not enjoy a life of success.

[4] Highland County Property Deeds, Vol. 30, Page 393, January 1862; Ibid., Vol. 33 Page 448, June 1865; Highland County Mortgage Record, Vol. 5, Page 356, June 1865; Highland County Probate Court Marriage Records, Vol. 7, Page 824. Available at Highland County, Ohio, courthouse. There are also records of C.R.'s family being in Greenfield prior to 1865. See David McBride, *Personal Property Taxpayers of Highland County, Ohio* (Hillsboro, Ohio: D. N. McBride, 1980); David McBride, *Common Pleas Court Records of Highland County, Ohio, 1805-1860* (Hillsboro, Ohio: Southern Ohio Genealogical Society, 1959), 249; Snider, *Same Spaceship,* 99; Ibid., 136; Highland County Common Probate Court, Case No. 1432, January 15, 1901. Available at Highland County, Ohio, courthouse.

[5] William T. Turner, *An Unusual Man for his Time: The Life of Peter Postell* (Hopkinsville, Kentucky: The Athenaeum Society, 2016), 2-7; *Hopkinsville Kentuckian,* "Richest Colored Man in Kentucky Passes Away," May 24, 1901; *Tacoma Daily Ledger,* "Rich Negro Dead," May 23, 1901; *Hopkinsville Kentuckian,* "Peter Postell's Will: Leaves Everything to his Widow, Pauline Postell," June 4, 1901.

---

[6] USDA, *List of Workers in Subjects Pertaining ot Agriculture in State Agricultural Colleges and Experiment Stations* (USDA, 1925).
[7] Kathleen L. Patterson, *It's Been a Wonderful Life: One Day at a Time* (Dayton, Ohio: Self-published, 2001).
[8] There were 221 properties listed and the range of real estate values was $100 to $40,000, with the mode being $1,000, so the numbers are slightly skewed owing to some large values (17 properties at $10,000 or higher). Despite this, there were only 50 people listed that had a higher property value than C.R. This placed C.R.'s real estate value in the top 22.6% of Greenfield residents at that time.
[9] As far as personal property value, C.R.'s value of $1,450 was slightly lower than the average of $1,631. The total number reported was 359, a higher total number than for real estate values since this included values for non-property owners as well. The range of these values was from $100 to $50,000, which once again, throws off the average owing to 11 values placed at $10,000 or more. The mode of personal property value was $100. This places C.R.'s personal property value in the top 22.8%.
[10] The number of Blacks that reported real estate ownership was low, only 18 total. The range of real estate values was $100 to $3,500 with an average of $836. The mode was only $400. C.R.'s real estate value was $3,500, which placed him first among Black property owners with the next closest value being at $2,000.
[11] A total of 35 Blacks reported the value of personal property. The range of personal property values was $100 to $1,450 with an average of $233. The mode for this was only $100.
[12] There were 11 entries that included a real estate value. The range of values was $600 to $3,500 with an average of $1,581. The mode for these values was $1,500.
[13] There were 20 total entries for personal property values with an average of $844. The range for these values was $100 to

---

$3,675. The mode was $100, $200, and $350, all relatively low values. Unfortunately, census records changed their format and later versions did not show the real estate and personal property values. This eliminated the use of these records to create a consistent index to wealth status for each decade.

[14] Including transfers of property deeds between family members through probate and other actions, this number increased to 57.

[15] Monroe N. Work, "The Negro in Business and the Professions," *Annals of the American Academy of Political and Social Science,* Vol. 140 (1928): 144.

[16] Charlotte Pack, *Time Travels: 200 Years of Highland County History Told Through Diaries, Letters, Stories, and Photos* (Fayetteville, Ohio: Chatfield Publishing Company, 2007), 182-183.

[17] Ibid., 183; 186; 185.

[18] Snider, *Same Spaceship,* 92; 91.

[19] Lisa Powell, "Wilberforce University has Place in History Books for its Influence on Black Culture, Education," *Dayton Daily News*, February 3, 2021.

[20] Ibid.

[21] Harold Schmidt, "State of Ohio on Relation of C.R. Patterson vs The Board of Education of the Incorporated Village of Greenfield, Ohio, and W. G. Moler as Superintendent," Greenfield Historical Society, http://www.greenfieldhistoricalsociety.org/PattersonvsBd.pdf (Accessed December 3, 2008).

[22] Ibid.

[23] Ibid.; Lapchick, *100 Pioneers,* 360; *Courier (Cleveland),* "Negro Family Made 'First' Cars," December 10, 1965.

[24] Frederick D. Patterson vertical file, Ohio State University Archives; Lapchick, *100 Pioneers,* 359; Ibid, 361; Bob Hunter, "Athletics Not Always an Open Field," *Columbus Dispatch* January 29, 1992.

---

[25] Dan P. McQuigg, *Days of Yore: The Men of Scarlet and Gray*, 2020; Robert J. Roman, *Ohio State Football: The Forgotten Dawn*, 2016.
[26] McQuigg, *Days of Yore*; Roman, *Ohio State Football*; The Lantern, "Football: O.S.U. vs. Kenyon," December 12, 1890; The Lantern, "Football," October 22, 1891.
[27] McQuigg, *Days of Yore*; Roman, *Ohio State Football*; *The Lantern*, "Athletics," April 14, 1892.
[28] McQuigg, *Days of Yore*; Roman, *Ohio State Football.*
[29] *Makio*, Ohio State University Yearbook, 1891: 68; Ibid., 1892: 58, 128, 53.
[30] McQuigg, *Days of Yore.*
[31] *The Lantern,* "Horton Anniversary," February 11, 1892; *The Lantern,* "Horton Anniversary," February 18, 1892; *The Lantern,* "Horton," April 25, 1890; *The Lantern,* "Horton," November 21, 1890.
[32] *The Lantern*, "The Joy at Home," June 16, 1892.
[33] *Makio*, Ohio State University Yearbook, 1891.
[34] *The Lantern,* "A Word From Pat," December 13, 1892. Ohio State University Student Newspaper.
[35] Michael De Bonis, "First Black Ohio State Graduate Dedicated his Life to Education," *WOSU NPR News*, June 2, 2022.
[36] *The Lantern,* September 20, 1892; *The Lantern,* "Our Alumni," January 23, 1895.
[37] Klise, *The County of Highland,* 428; Galbreath, *History of Ohio,* 243.
[38] Ibid.; Ohio State University, "ICY IOTA PSI History," Ohio State University, http://ques.org.ohio-state.edu/icy_iota_psi_history.htm (Accessed April 17, 2009); *The Lantern*, "Winners in Boxing Contests Announced," December 11, 1925.
[39] Ohio State University, *Register of Graduates and Members of the Ohio State University Association, 1878-1917* (Columbus, Ohio: Ohio State Unviersity, 1917), 251; Charles W. Napper,

---

"Occurrence of Carbonaceous Material in the Greenfield Member of the Monroe Formation," *The Knowledge Bank at O.S.U.* Vol. 16, No. 4 (1916); Charles W. Napper, "Concretionary Forms in the Greenfield Limestone," (paper presented for the Geological Section of the Ohio Academy of Science Meeting at Columbus, Ohio, on April 7, 1917); Ohio State University Association, *The Ohio State University Monthly* February (1913): 33.
[40] Williams, *History of Ross and Highland,* 424; *Greenfield Republican* (hereafter shown as *GR,* "A. M. E. Church," January 9, 1908.
[41] *GR,* "Business Men Close," December 2, 1915; *GR,* "A. M. E. Church Notes: Movement Started to Fix Up the Church Building – Women Organize to Assist," March 1, 1917; *GR,* "A. M. E. Distric Conference Here," May 6, 1926.
[42] Patrick McSherry, "The History of the 3rd Ohio Volunteer Infantry," website accessed May 15, 2022, https://www.spanamwar.com/3rdohio
[43] *GR,* "Proud Old Greenfield: Property Owners Come Bravely to the Battle Front Demanding Substantial Street Improvement," February 18, 1915; *GR,* Greenfield Commercial Club Advertisement, May 3, 1917; *GR,* "To Form Company: Colored Men of City Taking Steps to Form Company Similar to the Home Guards," May 16, 1918; *GR,* Patterson sponsored advertisement for Liberty Bonds, April 25, 1918.
[44] *GR*, "A Good Game," October 8, 1908.
[45] *Greenfield Independent,* "Addressed Colored Alumni," June 2, 1921; *GR,* "Emancipation Anniversary Will Be Observed by the Colored People of Greenfield and Vicinity," September 20, 1917; *GR,* "Emancipation Day Will Be Fittingly Observed in Greenfield Next Monday, September 22," September 18, 1919; *Greenfield Independent,* "Emancipation Day: Fifty-Eighth Anniversary of Lincoln's Proclamation Appropriately Celebrated," September 23, 1920.
[46] William Hartwell Parham, *An Official History of the Most Worshipful Grand Lodge Free and Accepted Masons for the*

State of Ohio (Columbus, Ohio: Grand Lodge for the State of Ohio, 1906), 41; Ibid., 65.
[47] Joseph Mason Andrew Cox, *Great Black Men of Masonry, 1723-1982* (Bronx, New York: Blue Diamond Press, 1982), 146; *GR,* "Will Lay Corner Stone," April 27, 1922.
[48] Klise, *The County of Highland,* 429; John Howard Burrows, "The Necessity of Myth: A History of the National Negro Business League, 1900-1945" (PhD diss., Auburn University, 1977), 56; Ibid., 36.
[49] Records of the National Negro Business League (microfilm), compiled by Kenneth M. Hamilton, (Bethesda, Maryland: University Publications of America, 1994). 6th Annual Meeting 1905:152.
[50] Ibid., 1907:62, 117.
[51] Records of the National Negro Business League (microfilm), compiled by Kenneth M. Hamilton, (Bethesda, Maryland: University Publications of America, 1994). 5th Annual Meeting 1905:24.
[52] Ibid., 1905:26.
[53] *The Hocking Sentinel* (Logan, Ohio), August 17, 1905.
[54] Ibid., 1905:26; Mifflin Wistar Gibbs, *Shadow of Light: An Autobiography with Reminiscences of the Last and Present Century* (Washington D. C., 1902), 322-323.
[55] Records of the National Negro Business League (microfilm), compiled by Kenneth M. Hamilton, (Bethesda, Maryland: University Publications of America, 1994).
[56] University of Massachusetts Amherst, http://scua.library.umass.edu/exhibits/dubois/page6.htm
[57] Ibid.
[58] *The Highland Weekly News*, "Delegates to the Republican County Convention," August 1, 1872; *The Highland Weekly News*, "Republican Central Committee of Highland County," August 22, 1872; *The Highland Weekly News*, "Republican County Convention," May 29, 1879;

---

[59] *The Lantern,* October 7, 1891. Ohio State University Student Newspaper; *The Lantern,* May 19, 1892. Klise, *The County of Highland,* 429.
[60] Ibid., 429; *GR,* "Patterson to Speak," February 20, 1902.
[61] *GR,* "Relation of Republican Party to the Negro Ably Discussed by Fred Patterson Before Foraker Club," March 13, 1902.
[62] *GR*, "Council," May 29, 1902.
[63] *The News Herald*, "Treasury Inspectors Certificate," September 10, 1896; *The News Herald*, May 2, 1901.
[64] Ohio State University Association, *Who's Who in the Ohio State University Association* (Columbus, Ohio: Ohio State University Association, 1912), 187; Naughton, "Former Slave," 1.
[65] Randolph C. Downes, *The Rise of Warren Gamaliel Harding, 1865-1920* (Columbus, Ohio: Ohio State University Press, 1970), 386; *GR,* "Patterson Candidate: Greenfield Man Agreed Upon as Republican Leader of his Race for Big Four Delegate," February 19, 1920; Downes, *Rise of Harding,* 387; Bob Hunter, "Athletics Not Always an Open Field," *Columbus Dispatch* January 29, 1992; *GR,* "Emancipation Day Will Be Fittingly Observed in Greenfield Next Monday, September 22," September 18, 1919.
[66] Downes, *Rise of Harding,* 387; *Cleveland Advocate,* "Ohioans Unite to Boost F. D. Patterson," March 13, 1920; William E. Myers, *Book of the Republican National Convention, Cleveland, Ohio, June 10th, 1924* (Cleveland, Ohio: Allied Printing, 1924).
[67] U. S. Committee on Industrial Arts and Expositions, *Celebration of the Semicentennial Anniversary of the Act of Emancipation* (Washington D. C.: Government Printing Office, 1914), 1. (Bill HR15733 presented to the 63rd United States Congress on May 27, 1914); Ibid., 6-7.

2

C.R. PATTERSON ENTERS THE CARRIAGE BUSINESS

The Carriage Industry in the United States

Up to the mid-19th century, the work of carriage making was accomplished by the efforts of several specialized artisans. These workers built the parts of a carriage by hand, and they were highly skilled at their craft. Many skilled workers were needed to build a carriage including woodworkers, blacksmiths, wheelwrights, trimmers, leather workers, and painters.[1] The quality of carriages was high which often made the prices of the vehicles too high for the average customer to purchase. As the 19th century moved forward the gradual development and use of machines to aid in making carriage parts allowed the prices to be reduced.

The introduction of machines in carriage making allowed specialty shops to produce specific products in mass quantities at cheaper prices.[2] Examples of this would include shops that specialized in building only wheels, axles, or springs for carriages. This allowed other carriage shops to assemble carriages without making all the parts. The smaller shops did not have to purchase expensive machinery to produce those parts, nor did they have to pay skilled craftsman to make them by hand. It was cheaper to buy these parts from the specialty shops in bulk.

As the use of machines expanded rapidly during the last quarter of the 19th century, larger factories began to produce carriages in high production numbers. These large factories purchased the machinery to produce all the specialized parts of a carriage and utilized assembly methods that allowed mass production of the vehicles. The use of machines enabled the rapid and consistent production of parts, which allowed carriages to be assembled quickly. One example of a large carriage factory was the Studebaker Brothers of South Bend, Indiana. In 1874, after just a few years of mechanizing their factory, they produced 11,050 carriages and employed 550 workers. As growth and development of the company continued, by 1895 they managed to turn out 75,000 vehicles yearly, producing one every 6 minutes, and employed around 1,900. The carriage industry was one of the first to utilize this type of mechanized vertical integration. This, in turn, allowed the production of carriages to become cheaper for the large factories and they were able to sell their products at a price much lower than before. In some of the larger factories in Cincinnati, the main center for carriage production in the United States, the price dropped so low that one could purchase "three for a hundred" dollars by 1900.[3] These prices made the carriage affordable for anyone with a horse, and it was no longer a luxury reserved only for the wealthy.

The carriages produced in the large factories were cheaper in price; often they were also viewed as being cheaper in quality. While most factories still managed to produce high quality products through mechanization, others used cheaper materials in conjunction with mechanization to keep their prices as low as possible to create high numbers of sales. This meant that the days of the smaller custom shop were not completely over. While these small shops did not mass-produce vehicles like the large factories, they participated in custom and batch production. Batch production is described as turning out "small groups of similar or identical goods to order or in anticipation of demand. It shares some of mass production's efficiency but, like custom work, fills specific orders…Together referred to as flexible production, both custom and batch manufacture shares the virtue of flexibility as a means of coping with fluctuating demand and the vagaries of fashion."[4]

Business continued for many small shops in the United States because of various factors. Although "industrial potential did provide new opportunities, the resilience of long-standing practices…particularly, the shop economy that dominated the first half of the century, remained strong." Granted, business was likely much slower for the small shop at this point, but those customers who wanted and could afford a custom-made vehicle could still purchase one from these shops. "In a way, one could argue that a mechanic's reputation for craftsmanship was a form of compensation. His skills and personal attention would assure the finest work and accommodate all demands at reasonable prices."[5] The skilled craftsman cared much more about the products that they produced and took pride in their products – while many times the factory workman was there for a paycheck. This is one of the reasons that many small carriage makers were able to stay in business even after the process of carriage building was mechanized.

By the turn of the century, there was room for all three types of manufacturers in the carriage industry: the specialty factory, the vertically integrated factory, and the small traditional shop. This diversity gave customers a choice of which type of vehicle they wanted to purchase. Another factor that led to the continued success of some smaller shops had a lot to do with location. Those shops in rural, agrarian settings typically did not have to compete with the large urban factories and could occupy a middle ground between the blacksmith that occasionally turned out a vehicle and the vertically integrated factories in the city.[6] These small shops continued to operate into the 20th century, although on a much smaller scale than the large urban manufacturers. They continued by filling the specific vehicle demand in their local rural market. The Patterson Company was in a rural community, and it continued to serve its customers in the same tradition as it had for several years.

The C.R. Patterson and Sons Company never turned to large-scale mechanization to produce its parts, but rather relied on hand tools and, later, small electrically powered tools to accomplish the task. The Pattersons preferred to remain a custom shop that produced high quality vehicles and provided expert, personalized service to its customers. According to historian Horace Greeley "some of the best examples of American carriage building are afforded by men who have risen from obscurity and poverty to wealth, success, and reputation by their own energy, industry, and intelligence."[7] C.R. Patterson overcame adversity to become a successful businessman in the carriage industry.

The Beginnings of the Patterson Company

The earliest records of the Patterson family show that they were mostly engaged in the blacksmith trade. According to the 1850 and 1860 United States Census, Charles, C.R.'s father, was a blacksmith in Greenfield. The 1860 Census also showed that William D. Patterson, C.R.'s brother, was an apprentice blacksmith in neighboring Fayette County, and another brother, John E. Patterson, had become a blacksmith by the time he enlisted for service during the Civil War. It was no surprise that C.R. worked as a blacksmith as well.

This was not uncommon for the time. Blacks frequently occupied the blacksmith trade during the antebellum period. This, along with other artisan trades such as carpentry and shoemaking, came "as a result of generations of support on the part of slave owners" and "these skills became concentrated almost exclusively in the hands of Blacks, whether slave or free."[8] Even during the post-Civil War period, owing to a lack of education, these were still popular trades for Blacks to enter. Blacks faced obstacles such as credit and legal difficulties when they tried to enter business in competition with Whites. Prejudice and other factors limited most Blacks to operating small business establishments, such as blacksmith or carpenter shops, and seldom were they able to start large businesses.

Not much is known about the earliest places where C.R. worked. By 1870, he served as a shop foreman at the Dines and Simpson Coach and Carriage Company in Greenfield.[9] He worked side by side with Whites, and several of these men worked under his supervision. During the time that C.R. was foreman, the firm of Dines and Simpson was considered successful, and the company received praise as outlined in an 1870 article in *The Highland Weekly News* (5/12/1870). The article stated "some of the finest carriages and buggies are manufactured here by the firm of Dines & Simpson and sold all over Southern Ohio and Kentucky. This firm turns out more fine work than any establishment (outside of Cincinnati) in Central or Southern Ohio. They have one of the finest carriage painters (Mr. Bush) in the State. They never fail to take the premium at our Fairs, over all other competitors."

Typical Dines and Simpson advertisement as shown in the Fayette County Herald on May 9, 1872.

This arrangement with Dines and Simpson apparently was not to his satisfaction, as he wanted to operate his own business. Owing to a lack of startup funds and prejudicial reasons, C.R. partnered with a White man, James (J.P.) Lowe, in 1873 and they started their own carriage company, known as Lowe and Patterson. This company "became noted for the expert craftsmanship that went into their products."[10]

C.R. Patterson claimed that his carriage business (or at least service) began in 1865. His partnership with J.P. Lowe began in 1873, but it is quite possible that he had been in other partnerships prior to this date with the Dines and Simpson Carriage Company and possibly others dating back to 1865. In a speech at the inaugural meeting of the National Negro Business League in 1900, C.R. spoke of how he had been in the carriage "business for fifty years in various partnerships" and "the older firm of Lowe and Patterson, once extensive, dwindled through business depression to such proportions that the formation of the present partnership was opportuned and consummated."[11] It is possible that C.R. had been in a partnership with Dines and Simpson, which would further explain the unquestioned acceptance of C.R. as a foreman at the time. Also, according to the obituary of Ed Dines, it mentions that Dines had been a partner with J.P. Lowe. There is a distinct possibility that Ed Dines, Joseph Simpson, Lowe, and C.R. had all been partners in the earlier company and then in 1873 Lowe and Patterson split off to create their own company. It is possible that Lowe and Patterson had bought out the Dines and Simpson shop since the two businesses worked out of the same location, where the later Patterson shop would also be located.

Even within the Lowe and Patterson Company, there were several changes of partnerships with only C.R. and J.P. Lowe being the constant partners. After 10 years in business together as Lowe and Patterson, a company name change to James P. Lowe and Company was announced in The Highland Weekly News on 1/10/1883. This appears to have been the inclusion of partners J.P. Custus and Charles Grassley. Apparently, Lowe was the major partner since his name is the only one included in the new name. The other partners did drop out of the business over the next decade, which left Lowe as the only remaining partner with C.R. in 1893 when C.R. bought out Lowe and took full control of the business. Despite the addition of more partners, the business apparently wasn't busy enough by the end of the year (1883) to carry them through the entire winter season. In early December, there was a notice in the newspaper indicating that the shops of J.P. Lowe & Co. would be closing until at least the beginning of the new year due to slowness of business.[12]

Information is lacking concerning the Lowe and Patterson and J.P. Lowe and Company during their time in business. Advertisements were not identified in newspapers of the period and there were few mentions of the company. However, the few mentions typically included praise for their work as a manufacturer of high-quality carriages. One reference to their company was a status of Greenfield carriage shops in 1880 that was mentioned in a Cincinnati newspaper. Why folks in Cincinnati, the "carriage capitol of the world" at the time, would care about smaller manufacturers in the small town of Greenfield is up for debate. However, the article indicated that for J.P. Lowe & Co., Mr. Lowe indicated that they were "doing more in the way of repairs than at any time since he has been in business. In connection with which he has new work enough to run at least half the season, and that the amount of work on hand will double that of last year at the same time in the season." The three other carriage shops that were visited in Greenfield had similar responses, indicating that the carriage industry was faring well in the community at that time.[13] Another newspaper mention of the company also occurred in 1880, but was not related to a positive outcome. Two carriage shops in Greenfield were burglarized in one night. The thief, or thieves, targeted specific high dollar items related to the carriage industry, which gave the impression that they knew exactly what items to target. The combined losses totaled approximately $200. It appears that the guilty party was never identified.[14]

Another negative event that happened to C.R. during this time did not appear to have any connection to the company but speaks to C.R.'s disposition. A revolver was pulled on C.R. in the office of the Harper House in Greenfield. The Harper House was an elegant hotel in town at that time. Apparently, there was some form of disagreement between C.R. and James Crocket, who according to the newspaper account "it seems seeks to make himself disagreeable, and Charles Patterson, one directly opposite in character, both colored. Crocket was under the influence of liquor and began abusing Patterson, who sought to absent himself from the room, but was detained at the point of a revolver until the proprietor took the weapon from Crocket." Crocket was arrested and was put under bond. Patterson did not want to press charges.[15]

According to Kate's handwritten history of the company, she stated, "My father, thinking of the future of his two young sons, acquired the business for himself." In 1893, C.R. bought out J.P. Lowe's share of the business and became sole owner. Many recent stories indicate that J.P. Lowe died and, with no sons to leave the company to, by default C.R. became owner of the company. This is incorrect as Lowe lived until February of 1928. Lowe left the company to become President of the local Home Building & Loan Company for many years. In fact, this would become an important point in the story as nearly half of the mortgages that the Pattersons received over the next couple of decades came from the HB&L Company bank. It is interesting that Lowe decided to leave in 1893 but was perhaps a result of the Financial Panic of 1893. Banking wasn't necessarily a more secure venture during a panic, but something prompted him to get out of the carriage business. Obviously, he and C.R. parted on good terms as evidenced by the continued relationship between the two. At one point, Lowe provided a mortgage to the Pattersons from the HB&L and, at the same time, one out of his own pocket.

According to the handwritten history of the company by C.R.'s youngest daughter, Kate, the original name of the newly formed company in 1893 was "C.R. Patterson, Carriage Builder". The newest partnership C.R. spoke of at the 1900 Negro Business League meeting occurred in 1899 when he added his sons Frederick and Samuel to the company.[16] At this time, the company name changed to C.R. Patterson and Sons Company.

C.R. went into the carriage business with the belief that he could compete with White businessmen. Like other prominent Black businessmen of the period, C.R. knew that by producing a product that "somebody else wants, whether it be a shoestring or a savings bank, the purchaser or patron will not trouble himself to ask who the seller is. Recognize this fundamental law of trade; add to it tact, good manners, a resolute will, a tireless capacity for work, and you will succeed in business." Although many Black leaders felt that "removing Negroes from politics and putting them into industry of business seemed both a justifiably humanitarian and politically wise course of action, it would certainly offer no threat to the real business community."[17] C.R. Patterson was not willing to just exist, he wanted his company to be noticed in the business community and be seen as a competitor.

C.R. had become quite the popular blacksmith in town. It has been said that he "could fashion almost any gadget of metal with his hammer and anvil. His sinewy arms were strong as iron bands, and he could build anything that rolled on wheels." A 1902 history stated that C.R. "is a mechanic of the very first order in his line, having no superior as a smith. He is besides an excellent businessman, sound in judgment and full of enterprise and push." His son Frederick, although being a college educated man, had this to say of his father at the 1901 meeting of the National Negro Business League: "A college education has nothing to do with making a successful businessman. Often [I] thought [my] father was wrong, but every time it was [I] who was wrong, and [my] father was right...[I]n carriage making, it takes a man of common sense, push, and hustle."[18] Evidently C.R. had become that man.

In that same 1901 meeting Frederick spoke of how large their business had become. He made it very clear that "their carriages are in demand only because of their superior quality." A 1914 article about the company stated that "from a small shop with one smith and one helper it has grown until 50,000 square feet of floor space with additional yards are inadequate to house this business." The article went on to say, "upon their character as men and their reputation for mechanical skill were laid the foundations of a business that has developed to country-wide extent."[19] With his good business sense, the failures of earlier partnerships were eliminated once C.R. took control of the company.

The C.R. Patterson and Sons Company became known for the high quality of its products. Even the sons, Frederick and Samuel, were recognized as being natural mechanics. "The sons learned the trade from the anvil on up and are practical men of actual experience." By 1902, the business was seen as being "far beyond the experiment stage. It is on a safe foundation, with capital behind it – and better still – a store of indomitable energy and unflagging perseverance."[20] The company became widely known for using good materials and standing behind its work, which made its vehicles popular across a range of states.

Patterson vehicles were extremely popular in the local area. According to Frederick in a 1903 *New York Times* article, "nearly 80 percent of the carriages built in his town came from his factory." The business did not remain only local. In 1910, when C.R. died, his obituary stated that "for many years [he] was a carriage builder and his factory has grown to the present large proportions and the carriages and buggies are sent into many states." His name was so well known in the carriage industry that in the month following his death, news of C.R.'s death was the first topic in national magazine *The Carriage Monthly* in May of 1910 under the "News in the Trade" section. Son Frederick had a lot to do with achieving widespread patronage. As he traveled to meetings of groups he belonged to, such as the National Negro Business League, he solicited business while he was there. One account in 1909 stated that he had been at Tuskegee Institute in Alabama for a meeting of the League, and by the time he returned, he had $5,000 worth of carriage orders.[21] These meetings were held in multiple cities and attracted Black businessmen from all over the nation. These men ordered carriages from the Patterson Company and had them delivered to their hometowns where they became rolling advertisements for the company.

Frederick was a firm believer in advertising to push the company's products. One article stated, "One good vehicle selling another, together with constant, consistent use of printer's ink, has caused this concern to have a remarkable growth." The company sent out monthly bulletins in the mail to the local area and created "handsome catalogues which have carried the name and fame of the Patterson vehicles throughout the length and breadth of Ohio, through Illinois, even past the portals of the 'White City,' [Chicago] down into the Blue Grass State and into sunny Tennessee."[22] By 1902, the Patterson Company placed weekly advertisements in local newspapers. They advertised in other areas as well. A staple of a vast majority of the Patterson advertisements was an image of Frederick pointing in the direction of something within the advertisement. This "pointing man" makes the Patterson advertisements easy to quickly recognize and stand out, especially in newspaper pages that were cluttered with advertisements for all types of businesses and products. A total of 1,202 Patterson advertisements were collected and analyzed during this study (see Appendix A for a breakdown of these advertisements, including the general contents of the ads). The company focused most of its advertisements on general consumers (90.5%), but they also targeted advertising toward select groups.

Frederick as "The Pointing Man," an image that was included in over 1,000 of the Patterson advertisements. (R.P.)

One interesting journal that the Pattersons advertised in was the *Journal of the National Medical Association*. Physicians were constantly making house calls during the period and therefore needed durable, high-quality carriages to take them from place to place. The Pattersons recognized this, and they directed advertisements towards physicians and made a special buggy for doctors. This journal was a bit of a "double dip" for the Pattersons. Black physicians founded the National Medical Association which operated out of Tuskegee Institute in Alabama. Black physicians had trouble publishing their work in White medical journals, so they began their own, the *Journal of the National Medical Association,* but clearly stated that it was open to both Black and White physicians. A Patterson ad in the journal stated, "a strictly Negro firm, yet we make more styles of physician's buggies and sell to more doctors than any other firm, White or Black, in the United States."[23] The Patterson Company was the first advertiser for the association and even placed ads in the annual meeting reports before the journal was ever published. Once the quarterly journal got its start, the Pattersons never missed an issue from 1909 into 1918. The Pattersons reached physicians through advertising in this journal and Black physicians, in turn, could patronize a Black owned business by purchasing from the Patterson Company.

Another group that was targeted by Patterson ads was the general Black community, but only in certain media. According to a 1914 article, the Pattersons "have always appreciated the business that has come to them through racial preference, but never have they permitted this to be the sole claim for patronage."[24] They did not want to provide products solely for Blacks; they wanted to compete and be recognized as being high-quality manufacturers by all consumers, White or Black. They avoided mentioning that they were "Negro owned" in local media, but in publications intended mostly for a Black audience, they almost always included this fact. The ads placed in the *Journal of the National Medical Association* contained this information, as did those published in *The Crisis*, a journal published by the

Portions of some of the Patterson advertisements that included mentions of them being a Negro owned company. (*JNMA* 1914 and 1915)

NAACP. All 73 advertisements collected from these two journals specified that the company was Negro owned.

An interesting observation of the Patterson ads between these two journals includes the selection of targeted audiences within the Black community according to economic class. While *The Crisis* was aimed at serving Blacks of all economic classes, the *Journal of the National Medical Association* was specific to Black physicians, which tended to be in an upper economic class. While still building carriages, the Patterson Company advertised in both journals simultaneously, but once the company shifted to producing the higher priced Patterson-Greenfield automobile, the advertisements stopped in *The Crisis*, but continued for the higher paid physicians. Although the Pattersons operated a Black-owned company, they still recognized that the average Black consumer at that time was of a lower income bracket and could not afford an automobile, so this group was no longer targeted in their advertisements.

Patterson advertisement in the *Journal of the National Medical Association* (1909), indicating that they were "A Strictly Negro Firm."

The company also aimed its products and services at farmers. Obviously, in a town like Greenfield, located in the middle of a farming community, the farmer was an important customer. Although farm machinery repair was included in many of the Patterson advertisements (344 ads [28.6%] mentioned other services besides the primary theme of the ad), 39 of the ads (3.3%) were directed specifically toward farmers. One Patterson ad discussed farming and stated, "it is our business to serve this army of workers." The company also placed several ads that supported American farmers and the importance of their work. Ads that stated, "in no business does individual effort bring larger returns than in the business of farming," and "business is business, but the business of successful farming is the greatest enterprise on earth."[25] The Pattersons placed advertisements in a statewide newspaper called *The Ohio Farmer* from time to time, which helped them reach more farmers across Ohio.

The Patterson Company often included testimonials in its advertisements. Sixty-eight ads (5.7%) included testimonials, ranging from brief statements to full-length letters from customers who had bought and were highly satisfied with the Patterson products. Many Patterson ads included the names, and sometimes places, of customers who had made purchases that week (see Appendix B for a sample of client locations mentioned in Patterson advertisements). From the 1,202 advertisements, a total of 699 customers were identified. Of these 699 identified customers, 168 (24%) did not specify a location. As expected, most customers came from Greenfield and other Ohio towns. The Greenfield total was 263 (37.6%) and other Ohio towns accounted for 216 (30.9%). Out of state customers totaled 50 (7.2%), while those from out of country amounted to 2 (0.3%). Sixteen different states were represented in the ads along with Puerto Rico. Most of the states were east of the Mississippi River, but some were as far west as Oklahoma and Texas. Advertisements provide only a small window into Patterson's total client base, but the percentages given likely reflect its normal distribution between locations.

The workers at the shop played an important role in the success of the company. The employees were a mix of both White and Black workers. The feelings toward the Patterson family must have been good considering Whites were willing to come and work for a Black owner. In a 1965 article, Kate, C.R.'s youngest daughter, stated "the great majority of our employees were White, and we never experienced any labor trouble." The Pattersons treated all employees fairly, regardless of race. As far as employment practices, a 1914 article stated that "their factory is operated on the single standard of merit and is open to all who have the determination to make good in any department."[26] During the 1907 meeting of the National Negro Business League, Frederick stated:

> In the operation of our plant we have found it difficult, at times, to get the kind of help that our standard product and the character of our trade demands. It is not a problem in my section [of the nation] to get colored and White men to work side by side; in some sections this is impossible – at least under present conditions. While it has always been the policy of our firm to employ colored artisans as far as possible, nevertheless we place a premium upon competency and skilled labor – which every rational manufacturer must do if he expects to meet the exacting competition that confronts him in America.[27]

On this point, though, it seems that their preferences were flexible. A series of "Help Wanted" ads that the company placed in the Black oriented journal, *The Crisis*, fluctuated on its requirements. In December of 1919, they advertised for a "first-class man. None other need apply." Only three months later in March, the ad changed to "colored men only," but by May it had changed to "colored preferred," but they would take any men as long as they were "young men of good habits, intelligent and fair education, and with desire and ambition to become experts." Frederick believed in teaching trades to Black men to help encourage industrial growth in the Black community. He told Black-owned companies that "those who are pioneers in various trades and various lines of business should lose no opportunity of teaching the same to colored apprentices, and thus promoting the industrial and commercial welfare of the race." A former White employee of the company stated in a 1992 article that Frederick was a "nice, kind man to work for."[28] To stay in business for so long utilizing a workforce of primarily White employees says a lot about the relations between the races at the company.

As the business developed over time, the number of employees fluctuated. In 1888, when the firm was still J. P. Lowe and Company, it employed only 10. After C.R. took full control of the company, the business grew, and he employed 35-50 men on average by 1900. The company employed 30 in 1910 and 35 by 1913. A 1914 article stated that the Patterson Company had "a large workforce of skilled mechanics with a yearly payroll in the ten-thousands" and were still "unable to promptly meet the demands made on the business." During the period 1915-1918, while building carriages and automobiles at the same time, the number of employees dwindled to 18 men and one woman by 1918. Even at this lower number of employees in 1918, this still placed them as having more employees than 71.8% of other companies in Ohio that year that were listed as manufacturers of carriages, wagons, and carriage materials.[29] Of 163 companies listed, only 46 (28.2%) had more employees than the Patterson Company, ranging from 21 to 494 employees (22 of these had 50 or more employees).[30] The number of employees fluctuated, but for the most part, it appears that the company typically employed around 30 workers.

Employing 35 to 50 workers around the turn of the century placed the Patterson Company as a major employer in Greenfield at the time. While TAPATCO and Waddell both employed large numbers, 250 and 100 respectively, the Patterson shop fell into the lineup as the next highest employers in the town. Both the W. I. Barr sawmill and the Columbian Manufacturing Company employed 30 workers at this time, which was comparable, but slightly lower than the Patterson Company's number of employees.[31] The company employed a significant number of residents in the town, and this may explain why it received considerable and favorable newspaper coverage, even though it was a Black-owned company. Their business was important to the town of Greenfield.

The Pattersons placed an advertisement in 1909 that stated, "we are going to have one of the largest industries in the United States." While displaying this confidence in their business, they did not manage to achieve this goal, but according to author Frank Mather in 1915, they did become the "largest plant owned by colored people in the United States." According to C.R. in a 1900 speech to the National Negro Business League, the company output had reached 400-600 vehicles per year. During this speech, he stated "while this by no means compares in magnitude with the large carriage concerns of the country, you will please observe that when it comes to a comparison with builders of strictly high-grade work it shows up fairly well."[32] Author Jack Salzman added that the company earned $75,000 annually at this time.[33]

Far from its humble beginnings as a small repair shop that built a few custom carriages, by 1900 the Patterson shop had become an establishment capable of both custom and batch production of vehicles. According to a Patterson advertisement in 1912, "the Patterson Buggy factory is simply a custom shop grown to large proportions, with every modern facility applied for the production of the best buggies and carriages that man can construct."[34] Many early articles, and the Pattersons themselves, referred to the Patterson shop as a factory. By today's definition, a factory mass-produced a standard line of products through means of mechanization. The Patterson shop did not do this, and therefore, would be considered a shop. They concentrated on custom and batch production, as well as offering many additional services on the side. Although the company had grown to significant proportions, they still prided themselves in

Photograph of Frederick standing at the front doorway of the Patterson shop, likely prior to 1915. (from *Courier* [Cleveland], December 10, 1965)

offering strictly custom-made vehicles, rather than mass-produced mail order carriages from a distant factory. Even with the standard vehicles that were typically kept in stock, a customer could request specific materials, accessories, or details and the Pattersons would construct it for them instead of pushing one on them that was already built. That was the advantage of the customer buying locally and having a more personal experience with the builder.

The shop went through several changes and modifications during the carriage building years. Like many small shops, they never quite felt that there was adequate space for working and storage. Not only were the vehicles themselves rather large, but "bulky, awkward items like shafts and poles, folding tops, wheels, and other impedimenta occupied a great deal of room. Similarly, while sufficient quantities of cloth, leather, paint, fasteners, and the like might occupy corners and closets, iron and lumber required dedicated storage space." With this in mind, the Patterson Company experienced several periods of expansion and new construction. The following illustrations show Sanborn maps of the period and changes for both the west and east buildings. An 1890 property deed stated that the original shop on the west side of North Washington Street had a total of 18,562 square feet. By 1900, business had increased greatly, and they expanded the shop to provide more room for workspace and as a storage repository. The Patterson Company purchased the property on the west side of North Washington Street and the total workspace increased to 30,000 square feet. A 1902 *Greenfield Republican* article stated, "as the years went by and business grew, larger quarters were found necessary, until now the factory and show rooms on Washington and Lafayette Streets are models of convenience and spaciousness." A May 1907 article in *The Carriage Monthly* provided news of the factory being enlarged to accommodate the large amount of business they were experiencing. It mentions that the Patterson Company is "the only carriage manufacturing company in the United States owned by colored men, and they are turning out over 300 new jobs per annum." In 1912, one more expansion of the shop occurred when the Pattersons constructed another three-story building that year on their property.[35] This building was the last to be built by the Patterson Company.

The Patterson Company also grew by buying out other carriage companies that were closing their doors. In 1910, the Pattersons bought out the Scioto Buggy Company of Columbus, Ohio, and the Troy Buggy Company of Troy, Ohio.[36] With these purchases, the Pattersons received all their stock of buggies and materials. The ability to make these purchases at such a low price allowed them to pass the savings on to their customers. These purchases came at a time when many carriage concerns were closing owing to the rise of the automobile industry. The Patterson Company was still hanging on and trying to make the best of the situation.

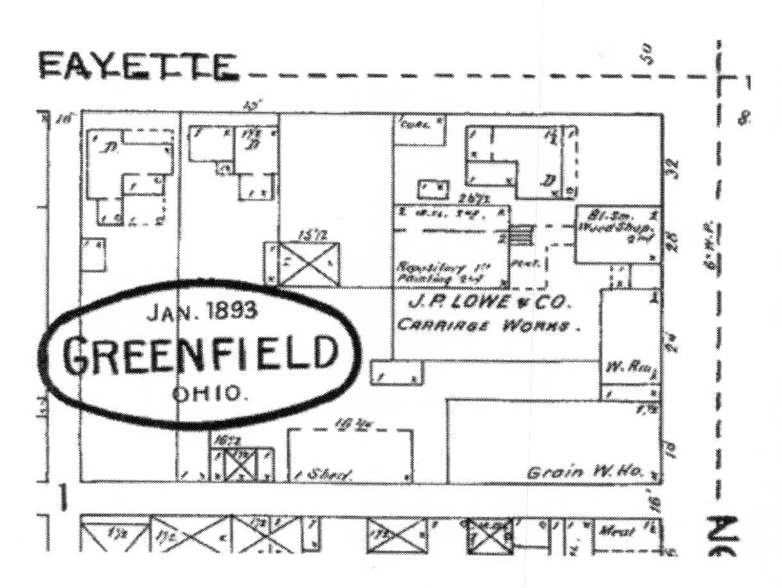

Sanborn Fire Insurance Map from 1893 showing the west building as it appeared under the partnership of C.R. Patterson and J. P. Lowe.

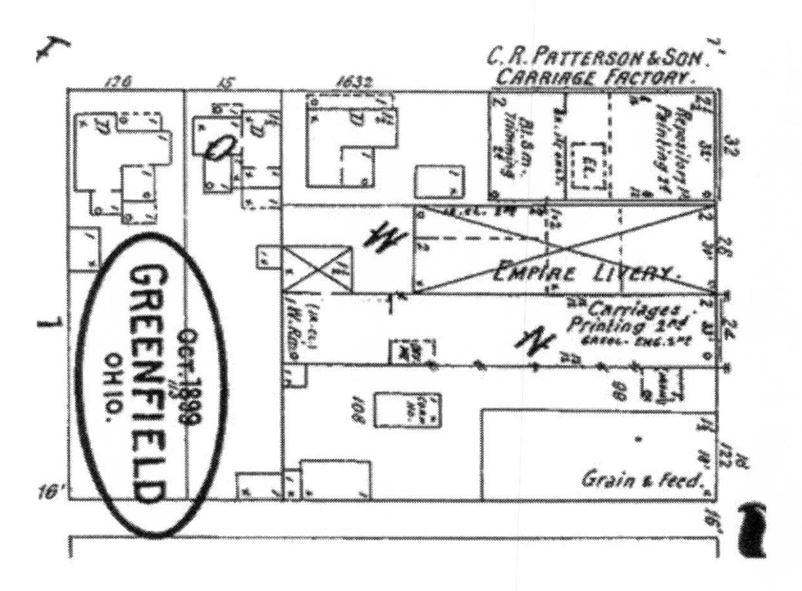

Sanborn map showing west building addition as of 1899.

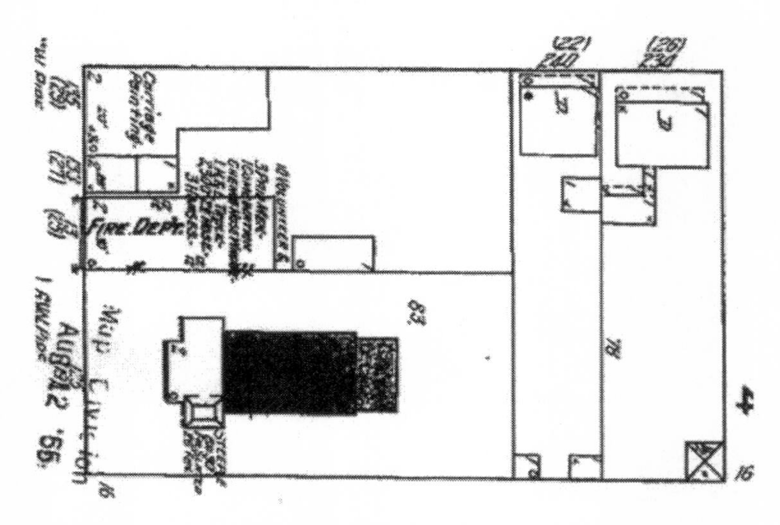

Sanborn map showing the east building as it appeared in 1910.

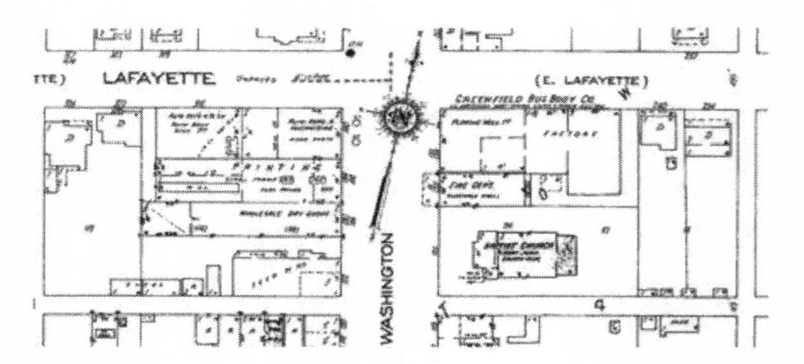

Sanborn map from 1927 showing both the east and west buildings and the additions and changes that had been made.

Photograph of the west building in 1899 or after. (R.P.)

Apparently, the rapid growth of the company caused some of the Patterson inventory to spill out onto the sidewalk at times, despite the large repositories that they claimed. A funny article appeared in *The Greenfield Republican* on June 18, 1914, stating that an odd edict had been made by Frederick. The article stated, "It isn't Fred's business to go about issuing edicts, but when he does the edicts have whiskers and command attention." Frederick said that the buggies that he can't keep in his repositories will line up in the street "at the curb where vehicles belong" and not on the sidewalks. "In other words, he's going to intrude nothing in the way of the pedestrian, and the pavement in front of the Patterson domain will be kept clean as a walrus tooth." The article continued, "he says he owes this not alone to himself, but to his neighbors in business and he's right – preeminently right." Some of the language in the article strikes a funny tone, and it is unclear why Fred would make such an edict – perhaps it came as the result of a complaint of the sidewalk being blocked and this was his method of public relations control, or if it was a bit of a passive-aggressive statement made to catch the attention of certain other businesses in town that tend to block sidewalks with their goods.

Not many details are known about the inside of the Patterson shops, but some information was gleaned from articles and advertisements. The Pattersons claimed to have the "most complete and thoroughly equipped custom repair shop in southern Ohio." They stated that not only did they have skilled workmen, but they also had the room and equipment to handle any problem at their "perfectly equipped plant." A 1914 article mentioned that the plant was powered by 8 electric motors that provided all-day power. A 1912 advertisement added that they had 10 men working in their blacksmith shop. Some tools in this shop included three forges, power drills, and an electric blower. In 1913, they advertised that they had added a gas brazing system that "does your work better and very much quicker. Come and see it work." A former employee stated in a 1992 interview that there had been "a bridge over the street connecting the two buildings" when he worked there in 1914 and 1915. He further said that five or six blacksmiths worked in the west building forging the metal parts while the wheels were assembled in the east building.[37]

While not much is known about the Patterson shops, it can be assumed that they had fully equipped blacksmith and woodworking shops with all the tools necessary for the purpose of building carriages. To better illustrate what a typical small shop would include, information was taken from Thomas Kinney's 2004 book, *The Carriage Trade,* which aids in visualizing how the Patterson shop may have been equipped. According to Kinney, four fundamental processes existed in carriage making: woodworking, metalworking, painting, and trimming. These departments are clearly labeled on the Sanborn maps of the Patterson property. In larger shops, each of these skills was kept in separate departments and independent of each other. This ensured a high level of skill in each department owing to the large amount of time required for workers to master more than one skill. Smaller rural shops, such as the Patterson Company, which were "far from the demands of the urban markets were more likely to train workmen capable of a broad range of tasks."[38] It is likely that several of the workmen at the Patterson Company worked on several different tasks throughout their employment with the company.

Each department would have its own special tools and environmental requirements. The blacksmith shop had to be well ventilated and have solid vibration-free floors to keep the hammer strikes as a dead blow instead of rebounding. Many times, the blacksmith shop was in a basement or on a dirt floor that would provide the proper sturdiness needed for anvil work. The room also needed sturdy racks and bins for the iron stock and finished parts that were being stored for later use. The blacksmith shop would be equipped with anvils and its associated hardies and other attachments, basic hammers, tongs, punches, grindstones, and vises, as well as the forges, power drills, and blowers mentioned previously that were known to be in the Patterson shop.[39]

The woodworking shop would be a spacious room with a large area free of obstructions. This room would have plenty of natural lighting and would be equipped with plenty of sawhorses and trestles. Tools would include basic woodworking tools such as saws (rip, crosscut, and bow), drawknives, spoke shaves, hand routers, and several types of planes (jack, smooth, joiner, rabbet, dado, and block).[40] It took several different types of tools for mortising, chiseling, paring, hollowing, and carving, so these were probably on hand as well. It is also likely that the Patterson shop kept the wheel making tools in this department as well since there is no distinctive labeling of a wheel shop on the Sanborn maps. There may have been some basic wheel making machinery in this department, such as a spoke tenoning machine, throating machine, and possibly a spoke driver. It took a lot of work and time to shape and tenon spokes, as well as boring hubs and chiseling out their mortises, so it is likely that the Patterson shop used at least some basic hand operated machines to accomplish this task. This would speed the process along and, at the same time, ensure a more consistent fit and interchangeability of parts.

The painting and trimming departments both required large spaces to work in as well. By the time the vehicle made it to this point, it was a rolling mechanical assembly, which would take up a lot of room, so space was usually an issue in each of these departments. Each would need lots of storage space for items such as paints, varnishes, leather, and other necessary items. The tools in these departments would include an assortment of brushes for paint and varnish, striping pencils, color grinders, tack hammers, needles, and stuffing irons.[41] The trimming room would also be equipped with cutting tables and implements to cut the leather and other materials with. Although it is unknown exactly what the Patterson Company held in its inventory of tools and equipment, it is almost certain that they would have each of these basic tools in its shop.

The Patterson Company believed in making money, but it did not try to do this by taking advantage of its customers. They felt that by using uniform prices that were based on a low-profit principle, the volume of the sales would make up the difference in profit. They kept prices low by eliminating the middleman and selling from their establishment. They even offered credit to those with references. They sold carriages on 4, 6, or 8-month notes with 6% interest. Their philosophy was to take care of the customer and they will return the favor. They provided services of convenience with no extra cost, such as providing plenty of hitching space outside while potential customers looked at their inventory, and they would also provide a loaner buggy for customers to use while theirs was being painted. One bonus was that they carried insurance on the customer's buggy while it was in for service or painting.[42] They did what they could to make the customer's interaction with the company more convenient and a better experience.

Another method that the company used to procure cash was their annual "at cost sales." During the first decade of the century, the company discovered that, during the winter months, business slowed enough that they did not have the operating funds to start production of their spring stock. The Pattersons explained in a 1910 advertisement that "our products have been of such uniformly high standard, our profits so small and our increase in our business so great, we have always been hard up. To get cash to run this business in the slow months, we were compelled to sell goods at factory cost."[43] The article went on to state "what we once did from necessity, we now do from choice. To get cash, we prefer to go to our customers." A 1908 advertisement stated that "our spreading, growing business demands more cash. How enormous the foreign demand has grown for our work. We can't swing this business unless we have cash."[44] This same advertisement stated that they had to sell $30,000 worth of buggies that were currently sitting in their repository, suggesting that the company carried a large stock of goods on hand.

Several sources have claimed that during the earlier years of business, likely during the 1890s, C.R. Patterson received a lot of financial help from Edward L. McClain, a prominent businessman of Greenfield.[45] McClain made his fortune by improving the horse collar pad. His improvement allowed the pad to be detached easily and the new shape allowed it to follow the contours of the horse's neck better than those that had previously been in use. McClain's company, The American Pad & Textile Company, had a capital stock of $1.25 million.[46] The detachable pads needed special steel hooks to hold the collar in place. According to one source, McClain "prepared a rough drawing of the steel hooks and the town blacksmith manufactured them for him."[47] Although it does not specify that this blacksmith was C.R., he was one of the few blacksmiths in town at that time and it would make sense that McClain would freely offer assistance to the man who helped him to achieve his wealthy status.

The Patterson shop did a lot more than build carriages during this period. As mentioned earlier, C.R. began as a blacksmith. While this was part of carriage building, his blacksmithing skills served other purposes as well. These included general blacksmithing services, which could be used for many tasks such as welding, creating forgings, and farm machinery repair. The Patterson Company advertised that it performed all sorts of farm machinery repair, including repairs to plows, planters, harvesters, and sickle grinding. Other services included obvious tasks such as carriage and wagon repair, carriage painting, as well as repairing and replacing upholstery. They also sold accessories for the horse-drawn trade: harness, rain aprons, side curtains, foot warmers, and lamps. The Pattersons made some of these items, while others were purchased elsewhere and stocked in the shop.

A 1902 article stated that "what won the confidence of the people was the fact that the Pattersons always did good work…the news spread that 'you get your money's worth at Pattersons' and the farmers got into the habit of coming into Pattersons and giving orders for buggies." A 1912 Patterson Company advertisement told its secret of a successful business. They stated that "the real test comes in holding good patronage. Personality won't do it; cheap goods, even at low prices, won't do it and a whole lot of talk won't do it. What is the secret? Simple indeed but must be persistently pursued. Give honest values, give dependable quality, go to your patrons personally with every assurance that their interests are always safeguarded."[48] Patterson Company advertisements and articles written about the company clearly show that it stood by this philosophy through the years. It always offered quality products and personal service to its patrons.

Despite offering quality products, friendly service, and reasonable prices, the C.R. Patterson and Sons Company, like so many others, eventually lost out to the automobile age. The carriage business slowed greatly in the second decade of the twentieth century and the competition with automobiles became too overwhelming for the company. Although they did not want to abandon the carriage business altogether, they knew that they had to change with the times or close the doors of the business. After 50 years in the carriage business, the C.R. Patterson and Sons Company made the decision to begin its shift to the automobile age.

Technical Details of the Patterson Carriages

C.R. Patterson spoke in front of the inaugural meeting of the National Negro Business League in 1900. He talked about the state of carriage manufacturing at the turn of the century. He stated that three types of carriage builders operated in the United States, and he gave a brief explanation of each group. He referred to the first group as "assemblers." This group bought all finished parts from different manufacturers and assembled a carriage from these. This took no "special knowledge of styles and no good judgment of quality." This group first asked, "what can I sell?" and then, once he received an order, "where can I buy it?"[49] They simply built the carriage from the parts that they had assembled and coated it with cheap paint to give the customer what they wanted.

The second group of builders that C.R. described were those that were "good examples of that American mechanical ingenuity that has astonished the world." They built the carriage from top to bottom on their own (with the aid of lots of machinery), but they did not care much about the quality of the vehicle. Their questions involved "how many?" and "how much?" can I get for the dollar when it came to materials used in their carriages. With the cheap materials and the high amount of production that was possible using machines, these large factories were able to sell the vehicles at a low price. By utilizing machines, this also eliminated the use of skilled workers that would require higher pay. This builder was more concerned with quantity, rather than quality.

The last group of builders produced high-grade work. They considered the quality of materials and craftsmanship from the beginning of construction to the completion. They only used the best materials they could find on the market. "The workmen are intelligent, competent, well-paid mechanics; the undertakers are designers of styles, expert judges of quality and men of high honor. The name plate alone is a guarantee of excellence." The vehicles built by these expert craftsmen may have cost more initially, but they saved people money and lots of trouble in the long run. The high quality ensured more durability and, therefore, less breakdowns and a longer time of service. The C.R. Patterson and Sons Company prided itself in belonging to this third group of builders.

A Patterson ad stated in 1912 that "our father made these high-grade buggies by hand a half century ago. He taught us the art and we have been maintaining the Patterson reputation by continuing to offer the same hand-made buggies in the same dependable quality." The Pattersons credit the quality of their vehicles to learning from other's mistakes. They carefully observed the faults and weaknesses in carriages that were brought into their repair shop and worked to eliminate and prevent those faults in their own carriages. Not only did they pay attention to carriages brought in for repair, but many times they also purchased some from competing manufacturers, mainly from the large factories, and examined them. Once finished with their examination and observation, they would offer them for sale to their customers at a discounted price.[50] By paying attention to detail and observing those components and materials that did or did not work, the Patterson Company was able to build carriages that would last.

Another matter of pride that the Patterson Company held revolved around its buggy guarantee. As early as 1905, the guarantee was for two years from the date of purchase. Unlike the buggies that were purchased from distant factories, usually through middleman dealers, the Pattersons could make repairs close to home, which resulted in a quicker return of the vehicle to the customer. They stated "the reason why we are able to make our guarantee good as gold is because we make the buggy we sell you. Any defect, any mishap can be remedied almost instantly. We take pride in the fact that we can relieve our customers from any long waits or from any inconvenience." If a customer had to wait on a distant factory to make repairs, it was possible for them to wait a week or more (transportation time included), but with the Pattersons, who operated locally, you only had to wait "an hour or two – that's all."[51] This quick turnaround would have been attractive to many residents of the area.

The Patterson Company also prided itself in its high quality of craftsmanship and its use of high-grade materials in the vehicles that they built. Although the carriages made at the large factories were much cheaper because of the use of advanced assembly methods and machines, the quality of these vehicles were usually at a lower standard. In regards to this, the Pattersons stated "we have never tried to make a buggy at $30 or $40, but our aim has always been to make the very best buggy that could be built out of high-grade materials and put together by real buggy builders who learned the art of hand construction long before cheap, machine-made buggies were thought of."[52] Although the Patterson buggies were high quality handcrafted vehicles, the low price of the factory-made buggies was difficult to compete with.

One of the most important parts of the buggy is the body. The Patterson Company took extra care in constructing its bodies so that they would last longer and be more durable. According to a Patterson ad, a buggy body weighed only 30 pounds, yet could carry a load of up to 600 pounds if built correctly. A good, strong wood had to be used for the body to achieve this type of strength and carrying capacity. The Patterson Company used rigid ash for the frames and "the panels must be of clear, solid poplar, a very light wood, but close grained with lots of strength. These panels must fit the frame exactly to which they are glued, screwed, and plugged. The front floor must be good wood to stand the wear of your feet. Every part must be further strengthened with suitable all wrought irons and no castings." The timber used for these bodies was stored for three years in the Patterson yards to ensure proper seasoning before being used in vehicle construction. To ensure that the buggy body was properly built, a superintendent would perform a rigid inspection of each finished buggy before any paint was applied.[53] A properly built body with solid materials kept the buggy in service longer.

Another important wooden piece of a buggy were the shafts. These are the attachment arms that created the connection between the horse and the buggy. The Patterson shaft features included "black hickory timber, the toughest stuff that grows; hand-welded and hand-shaped; heavy wrought irons; trimmed with genuine harness leather; flat straps and genuine patent dash leather points; painted with the same care as any part of the Patterson Buggy." These had to be very strong, yet flexible enough to not break easily under stress of an accident or a horse's fall. Also, important to construct using a high-quality wood were the wheels. The Pattersons used rock elm hubs, white hickory 15/16" barrel spokes, and a ¾" wide white hickory rim. The spokes had two screws placed at each spoke end to help prevent splitting and the metal tire was applied by hand with a screw between each spoke for security.[54]

Another option for tires, instead of using the bare metal rim, was the installation of a rubber tire. The Patterson Company made its own special brand of tire that was larger than the traditional size. A traditional tire size was 7/8", while the Patterson tire had a crown that was a full one inch but was still mounted on a 7/8" rim. This would allow for approximately "$6 more wear for the price," and since the Patterson Company owned the molds, they could still sell them at the same price as the other brands. These tires sold in large numbers and by June of 1914, the Pattersons claimed that they had sold over $10,000 worth of these tires and "not one defective set. This tire has solved our

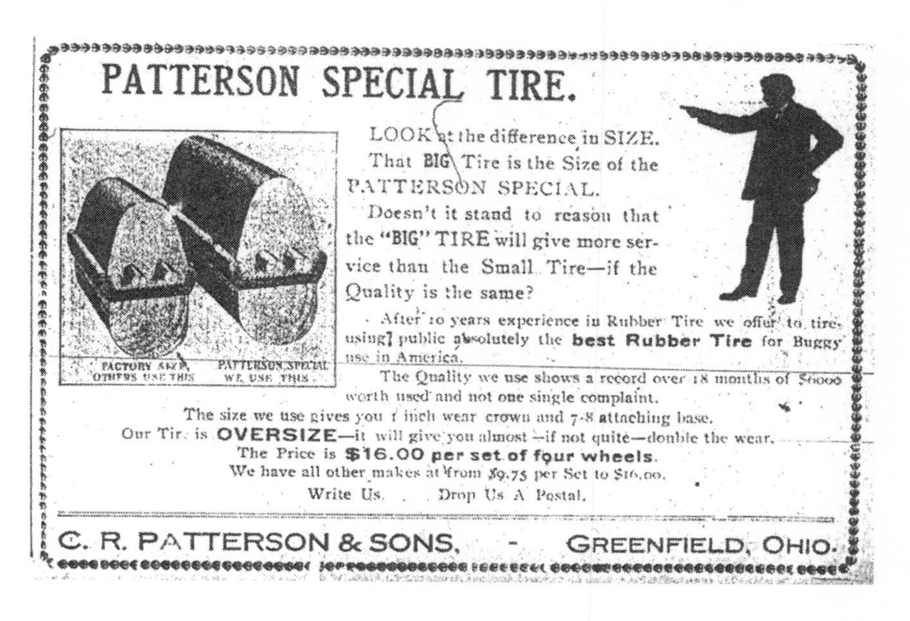

Advertisement showing illustration of the Patterson tire size in relation to the traditional tire sized tire. (*Hillsboro Dispatch* March 4, 1913)

troubles and it will solve yours too." One farmer wrote of the Patterson tire, "it's an insult to the intelligence and business judgment of the farmer to suppose that a commodity carrying the extra value of your special tire will not be demanded almost exclusively."[55] Although the Pattersons made this tire themselves, it was not placed on all their vehicles, as customers could still request any brand that they desired.

Other parts that the Pattersons used in their carriages included the exclusive use of 38" long elliptic springs for suspension. Most manufacturers used a traditional 36" length, but the Pattersons claimed that "this extra length gives you greater service and far more comfort." Strength, durability, and comfort were the keys to a high-quality Patterson buggy. Other parts of importance that required the use of good materials to ensure proper strength were the reaches. These were the long parts that stretched from the fifth wheel all the way back to the rear axle. The Pattersons used second growth hickory that was secured by a heavy bevel edged strip of iron to construct their reaches. This gave the buggy strength to resist torsional forces on the body.[56]

According to the Pattersons, their buggies contained high quality craftsmanship and good materials. They claimed to only use genuine leather, all wrought iron, and the best quality wood for their vehicles. To show off their operations, they would invite the public to come and see their work in progress and inspect the quality of their work at any time. They encouraged the public to "see the cloth on the bolt, and the leather in the hide" to experience how a custom handmade buggy shop operated and constructed the vehicles utilizing raw materials. A 1913 ad stated "Mr. Vehicle User, take this opportunity to visit one of the largest custom factories that remain. See for yourself. Get posted. Inspect our materials and watch us work. Then you will appreciate Patterson quality." The company took a lot of pride in its lasting quality even when most had turned to machine-made products. These values even extended into the nameplates they placed on the vehicles. "The Patterson nameplate stands for years of honest toil and earnest endeavor to afford you the very best buggy value. It represents half a century of thoughtful effort devoted to your best service."[57]

A 1916 article that highlighted top businesses in Highland County stated, "ever since this company first opened its doors it has been one of the prominent concerns in this section of the country and has been an important factor in the scheme of affairs in the city in which it is located." It continued:

> [The Pattersons] are among Greenfield's best citizens and we do not know of more popular business men in this section. They have always been fair and honest employers and straightforward businessmen, and most desirable and popular citizens. During the years that he and his associates have been in business there they have established a reputation as progressive businessmen who would rather retire than to be subjected to the least criticism in their dealings with the public. They have won

> for the firm of C.R. Patterson and Sons a warm place in the hearts of the Greenfield and Highland County people and have aided in making a more prosperous city...the Dispatch is proud to point out the C.R. Patterson and Sons Company as one of the foremost concerns of the middle western states and to direct your attention to the commendable straightforward and latter day business methods of the Messrs Patterson that have won such a wonderful success.[58]

The C.R. Patterson and Sons Company built several types of vehicles. It is difficult to determine just how many styles of vehicles that the Patterson Company built since only certain ones were included in its ads. It is known that the company sent out 100-plus page catalogs of its carriage styles twice per year. The shop specialized in custom made vehicles with each vehicle individualized based on the needs of the customer but was based on a standard design of each style. Only one of these catalogs is known to exist and a copy was provided to the author by descendants of the Patterson family. The ca. 1913 catalog is 102-pages in length and full of images and details of many of their standard designs, all of which could be customized. The catalog includes carriages, buggies, drays, special purpose vehicles for delivery of goods and people, and even includes an extensive line of Patterson-made harness and bridles. Replacement parts are available in the catalog as well as accessories for the vehicles they made or for the driver, such as gloves, aprons, and even rifles or shotguns. Accessories for horses, such as fly nets, are also included.

While the above details are important to consider, there are other details included in the catalog that are invaluable to the Patterson story and their way of conducting their business. Much of this information comes in the form of several pages of narrative descriptions of their business practices, including their two-year buggy guarantee, their method of eliminating the middleman and selling direct to the customer, along with their reasons why, and even a defense of accusations from competing manufacturers. There are also detailed descriptions and diagrams of their crating and shipping procedures, along with shipping costs to multiple cities across the United States. These narrative details are too important to the story to not include here and are presented in the images on the following pages (the following images courtesy of R.P.). Reading the statements in the Patterson's own words captures their attitudes and approach much better than any author can paraphrase and it eliminates the risk of losing important parts in the translation.

QUESTIONS AND ANSWERS

Do you change the color of painting from the regular colors described for different styles in catalogue? Yes, we are glad to change the color of painting of gear or even body on any of our carriages, surreys, phaetons, stanhopes, traps, or top buggies, but we cannot make such changes and ship as promptly as when painted regularly.

Do you allow special discount when more than one vehicle or harness is ordered? No, we do not.

How shall I send money? Either bank draft, post-office money order, express money order, or registered letter.

Do you furnish side curtains and storm aprons with all top vehicles? Yes, we do.

How many spokes in your wheels? Banded hub wheels have 14 spokes, Sarven patent, 16.

What is meant by leather quarter top. Our leather quarter tops have the strength and appearance of a full leather top; side quarters and backstays are leather; roof, back curtain and side curtains are rubber.

Do you furnish boots on all top buggies? All our Piano and Corning body top buggies have boots to cover the back of body.

Do you furnish four-bow top if desired where cut shows three-bow? Yes, we furnish four-bow top when wanted. No additional charge.

Are your tires bolted between each spoke? Yes, they are.

What kind of oil is the best to use on spindles? Either good castor oil or carriage axle grease.

Do poles have neck-yokes and whiffle-trees? Yes, they do. We also furnish stay straps with all poles.

Are all your nickel harness trimmings nickeled on white metal? Yes, our nickel harness trimmings are nickeled on white metal, and the nickeling will not chip off and show black specks as it does when nickeled on iron. Our brass trimmings are solid brass, finely polished.

What is meant by leather trimmings? By leather trimmings we mean the cushions and backs are covered with leather instead of cloth.

Do you furnish breast straps with pole? No, they belong with harness.

Will you make changes in your regular styles? We want to please our customers, and we are always glad to quote prices on any changes desired.

Are the dashes on your road wagons and spring wagons leather? Not unless specified. We find solid wood far more durable.

Do you furnish either pole or shaft with any of your vehicles? On buggies priced with shafts it is extra for pole instead of shafts. When priced with either pole or shafts we send what is ordered.

How do you measure width of track? Track is measured from outside to outside of tire along the ground. The regular widths are 4 feet 8 inches and 5 feet 1 inch. Bodies are the same width for both tracks.

Do you furnish your different style vehicles on any other width track than 4 feet 8 inches and 5 feet 1 inch? Yes, we can furnish any width track desired, but we do not carry anything but the 4 feet 8 inches and 5 feet 1 inch in stock, and when a different width track is wanted we cannot ship promptly.

Do you pay freight? No, we do not. Prices quoted in our catalogue are for goods crated on board cars at Greenfield. If for any reason you are not pleased, and what you order is returned, then we pay freight both ways.

Do you furnish storm apron with carriages and road wagons without tops? We do not furnish storm aprons with open vehicles unless ordered extra at $1.50.

How to Take Care of Vehicles

Wash new vehicles in clear, cold water, and wipe dry with chamois skin or soft cloth, before using to harden and set the varnish. Never use soap or hot water.

Do not leave the mud to dry on the vehicle—it kills the lustre and causes the paint to come off.

Never wash or allow a vehicle to dry in the sun.

Wash vehicles often, as dry or frozen mud damages paint.

To preserve paint keep vehicle in a dry barn, away from manure.

Vehicles allowed to stand out of doors constantly will lose their lustre.

Any wheel to give good service, no matter how well it is made, must be kept in proper dish and tires kept tight by the owner of the vehicle.

Clean the axles and boxes, and oil frequently.

Keep all nuts tight.

Use duster on trimmings only—never on varnish work.

(from 1913 Patterson catalog, courtesy of R.P.)

C. R. PATTERSON & SONS

MANUFACTURERS OF

FINE VEHICLES AND HARNESS

Catalogue
Number
Twenty

Office Factory Repository
Greenfield, Ohio
U. S. A.

We have the pleasure of presenting our new catalogue, and this book takes the place and revokes all catalogues and lists heretofore issued. If you have in your possession one of our catalogues or lists, please destroy it and accept this instead. The prices named herein are for the goods crated for shipment and delivered F. O. B. cars, Greenfield, Ohio.

Our Business is the Manufacture of Vehicles, which we sell direct to the consumer at wholesale prices. All the vehicles shown in this book are our own manufacture. Any purchaser dealing with us can be assured that the transaction is direct with the maker and that he is paying but one profit—the manufacturer's profit, and a reasonable profit—not two or three profits, as he is obliged to do when buying from a dealer or jobber, and 99 out of 100 mail order houses.

Our factory, office and repository is located on the corners of Washington and Lafayette streets. If you should visit our city, we cordially invite you to call and inspect our work in person. If that is not possible, however, by means of this book you can buy as safely and satisfactorily as though you were on the ground.

Terms of Sale—All prices quoted in this book are on a cash basis. Selling on such terms enables us to make lower prices on our work, as we are not obliged to add to our cost figures a large per cent to cover bad debts, which all manufacturers, jobbers and dealers who sell in the ordinary way are obliged to do.

Our Catalogue Gives Two Prices on Vehicles, one a "C. O. D." price, the other a "Cash with Order" price. The Cash with Order Price is less than the C. O. D. price and we allow this discount to our customers who remit in advance for two reasons. In the first place, when cash is sent with the order we save cost of making C. O. D. collections, which in many cases amounts to considerable. Then, again, we are often-times imposed upon by unscrupulous persons who order goods sent C. O. D. and who have no intention of taking them from the depot, in which case we are obliged to reship, and lose the freight.

When cash is sent with order, the goods are shipped to your address at shipping station you name and the day following the shipment receipted bill and bill of lading is mailed direct to your postoffice.

When Goods Are Sent C. O. D. we ship them to ourselves, to your freight station, and make draft on you through any Bank or Express Company you may name. The draft is attached the bill of lading, endorsed to you. Upon payment of the draft the bill of lading will be surrendered to you, and same should be presented to the Railroad or Express Agent, who will deliver the goods upon payment of transportation charges.

Satisfaction Guaranteed—You Run No Risk. We guarantee everything we sell just as represented and satisfactory, and if a purchaser finds on receipt of the goods that the work does not bear out all of our representations and is not satisfactory, he has the privilege of returning and all money will be refunded. This applies not only to shipments which have been paid for in advance, but also C. O. D. shipments which have been accepted and paid for. You absolutely run no risk in buying from us. We take all the risk. This proposition is certainly a liberal one. We make it, however, knowing full well that the work will be in every particular satisfactory.

Our Customers Have the Option of buying on either plan. We recommend, however, the "Cash with Order" plan. It saves the purchaser considerable money and very often trouble and annoyance, particularly

(from 1913 Patterson catalog, courtesy of R.P.)

when the freight or express office is located some distance from the residence of customer.

The Goods on Either Plan net us about the same. On the "Cash with Order" basis, the customer gets the benefit of the discount; on the "C. O. D." basis, the Bank or Express Company making the collection.

How to Remit—All remittances should be made payable to "C. R. Patterson & Sons," either by bank draft, postoffice or express money order, or registered letter. (It is not safe to send money in a letter without registering it, and we will not be responsible for remittances so sent.)

Safe Delivery—As illustrated and explained on page 65 we crate all our vehicles for shipment in the most secure manner and there is no question but that they will carry through to destination in good shape. It is to our interest to be very careful for it is our desire that the work reach our customers in good shape.

Our Responsibility—We refer you to the letter of our bankers, The Highland County Bank, published on page 2. Also to any other bank or mercantile house in this city. The mercantile agencies of R. G. Dunn & Co. and Bradstreet may also be consulted. In this connection we wish to state that we own our own plant (one of the largest in the country), and are able and more than willing to carry out every contract or agreement into which we enter.

Prices—The prices named in this book are the lowest figures. Under no circumstances can we make any further reductions or discounts. They are based on the cost of manufacturing with one profit, the manufacturer's profit only, added. In dealing with us you pay but one profit, and a reasonable profit; not two or more as is always the case when buying through a middleman or supply house. We know full well that there are cheaper vehicles and harness offered by our competitors in the mail order business, by dealers and jobbers, and if a person is simply going to buy and take price only into account and not consider quality, some other firm perhaps can do better for you. If, on the other hand, the purchaser will consider quality of material, finish and general construction, it will be found that our prices in reality are lower than any concern in the country.

Quality—It is impossible for us to give a detailed description of the quality of material which enters into our different vehicles, owing to the fact that there is such a great number of different kinds of goods used. As far as space will permit, we give a description of the material on page 90 of this book. It is not complete, however, in many ways, and any further information wanted we will be glad to give if our customers will write us. We are only too glad to give this information, for we believe if our customers can understand the quality of goods which we are using in the construction of our vehicles, it would work very much to our advantage.

Facilities for Manufacturing—Our factory, as before explained, is located on Washington Street, and is one of the largest, best equipped and modern factories for the economical production of superior vehicles in the country. We have all the latest improved machinery and tools necessary for the manufacture of these goods, and employ only skilled mechanics in every department.

Shipments—As a rule we are able to make prompt shipment. We have at our factory a large repository in which we carry a supply of finished vehicles. We also have in process of manufacture in our factory a very large stock of vehicles of different kinds. There are a great many variations in the way in which vehicles may be made, as will be observed by reading the descriptions, and in some cases there may be delay of a week or ten days. But if it will take longer that that time we will notify you on receipt of order, and if you cannot wait, we will refund any money which has been paid.

Misrepresentation of Facts—Our plan of selling Direct to the Consumer does not please the retail Dealer or Jobber in many instances, and to combat us and effect sales, misrepresentations which often reflect on us seriously are made. We, therefore, feel justified in explaining matters which are not generally understood.

A prospective customer is often told that our guarantee is worthless; that we are simply after his money and that when we get it he can look out for himself. There is absolutely no truth in this statement. The man does not live who can point to a single instance in which we have not, in every particular, done just as we agreed, and in fact a good deal more in many cases. Then again, the warranty which we put on our goods means a hundred times more than any dealer can give. Any guarantee which you receive from a dealer is second-hand; from the manufacturer to him, and from him to you, while ours is direct to you.

Another scheme that is frequently worked. Very often when a customer indicates to a dealer his determination to purchase elsewhere he is told that on account of friendship, or for other reasons, he can have the goods at cost, or at a small advance over cost, and shows the customer an invoice from the manufacture. The invoice which he discloses is simply one made out by the manufacturer for the purpose, with a sufficient price added over the dealer's price to afford a good profit. This scheme is worked frequently.

(from 1913 Patterson catalog, courtesy of R.P.)

Still another argument which is used is the fact that a prospective customer in buying from us is obliged to pay the freight. That is true. The dealer also, however, has to pay the freight and adds the amount to the cost of his goods before quoting you a price. The dealer pays exactly the same freight that you do, but in any event you may rest assured that the consumer always pays the freight. He may pay it in one way or the other, but just the same he pays it.

This, and many other arguments, are advanced by dealers all because of the fact that they know it is impossible for them to compete with us on an equal quality of work.

We have no fight with the dealers and do not desire to have one. We claim the right, however, to conduct our business as we see fit and should not have touched upon this subject had it not been that in many cases we have been grossly misrepresented.

IN ORDERING

It is necessary that you sign your name plainly, giving your postoffice address, county and state, also shipping point, if different from postoffice, and county in which shipping station is located. We must also have certain information so as to determine just what you want, as follows:

First—Give the number of the article which you have selected also the page on which it is shown in the Catalogue.

Second—In specifying for vehicles give the width of track wanted. (Carriages, surreys and phaetons are made either 4 feet 8 inches or 5 feet 1 inch track.) Buggies either 4 feet 8 inches or 5 feet 1 inch, and some buggies 4 feet 4 inches. It is absolutely necessary that we have this information, as we cannot fill an order until we have it.

Third—Kind of wheels. We furnish as per description, either Sarven patent or compressed band hub wheels, and on most vehicles furnish different width of treads. Always state which is wanted, and when more than one height of wheels is specified in the description, give the height you want.

Fourth—On most piano-box buggies we furnish different width bodies. If choice is allowed, always specify width you want. If Corning body is wanted it must be so specified.

Fifth—State whether the seat and back are to be trimmed in cloth or leather. (If more than one colored cloth is used, state color you wish.) We also give choice of different designs of seat trimmings, as shown on page 6? State which kind is wanted If not specified, we send regular trimming as follows: On buggies and road wagons we send style B, on carriages, surreys and phaetons we furnish style F.

Sixth—On vehicles which are made with extension or falling tops, state kind wanted; whether full leather with rubber side curtains, leather quarter or full rubber. Also state (on buggies) whether 3 or 4 bow top is wanted. On some top buggies we can also furnish, if desired and specified, handy top, without additional charge.

Seventh—State if shafts or pole, or both are wanted. All vehicles are quoted with shafts only.

Eighth—We paint the bodies of all vehicles a jet black; the gears and wheels of most vehicles Brewster green, regularly. If you wish to vary this, color of paint must be specified.

When goods are shipped C. O. D., please give us the name and location of the bank or express office through which you wish us to send draft for collection.

HOW WE CRATE VEHICLES FOR SHIPMENT

The illustration below will give our customers a good idea of the substantial manner in which we crate vehicles for shipment, and they may be assured that unless because of some accident, the vehicle will reach

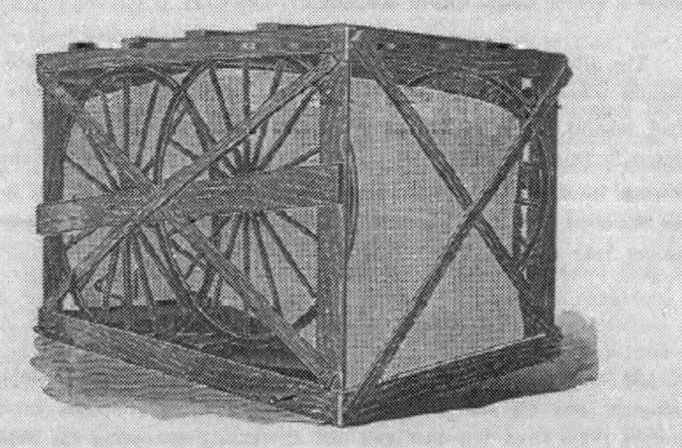

OUR 54 INCH CRATE

destination in good condition in every case. All vehicles are covered with heavy waterproof paper cover, to protect them from dust and dirt. Shafts and poles are carefully wrapped, which insures their carrying through in good condition.

(from 1913 Patterson catalog, courtesy of R.P.)

All carriages, surreys and phaetons, both canopy and extension tops, are crated in this size crate. Top buggies, also, when they are shipped with the top up, that is not crushed. Physician's closed top vehicles are shipped in crates measuring about 60 inches, as the top on these cannot be crushed, and are stationary. We advise shipment of buggies in 54-inch crates to points not too far distant. The freight will be a little more, but the top will be received in better condition, certainly repaying the purchaser for the extra freight paid.

OUR 34 INCH CRATE

The illustration above shows the manner in which we crate top buggies when the top is crushed flat; also open buggies, runabouts and road wagons. To far distant points we recommend top buggies to be crated in this size crate. If no instructions are given as to manner of crating, we always use our best judgment, endeavoring, of course, to get the shipment through in the best possible shape, and at the lowest possible freight rate.

UNCRATING AND SETTING UP VEHICLES

As explained all vehicles are securely crated with strong lumber in shipping, and all are covered with heavy waterproof paper cover to protect them from the dust and dirt in transit, excepting the cheapest road wagons and carts.

It does not require a skilled mechanic to set up a vehicle on arrival at the station. Anyone who can use a hammer or hatchet can do it in a very short time, and with but very little trouble. Every vehicle is put together before it leaves our factory, and all parts will go together without the least difficulty. It is very necessary and important to see that the spindles of the axles and the boxes in the hubs are clean and free from any foreign substance before the vehicle is run. Any paint, lead or varnish that may be adhering to these parts should be scraped off with a knife or other sharp instrument, and afterwards sandpapered until they are perfectly smooth and clean. Before running, of course, the spindles should be oiled with good castor oil or carriage oil. Do not, under any circumstances, use wagon grease, it is not intended for use on light pleasure vehicles.

Care of Vehicles—We wish to caution our customers not to allow mud to dry on a newly painted vehicle. In fact, on a vehicle that has been used for some time. Great damage is likely to result because of the varnish losing its luster.

It is well, also, if possible, to keep vehicles in as dry a place as possible. Dampness does not do the varnish any good and in some instances, much harm. In washing vehicles, nothing but clear, cold water should be used with sponge or chamois. Under no circumstances use warm water.

We cannot be responsible for the painting on a vehicle which is exposed to the action of ammonia. Ammonia will seriously injure the varnish and paint in a very short time.

Stable refuse is a constant source of ammonia. Therefore keep your vehicle **as far as possible** from it.

Guarantee Two Years—Our vehicles and harness are fully guaranteed and if within two years from date of sale any imperfection or breakage should occur, write us and we will take steps at once to adjust the matter in a satisfactory way.

Rubber Tires Guaranteed One Year—Our guarantee on rubber tires is for one year, the same length of time the rubber manufacturers guarantee to us.

We Do Not Pay Repair Bills Unless Price Has Been Previously Agreed Upon With Customer—Our customers must be very careful and keep the tires tight on all wheels. This is absolutely necessary to preserve the life of the wheel. No wheel will stand, no matter what grade or kind, unless the tire is kept tight.

(from 1913 Patterson catalog, courtesy of R.P.)

DESCRIPTION OF MATERIAL

In considering prices of our goods we must ask the purchaser to take quality, workmanship and finish into account. We fully realize that our competitors, catalogue houses, dealers and jobbers in many instances make lower prices on work which on paper looks like ours. There is a vast difference in the quality, however, and considering everything, our work will really be found the cheapest. The description of material following is not as complete as we would like to give. Any further information wanted, however, will be cheerfully furnished on application.

Wheels—On page 62 we show cuts of both Sarven patent and compressed band hub wheels, either of which we will use on most of our vehicles. We guarantee every wheel to stand, if tires are kept tight. This is very necessary. If tires are allowed to get loose, no wheel will stand, no matter what kind or grade. The spokes and rims of all wheels are made of selected second growth hickory, the hubs of thoroughly seasoned rock elm. The workmanship is the very best. In driving spokes into the hubs, the tenons are first compressed and scarified, then heated, thus insuring a complete dove-tail. The tires on all wheels are of the very best quality of steel, round edge, and are put on the wheels hot, by hand, with the greatest care; not cold, by machine, as a great many of the cheaper manufacturers do.

Shafts—All shafts are made of the best quality selected hickory. The heel irons are of steel and sufficiently heavy to stand any ordinary usage, and are securely bolted to the shafts. On all shafts we furnish japanned tips, harness leather straps and we trim with genuine enameled leather.

Gears—As stated in the description of the different vehicles the axles on all surreys, carriages, phaetons and certain buggies are made dust proof and have grooves cut in the spindles in such a manner that the axles are SELF-OILING. Axles are all steel. Springs are made of the best quality open hearth steel, ground full bright, and every spring is tempered in oil and thoroughly tested. On every vehicle we use a wrought fifth-wheel. The iron work on all gears, such as braces, etc., is refined iron. All clips, bolts, etc., are of the best quality Norway iron. Our gear woods are all hickory and strictly second growth. All axle caps are firmly cemented to the axles.

Bodies—We take particular pains in making all bodies, fully realizing that unless they are well made, they will not stand in some climates. The sills are of second growth ash; all panels of yellow poplar, screwed from inside and outside, securely glued and plugged. On phaetons, stanhopes and surreys, that portion of the sills which is exposed to view we cover with enameled leather, over which we put a metal tramp plate to protect same when getting in or out of the vehicle.

Tops—Our tops are made in the most approved manner. We use only strictly second growth ash bows. The leather in the highest grade vehicles is strictly No. 1 machine buffed, both top and trimming. In the lower priced work, it is No. 2 machine buffed. In vehicles built with rubber tops, or the side curtains on other tops, we use only genuine rubber, in weight from 30-oz. down to 22-oz., depending on the price of the vehicle. The cloth head lining in all tops is wool, indigo dye, and we guarantee it. All shifting rails and top joints are made of the best quality wrought iron.

Seat Trimming—On our highest grade vehicles when the cushion and back are leather trimmed, we use strictly No. 1 leather; when trimmed in cloth, 16-oz. all wool Slater Woolen Mills cloth, indigo dye, which will not fade. On our lower priced vehicles, when trimmed in leather, we use machine buffed, and when trimmed in cloth, Slater Woolen Mills make, 14-oz. weight, all wool, fast color, guaranteed not to fade.

Dashes—All dashes are made with steel frames covered with patent genuine enameled leather, and are attached to the bodies in such a manner that they will not rattle nor shake loose.

Painting—This is, as is well known, one of the most important things to consider in manufacturing vehicles. We give the matter most careful attention, and we are able to say to our customers without hesitation that our paint will stay where it is put in every instance; provided, of course, it is not exposed to the action of ammonia, and the vehicle is washed properly.

All gears and wheels are painted on the "Rubbed Lead System," the best process known for painting vehicles. Between each coat we give sufficient time for the paint to thoroughly set. Our rubbing and finishing varnishes are of the best quality that we can buy, and are rubbed down with pumice stone after hardening. Painting of the body of the vehicle is somewhat different. It is first given a coat of oil primer, then six coats of filler, following two coats of color, then two coats of rubbing varnish (which coats are rubbed down with pumice stone) last comes the finishing varnish.

The wheels and gears of all our vehicles are striped in an appropriate manner, and the moldings of the body in some cases, where it adds to the appearance, are also striped to correspond with the gear.

(from 1913 Patterson catalog, courtesy of R.P.)

FREIGHT RATES

We publish below a list of freight rates to many of the principal towns and junction points in the United States. Your shipping point may not be given, but it is probably near some town named, so that you can closely calculate the amount. We give two rates on medium weight top buggies, crated in 34 and 54 inch crates. As explained we can ship top buggies in 34 inch crates, by crushing the top. We can also ship open buggies and runabouts in 34 inch crates. All carriages, surreys and phaetons take a 54 inch crate, and top buggies, when crated with top up. Closed Top Stanhope and Station wagons, take special crates, and freight rates will be quoted upon application.

The rates given below are PER HUNDRED POUNDS:

The rates as named are correct as of today, but the railroads change their rates frequently, and we therefore cannot guarantee the amounts. If desired, at time shipment is made, we will procure exact rate on any vehicle and advise you.

The freight on vehicles may seem high, but the purchaser must bear in mind that he always pays the freight, whether buying direct from the factory or from a dealer or jobber. Firms quoting prices delivered, add an amount to cover the freight, and dealers and jobbers in fixing their selling prices, also add the freight charge.

	Crate 34 in.	Crate 54 in.
Alabama		
Birmingham	$1 16	$1 69½
Decatur	1 16	1 69½
Dothan	1 73	2 55
Greensboro	1 63	2 40
Huntsville	1 06	1 54½
Mobile	98	1 47
Montgomery	1 35	1 98
Talladega	1 25	1 83
Tuskegee	1 80	1 66
Tuscumbia	1 18	1 03½
Uniontown	1 69	2 49½
Arizona		
Phoenix	4 80	6 40
Flagstaff	4 80	6 40
Arkansas		
Pine Bluff	2 05½	2 74
Jonesboro	1 60½	2 26
Little Rock	2 05½	2 74
California		
Red Bluff	5 40	7 20
San Francisco	5 40	7 20
Los Angeles	5 40	7 20
Colorado		
Denver	3 11	4 15
Leadville	4 61	6 15
Trinidad	3 11	4 15
Connecticut		
Hartford	1 02½	1 37
New Haven	1 02½	1 37
Delaware		
Dover	90½	1 20
Wilmington	29	1 19
District of Columbia		
Washington	87½	1 17
Florida		
Jacksonville	1 22	1 78½
Kissimmee	1 96	2 89½
Pensacola	98	1 47
Tallahassee	1 74	2 56½
Tampa	1 72	2 53½
Miami	1 88	2 77½
Key West	2 19	3 24
Georgia		
Albany	1 50	2 20½
Atlanta	1 25	1 83
Augusta	1 30	1 90½
Brunswick	1 22	1 78½
Macon	1 30	1 90½
Savannah	1 22	1 78½
Valdosta	1 70	2 50½
Waycross	1 62	2 38½
Bainbridge	1 59	2 34
Columbus	1 30	1 90½
Idaho		
Boise	4 80	6 40
Coeur d' Alene	4 80	6 40
Pocatello	4 80	6 40
Illinois		
Cairo	70½	94
Chicago	63	84
Peoria	64½	86
Quincy	68	91
Springfield	60¾	80
E. St. Louis	68	91
Indiana		
Logansport	55½	74
Evansville	66	88
Fort Wayne	52½	70
Indianapolis	49½	66
Madison	51	68
Muncie	49½	66
Richmond	43	57
Terre Haute	63	84
Iowa		
Des Moines	1 43	1 91
Council Bluff	1 50½	2 01
Sioux City	1 88	2 51
Dubuque	88½	1 18
Burlington	88½	1 18
Kansas		
Atchison	1 50½	2 01
Kansas City	1 50½	2 01
Topeka	2 01½	2 69
Wichita	2 47½	3 30
Dodge City	2 88½	3 85
Colby	2 87	3 83
Kentucky		
Ashland	41	55
Lexington	55	73
Louisville	52½	70
Owensboro	73½	98
Paducah	73½	98
Bowling Green	80	1 15½
Somerset	78	1 12½
Louisiana		
Shreveport	2 35½	3 14
Alexandria	2 35½	3 14
Lake Charles	2 58	3 44
Baton Rouge	98	1 47
New Orleans	98	1 47
Monroe	2 35½	3 14
Maine		
Caribou	1 67	2 23
Bangor	1 02½	1 37
Augusta	1 02½	1 37
Portland	1 02½	1 37
Maryland		
Hagerstown	87½	1 17
Baltimore	87½	1 17
Massachusetts		
Boston	1 02½	1 37
Springfield	1 02½	1 37
Pittsfield	1 02½	1 37
Michigan		
Cheboygan	94½	1 26
Traverse City	82½	1 10
Saginaw	64½	86
Grand Rapids	64½	86
Lansing	63	84
Detroit	55½	74
Coldwater	61½	82
Minnesota		
Duluth	1 39½	1 86
Minneapolis	1 39½	1 86
Winona	1 30½	1 74
Marshall	1 84½	2 46
Mississippi		
Holly Springs	1 34	1 96½
Columbus	1 50½	2 14
Mound Bayou	1 62	2 38½
Jackson	1 33	1 95
Meridan	1 33	1 95
Natchez	98	1 47
Biloxi	1 37	2 01
Missouri		
Chillicothe	1 37	1 83
St. Louis	68	91
Jefferson City	1 35½	1 81
Joplin	1 79	2 39
Kansas City	1 50½	2 01
West Plains	1 80½	2 41
Montana		
Helena	4 80	6 40
Miles City	4 80	6 40
Kalispell	4 80	6 40
Nebraska		
Lincoln	1 65½	2 21
Alliance	3 00½	4 15
Nevada		
Elko	4 50	6 06
New Hampshire		
Colebrook	1 02½	1 37
Concord	1 02½	1 37
Portsmouth	1 02½	1 37
New Jersey		
Newark	92	1 23
Trenton	92	1 23
Vineland	94	1 26
New York		
New York City	92	1 23
Albany	88½	1 18
St. Albans	96	1 28
Utica	83	1 11
Ogdensburg	1 02½	1 37
Buffalo	63	84
Jamestown	63	84
Binghamton	73½	98
North Carolina		
Wilmington	1 09	1 59
New Bern	98½	1 45½
Raleigh	1 13½	1 68
Durham	1 13½	1 66
Win'n Salem	1 13½	1 68
Reidsville	94½	1 39½
Asheville	1 26	1 84½
Charlotte	1 20½	1 75½
Fayetteville	1 20½	1 78½
North Dakota		
Fargo	2 40	3 20
Bismark	2 89½	3 86
Grand Forks	2 40	3 20
Ohio		
Cincinnati	27	37
Dayton	29½	39
Lima	37½	50
Defiance	52½	70
Toledo	49½	66
Tiffin	47½	63
Marion	41½	55
Columbus	25½	34
Chillicothe	11½	15
Trenton	29½	39
Zanesville	37½	50
E. Liverpool	61½	82
Canton	51	68
Ashtabula	58	77
Oklahoma		
Boley	2 43	3 24
Muskogee	2 28	3 04
Ardmore	2 58	3 44
Guthrie	2 43	3 24
El Reno	2 43	3 24
Oregon		
Portland	5 40	7 20
Salem	5 40	7 20
Grants Pass	5 40	7 20
Pennsylvania		
Philadelphia	89	1 19
Scranton	89	1 19
Harrisburg	87½	1 17
Erie	63	84
Pittsburg	69	96
Oil City	63	84
Williamsport	87½	1 17
Rhode Island		
Newport	1 02½	1 37
Providence	1 02½	1 37
South Carolina		
Spartanburg	1 38½	2 02½
Columbia	1 27½	1 89
Aiken	1 28½	1 96½
Beaufort	1 13½	1 68
Charleston	1 22	1 78½
Florence	1 32½	1 96½
Sumter	1 34½	1 99½
Georgetown	1 17½	1 74
South Dakota		
Sioux Falls	1 92½	2 57
Watertown	2 07	2 76
Aberdeen	2 34	3 12
Pierre	2 46	3 28
Tennessee		
Memphis	75	1 12½
Jackson	1 26	1 84½
Union City	1 06	1 54½
Clarksville	80	1 15½
Columbia	1 05	1 55
Nashville	80	1 15½
Chattanooga	1 03	1 59
Knoxville	1 02	1 50
Bristol	1 27	1 86
Texas		
Galveston	2 68½	3 58
Houston	2 68½	3 58
Beaumont	2 68½	3 55
Temple	2 68½	3 58
Waco	2 68½	3 58
Dallas	2 68½	3 58
Brownsville	3 15	4 20
Austin	2 68½	3 58
Vernon	2 68½	3 58
Washburn	2 83½	3 78
El Paso	2 86½	3 82
Delrio	2 86½	3 82
Laredo	2 86½	3 82
San Angelo	2 68½	3 58
Palestine	2 68½	3 58
Utah		
Salt Lake City	4 08½	5 45
Mt. Pleasant	4 40	5 86
Vermont		
Montpelier	1 02½	1 37
Rutland	1 02½	1 37
Battleboro	1 02½	1 37
Virginia		
Norfolk	87½	1 17
Cape Charles	90½	1 20
Emporia	1 01	1 47
Danville	95	1 38
Roanoke	87½	1 17
Nottoway	87½	1 17
Richmond	87½	1 17
Winchester	87½	1 17
Culpepper	1 17½	1 57
Covington	87½	1 17
Washington		
Seattle	5 40	7 20
Aberden	5 40	7 20
Olympia	5 40	7 20
Walla Walla	4 80	6 40
Spokane	4 80	6 40
West Virginia		
Grafton	67½	90
Wheeling	51	68
Parkersburg	41½	55
Kenova	41½	55
Charleston	55½	74
Bluefield	1 05½	1 41
Clarksburg	67½	90
Wisconsin		
Madison	1 30½	1 74
LaCrosse	1 30½	1 74
Eau Claire	1 39½	1 86
Oshkosh	1 30½	1 74
Wyoming		
Granger	4 80	6 40

Remember the User always pays the Freight. We don't "Even-Up" on our Customers.

(from 1913 Patterson catalog, courtesy of R.P.)

Some details were gleaned from advertisements and the ca. 1913 catalog about some of its carriages. Vehicles with the most information were its highest selling vehicles; the No. 1 End Spring Piano Buggy, the No. 4 "The Greenfield" Piano Buggy, and the No. 176 Patterson Perfect Storm / Winter Buggy. The No. 1 follows the specifications listed above in the discussion covering the Patterson materials and craftsmanship in their custom and hand-made vehicles. Although, the No. 1 was their best vehicle with the highest qualities, it appears that the same standards applied to all their vehicles, with the only differences being in trimmings and other smaller details. More specific details and discussion will be provided in the following pages about the No. 176 and a couple of their special purpose vehicles and will be used as a case study of Patterson products. Other of their top selling vehicles have their catalog images and details shown but are not included in the case study.

The Patterson Line of Custom Made Buggies

We show you unquestionably the best buggy values in the United States. Even if you are willing to pay $25.00 more than our price you will then not be able to duplicate these values anywhere.

The selection of any one of these styles by the buyer is the choice of wisdom. The continued growing demand from every section of the United States attests undoubtedly the fact that in each vehicle we are giving the using public exactly the Quality Value it has been looking for but could not find.

Read carefully the description that follows. Better still, write us for description in detail, such as cannot be given in the most elaborate catalogue, and you will secure perfect satisfaction in your next buggy purchase.

No. 4. "The Greenfield" Piano Buggy.

Shown on Next Page

Many years ago, to supply the FARMER TRADE of this community, we designed and perfected our No. 4 "The Greenfield" Piano Buggy. Everything necessary to an absolutely dependable service-giving vehicle was incorporated into its construction. Every weak point of every make and style of buggy that came to our large repair department was studied closely with the purpose of eliminating it from our No. 4 "The Greenfield." We studied why top seams ripped; why body corners opened and panels split; why the shafts worked loose and broke; why gears and wheels gave way; why the draught was heavy and why the springs were either too weak or too stiff. All of this from the view-point of the repair man, in daily contact with dozens of vehicles brought for repair, we studied and learned. Rich with experience, informed by the mistakes of hundreds of other makes of vehicles we designed in quality of materials and workmanship our **custom made** No. 4 "The Greenfield" Piano Buggy, and endowed it with every possible wearing feature of merit known to buggy building.

Not a penny have we tried to save in the construction of this buggy where it could effect the wearing quality. Not a dime have we spent in talking points and useless selling features. But, the quality secure and the price unburdened by useless nick-nacks, we give our patrons today absolutely the most dependable service-giving, common sense, reasonably priced buggy on the market. Hundreds of the No. 4 "The Greenfield" have been in constant use for 10 or 15 years, and in every state in the Union; and the best possible evidence of the unequalled merit of this buggy is the fact that liveries prefer our No. 4 "The Greenfield" to buggies **that cost $50.00 more than our price.**

In our judgment, on the next page we show the best combination of brains, material and labor from an economic standpoint known to the vehicle building industry. "No. 4, The Greenfield Piano Buggy" is the practical embodiment of cost and service in its truest combination. There isn't a single element of expense incurred in the construction, no waste energy in furbelows. Service is the keynote with No. 4 "Greenfield." No weak spot is permitted to remain under any circumstances. Whatever is necessary to the strength of wheel or body, to the service of gear or endurance of shaft, to the wearing quality of trimmings and top, in other words whatever will contribute to its solid merit, must be added. **But not one cent to waste.**

(from 1913 Patterson catalog, courtesy of R.P.)

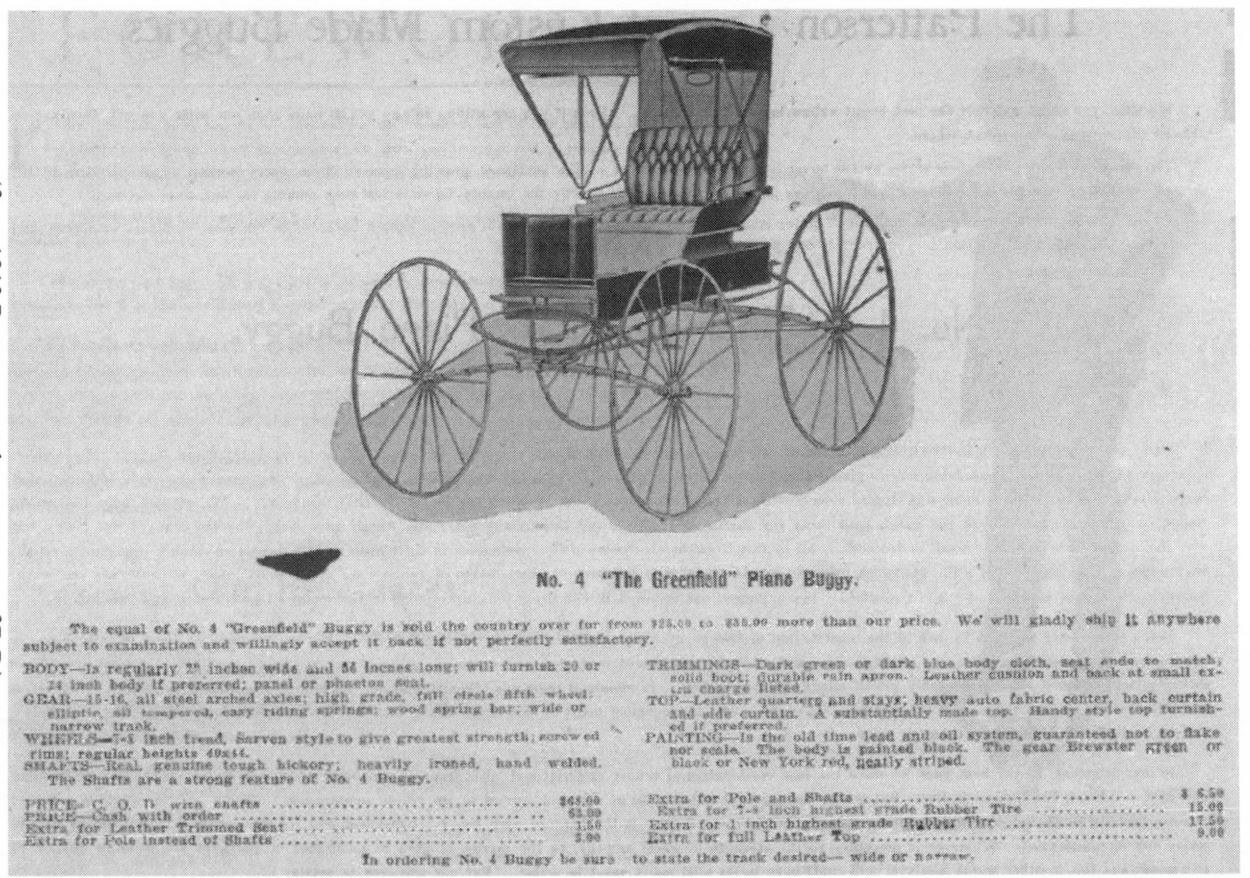

No. 4 "The Greenfield" Piano Buggy.

The equal of No. 4 "Greenfield" Buggy is sold the country over for from $25.00 to $35.00 more than our price. We will gladly ship it anywhere subject to examination and willingly accept it back if not perfectly satisfactory.

BODY—Is regularly 23 inches wide and 56 inches long; will furnish 20 or 24 inch body if preferred; panel or phaeton seat.
GEAR—15-16, all steel arched axles; high grade, full circle fifth wheel; elliptic, all tempered, easy riding springs; wood spring bar; wide or narrow track.
WHEELS—7-8 inch tread, Sarven style to give greatest strength; screwed rims; regular heights 40x44.
SHAFTS—Real, genuine tough hickory; heavily ironed, hand welded. The Shafts are a strong feature of No. 4 Buggy.

TRIMMINGS—Dark green or dark blue body cloth, seat ends to match; solid boot; durable rain apron. Leather cushion and back at small extra charge listed.
TOP—Leather quarters and stays; heavy auto fabric center, back curtain and side curtain. A substantially made top. Handy style top furnished if preferred.
PAINTING—Is the old time lead and oil system, guaranteed not to flake nor scale. The body is painted black. The gear Brewster green or black or New York red, neatly striped.

PRICE—C. O. D. with shafts $65.00
PRICE—Cash with order 63.00
Extra for Leather Trimmed Seat 1.50
Extra for Pole instead of Shafts 3.00

Extra for Pole and Shafts .... $ 6.50
Extra for 7-8 inch highest grade Rubber Tire 15.00
Extra for 1 inch highest grade Rubber Tire 17.50
Extra for full Leather Top 9.00

In ordering No. 4 Buggy be sure to state the track desired— wide or narrow.

(from 1913 Patterson catalog, courtesy of R.P.)

No. 1. AA. PIANO BUGGY

COMPLETE WITH ONE INCH RUBBER TIRE—BEST POSSIBLE.

This shows our highest grade piano buggy and you will find no buggy anywhere of higher quality. It is a combination of the very highest materials, most painstaking workmanship, and very latest style.

We haven't the space to give as complete description of this buggy as we would like; upon application however, we will enter into detail exhaustively and request a comparison with any joy that sells from $25.00 to $35.00 more than our price.

We will send this job anywhere with the understanding that it is to be found all we claim and perfectly satisfactory, or we will take it back at our expense both ways.

BODY—Is 22 inches wide and 56 inches long; will furnish body 20 or 24 inches wide if preferred; seat has solid back.

GEAR—Axles are slightly arched; they are 15-16 inch dust proof double collar, high grade steel; elliptic springs; regular style body loops with wood bar; will furnish Bailey loops if preferred roller rub irons; rubber covered steps; Bradley or standard couplers; wrought Wilcox fifth wheel,

WHEELS—Sarven Patent, 7-8 inch, with 7-8 inch rubber tires; will furnish Banded Wood Hub wheels if preferred; screwed rims.

TRIMMINGS—Dark green all wool 14 oz. broadcloth or H. H. trimmin leather; will trim with fine whipcord or dark blue broadcloth if pre ferred; ends of seat are trimmed to match the cushion and back both cushion and back are upholstered with hair and springs.

TOP—Is full leather, with leather covered bows and leather side cur tains; fleece lined apron; highest grade leather used.

PAINTING—Body black; gear dark Brewster green, neatly striped; wi furnish New York red if prefe red to Brewster green.

PRICE—C. O. D. with 7-8 inch best quality Rubber Tires and Shafts .. $125.00

PRICE—Cash with order 120.00

This buggy appeals to the user who desires his vehicle to be expressive of the very best taste.

(from 1913 Patterson catalog, courtesy of R.P.)

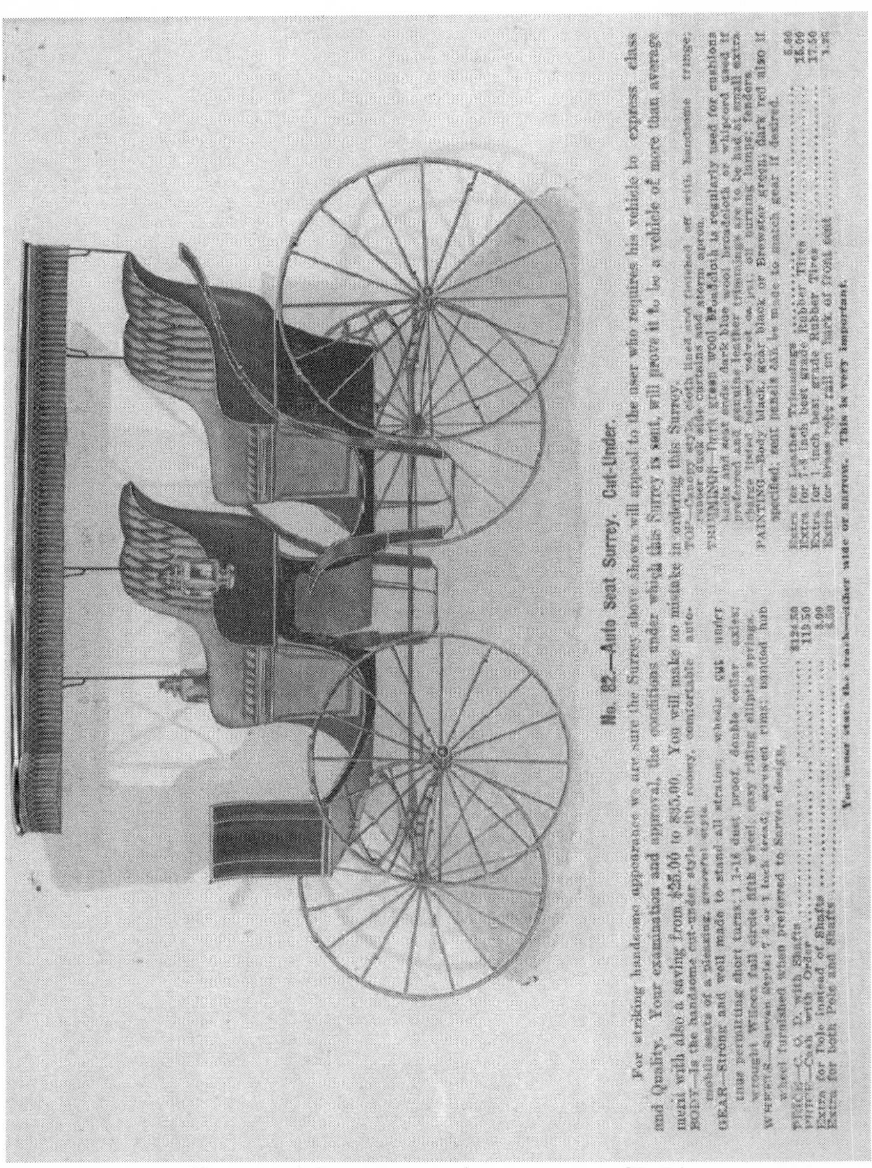

(from 1913 Patterson catalog, courtesy of R.P.)

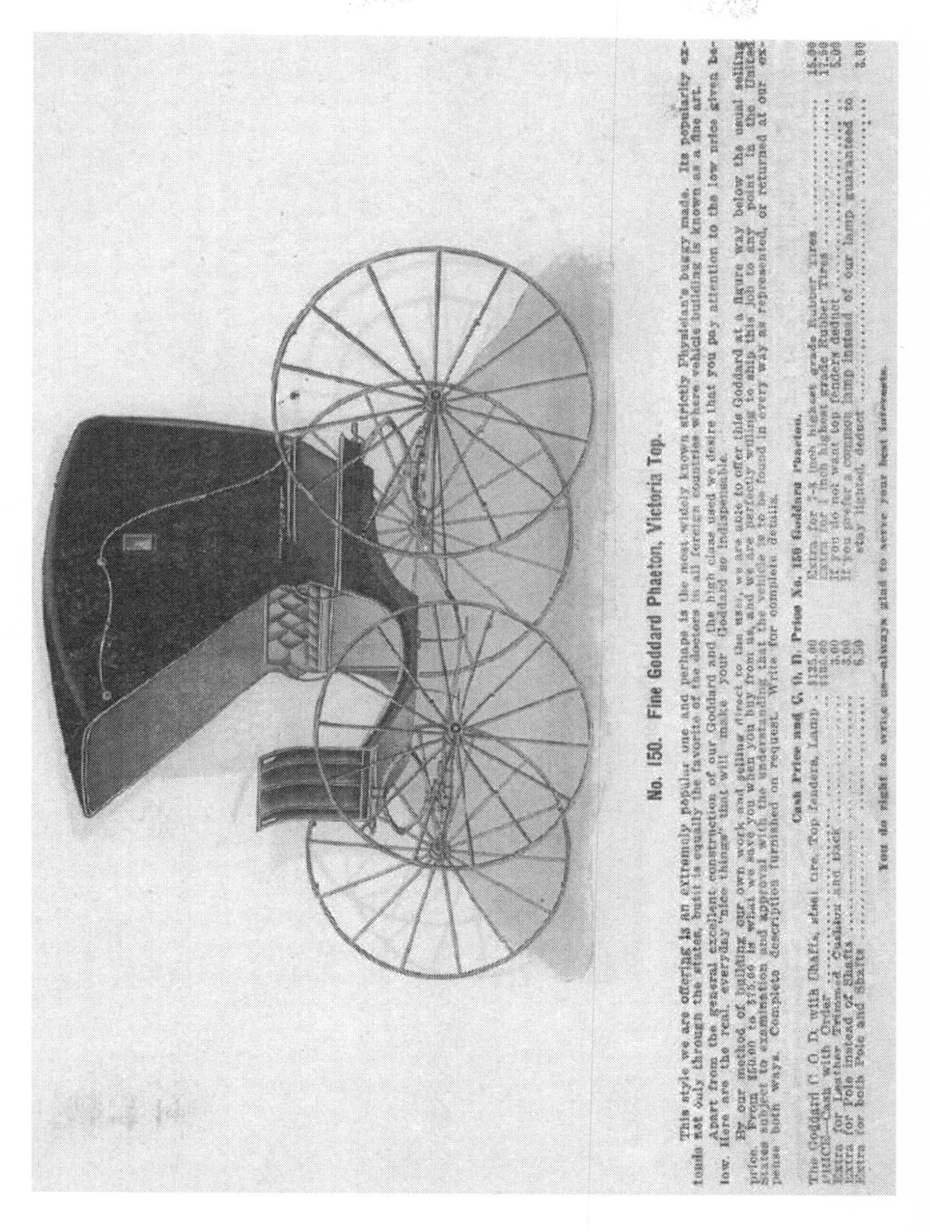

No. 150. Fine Goddard Phaeton, Victoria Top.

This style we are offering is an extremely popular one and perhaps is the most widely known strictly Physician's buggy made. Its popularity extends not only through the states, but it is equally the favorite of the doctors in all foreign countries where vehicle building is known as a fine art.

Apart from the general excellent construction of our Goddard and the high class used we desire that you pay attention to the low price given below. Here are the real, everyday "nice things" that will make your Goddard so indispensable.

By our method of building our own work and selling direct to the user, we are able to offer this Goddard at a figure way below the usual selling price. From $50.00 to $75.00 is what we save you when you buy from us, and we are perfectly willing to ship this job to any point in the United States subject to examination and approval with the understanding that the vehicle is to be found in every way as represented, or returned at our expense both ways. Complete description furnished on request. Write for complete details.

Cash Price and C. O. D. Price No. 150 Goddard Phaeton.

The Goddard C. O. D. with Shafts, steel tire, Top fenders, Lamp	$125.00	Extra for 7-8 inch highest grade Rubber Tires	15.00
PRICE—Cash with Order	$1[illegible]0.00	Extra for 1 inch highest grade Rubber Tires	17.50
Extra for Leather Trimmed Cushion and Back	3.00	If you do not want top fenders deduct	5.00
Extra for Pole instead of Shafts	3.00	If you prefer a common lamp instead of our lamp guaranteed to stay lighted, deduct	3.00
Extra for both Pole and Shafts	6.50		

You do right to write us—always glad to serve your best interests.

(from 1913 Patterson catalog, courtesy of R.P.)

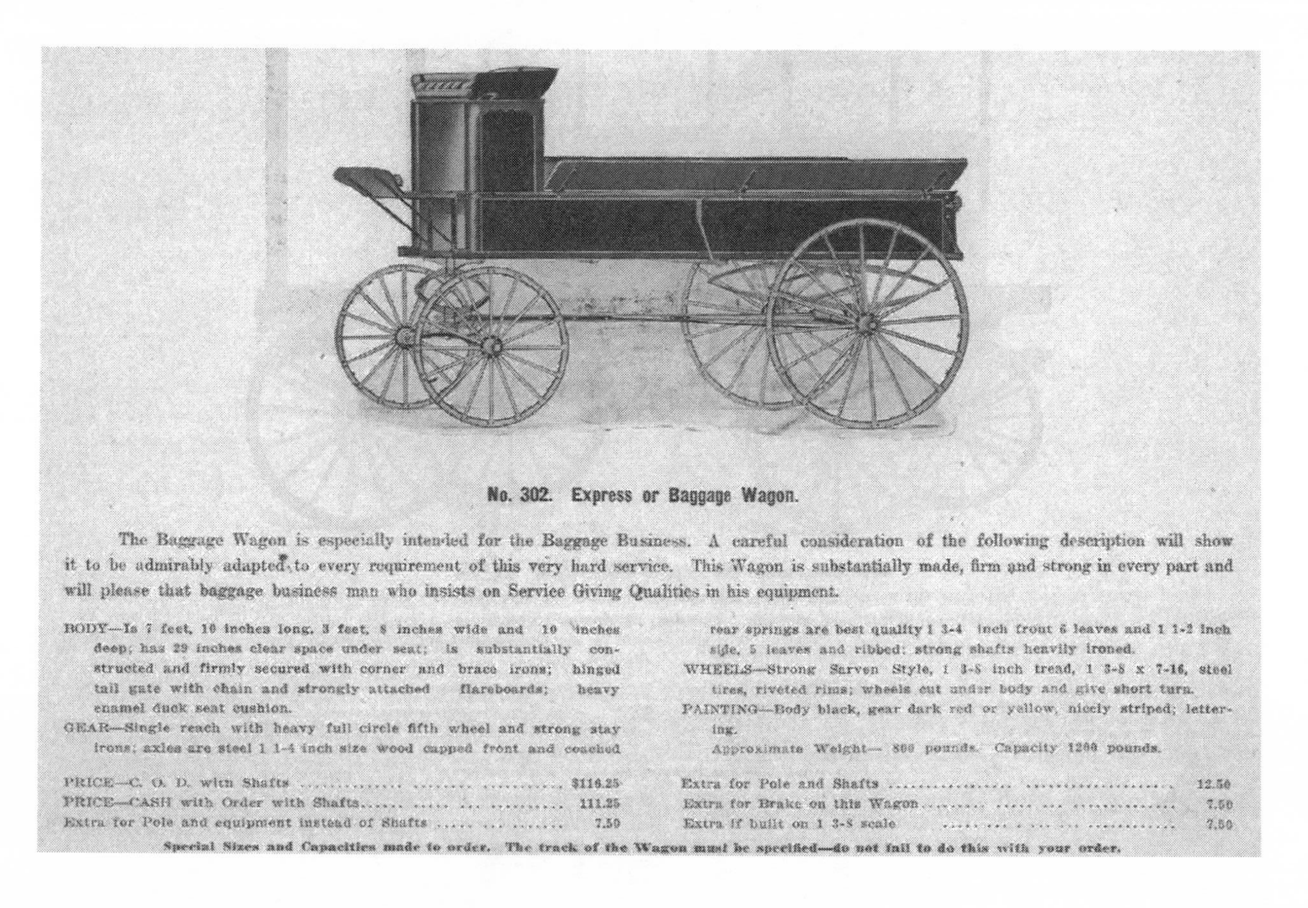

No. 302. Express or Baggage Wagon.

The Baggage Wagon is especially intended for the Baggage Business. A careful consideration of the following description will show it to be admirably adapted to every requirement of this very hard service. This Wagon is substantially made, firm and strong in every part and will please that baggage business man who insists on Service Giving Qualities in his equipment.

BODY—Is 7 feet, 10 inches long, 3 feet, 8 inches wide and 10 inches deep; has 29 inches clear space under seat; is substantially constructed and firmly secured with corner and brace irons; hinged tail gate with chain and strongly attached flareboards; heavy enamel duck seat cushion.

GEAR—Single reach with heavy full circle fifth wheel and strong stay irons; axles are steel 1 1-4 inch size wood capped front and coached rear springs are best quality 1 3-4 inch front 6 leaves and 1 1-2 inch side, 5 leaves and ribbed; strong shafts heavily ironed.

WHEELS—Strong Sarven Style, 1 3-8 inch tread, 1 3-8 x 7-16, steel tires, riveted rims; wheels cut under body and give short turn.

PAINTING—Body black, gear dark red or yellow, nicely striped; lettering.

Approximate Weight— 800 pounds. Capacity 1200 pounds.

PRICE—C. O. D. with Shafts	$116.25
PRICE—CASH with Order with Shafts	111.25
Extra for Pole and equipment instead of Shafts	7.50
Extra for Pole and Shafts	12.50
Extra for Brake on this Wagon	7.50
Extra if built on 1 3-8 scale	7.50

Special Sizes and Capacities made to order. The track of the Wagon must be specified—do not fail to do this with your order.

(from 1913 Patterson catalog, courtesy of R.P.)

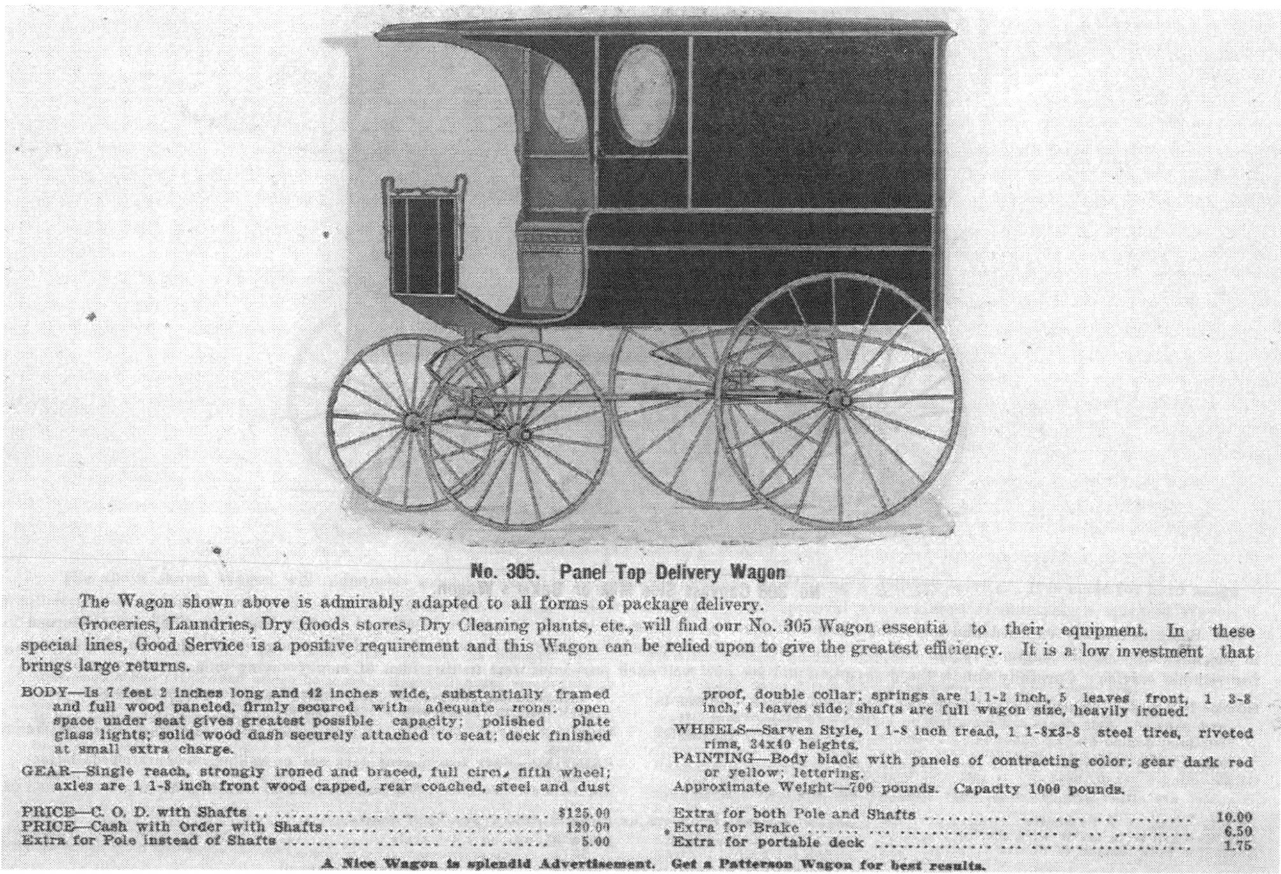

No. 305. Panel Top Delivery Wagon

The Wagon shown above is admirably adapted to all forms of package delivery.

Groceries, Laundries, Dry Goods stores, Dry Cleaning plants, etc., will find our No. 305 Wagon essentia to their equipment. In these special lines, Good Service is a positive requirement and this Wagon can be relied upon to give the greatest efficiency. It is a low investment that brings large returns.

BODY—Is 7 feet 2 inches long and 42 inches wide, substantially framed and full wood paneled, firmly secured with adequate irons; open space under seat gives greatest possible capacity; polished plate glass lights; solid wood dash securely attached to seat; deck finished at small extra charge.

GEAR—Single reach, strongly ironed and braced, full circle fifth wheel; axles are 1 1-8 inch front wood capped, rear coached, steel and dust proof, double collar; springs are 1 1-2 inch, 5 leaves front, 1 3-8 inch, 4 leaves side; shafts are full wagon size, heavily ironed.

WHEELS—Sarven Style, 1 1-8 inch tread, 1 1-8x3-8 steel tires, riveted rims, 34x40 heights.

PAINTING—Body black with panels of contracting color; gear dark red or yellow; lettering.

Approximate Weight—700 pounds. Capacity 1000 pounds.

PRICE—C. O. D. with Shafts	$125.00
PRICE—Cash with Order with Shafts	120.00
Extra for Pole instead of Shafts	5.00
Extra for both Pole and Shafts	10.00
Extra for Brake	6.50
Extra for portable deck	1.75

A Nice Wagon is splendid Advertisement. Get a Patterson Wagon for best results.

(from 1913 Patterson catalog, courtesy of R.P.)

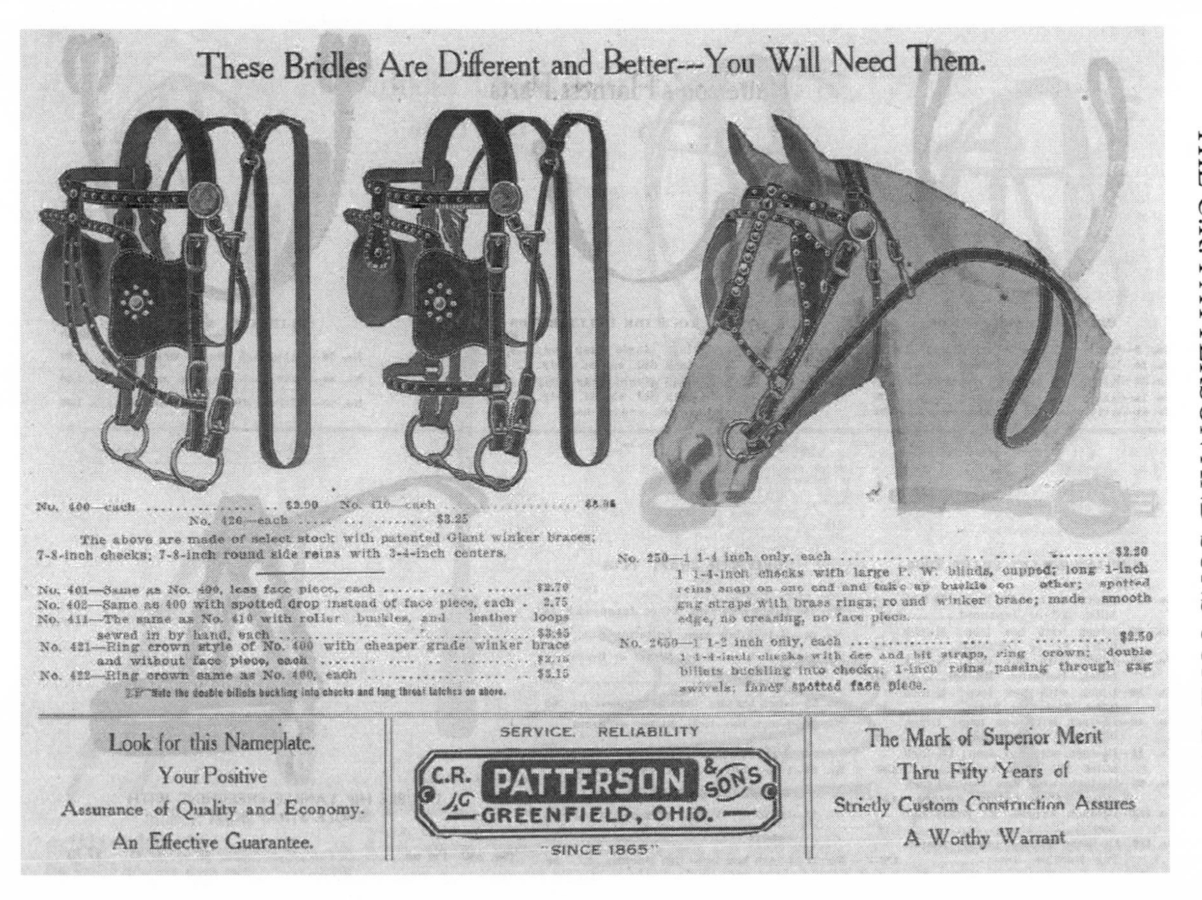

These Bridles Are Different and Better---You Will Need Them.

No. 400—each .... $2.90   No. 410—each .... [illegible]

No. 420—each $3.25

The above are made of select stock with patented Giant winker braces; 7-8-inch checks; 7-8-inch round side reins with 3-4-inch centers.

No. 401—Same as No. 400, less face piece, each $2.70
No. 402—Same as 400 with spotted drop instead of face piece, each . 2.75
No. 411—The same as No. 410 with roller buckles, and leather loops sewed in by hand, each $3.45
No. 421—Ring crown style of No. 400 with cheaper grade winker brace and without face piece, each .... [illegible]
No. 422—Ring crown same as No. 400, each $3.15

Note the double billets buckling into checks and long throat latches on above.

No. 250—1 1-4 inch only, each $2.20
1 1-4-inch checks with large P. W. blinds, cupped; long 1-inch reins snap on one end and take up buckle on other; spotted gag straps with brass rings; round winker brace; made smooth edge, no creasing, no face piece.

No. 2650—1 1-2 inch only, each $2.50
1 1-4-inch checks with dee and bit straps, ring crown; double billets buckling into checks; 1-inch reins passing through gag swivels; fancy spotted face piece.

Look for this Nameplate.
Your Positive
Assurance of Quality and Economy.
An Effective Guarantee.

SERVICE. RELIABILITY
C.R. PATTERSON & SONS
GREENFIELD, OHIO.
"SINCE 1865"

The Mark of Superior Merit
Thru Fifty Years of
Strictly Custom Construction Assures
A Worthy Warrant

(from 1913 Patterson catalog, courtesy of R.P.)

No. 176 Patterson Perfect Storm / Winter Buggy

The Patterson Company first introduced the No. 176 Storm Buggy in December of 1908. Although company representatives immediately claimed that this was the "perfect" storm buggy, they continually made improvements to its design. Several patents were applied for concerning this vehicle, especially for door devices. Unlike the open cab type design of most carriages and buggies of the period, the storm buggy included an enclosed passenger cab, which made the vehicle especially attractive during winter months and stormy weather. For this reason, the Patterson ad campaign constantly reminded readers that with a storm buggy, you get "comfort, health, and pleasure. Protect you and your family."[59] This vehicle quickly became popular among those who traveled in bad conditions.

The fully enclosed buggy cab added weight to a carriage. The top, doors, and glass made the vehicle heavy. Despite this, the Pattersons claimed to have the "lightest winter buggy on the market, while others drive a lumbering horse killing buggy." To achieve this, the company had to use the best light, yet strong, materials available. They stated, "a winter buggy requires the most careful construction, and we certainly give close attention to every little detail." In reference to quality, they also stated "think of it. Not a factory buggy, not one just slapped together, but a real genuine custom quality winter buggy at a lower price."[60] They wanted to construct the best winter buggy on the market yet keep it at an affordable price so that everyone could enjoy the comfort offered by the buggy.

To create a quality winter buggy, there were several obstacles to overcome that others had encountered when building one. First was the issue of weight, which has already been discussed. Another involved weatherizing the buggy cab. The Pattersons solved this by various means such as setting the glass in rubber grooves and making a better door.[61] This also eliminated rattles in the vehicle and moisture from seeping in. Another problem involved the door as well. As buggies would travel through the mud, sleet, and snow, the tracks on the outside would get clogged up and freeze. This would create a dangerous situation, not to mention an inconvenience. The passengers could be trapped inside the vehicle in case of an emergency. The Pattersons solved this problem by creating and patenting a door device that placed the slides on the inside of the vehicle, leaving nothing to get clogged or frozen. This device would make the Patterson Perfect Winter Buggy a success.

This door device was not perfected immediately though. As early as 1909, the Pattersons had created a door that had a pending patent.[62] In 1911, they received a patent (U. S. Patent # 983,992, approved February 14, 1911) for a different door device that pivoted on a rod and sealed tightly against the doorframe. This, apparently, was still not to their liking because another patent was received the following year, in 1912, for a sliding style door (U. S. Patent # 1,029,288, approved June 11, 1912). This door opened and closed in a manner like the side door of a modern van. It slid shut and then set into the doorframe and was locked into position. As mentioned previously, placing all exposed track on the inside of the vehicle eliminated the problems with slides becoming clogged and freezing.

This sliding door solved the problems of winter buggy design. This patent was so good that many competitors conceded it was "the best on the market." A December 6, 1911, letter from Charles W. Napper to W. O. Thompson, President of the Ohio State University, stated "we have designed and patented the only Perfect Winter Buggy in this country. The comfort and convenience we have thus afforded country life in the severest weather, is an achievement of which you will grant that we justly can be proud. This vehicle has been adopted by all the large carriage concerns on a royalty basis, and that is the positive proof of its superiority." The following illustration shows an October 1911 advertisement that listed 21 companies that would be using the patent and stated that there were "twice over as many more." The Patterson Company placed a block on the patent use by the other companies for a waiting period of two years. The other companies could not use it until 1913 and, in the meantime, the Pattersons were using it exclusively.[63] By this time, the Patterson name had spread across several states, but this patent and their Patterson Perfect Winter Buggy became their real claim of success.

The Winter Buggy compared favorably to its competitors. When comparing the Patterson buggy to the winter buggies of both the Ohio Carriage Manufacturing Company and the Columbus Carriage and Harness Company, both of nearby Columbus, Ohio, to the Patterson version, there are few noticeable differences.[64] The size was very similar for all three vehicles as well as upholstery, material types, and basic construction. Of course, the Patterson vehicle had their standard 38" springs rather than the 36" used on both other vehicles. The vehicle from the Ohio Carriage Manufacturing Company was virtually identical to the Patterson Winter Buggy. This was one of the companies listed that paid royalties to the Patterson Company for the sliding door patent, so the door and body shape were quite similar, if not the same, to the Patterson vehicle. The price of the vehicle was also essentially the same at $85-$87.50. Perhaps the biggest differences were in the vehicle from the Columbus Carriage and Harness Company. This vehicle used roll up curtains for the doors that provided no visibility through them. Despite this obvious inferiority compared to the other two buggies, the price from this company was much higher, as well, priced at $140. This is not enough data to suggest that the Patterson Winter Buggy was superior to others, but the use of the door patent among several top companies around Ohio and surrounding States does suggest that at least that one part was acknowledged to be superior.

Once the Patterson Company introduced the Winter Buggy, it became a dominant focus of its advertising. The company typically ran ads specific to the Winter Buggy from October until around March of each year from 1910 to 1917. Of the 609 advertisements collected that specifically mentioned a style of carriage, 22% of these were for the Winter Buggy. The only other vehicle that received double-digit percentages for advertisements was the No. 1 End Spring Piano Buggy (11%). Most of the ads were spread out among almost 30 different styles of vehicles. Many of the later ads (1914-1917) focused on how the Winter Buggy could help to save a person's automobile. It was better equipped to take the abuse of winter driving than an automobile and could supplement the automobile in many ways.[65] The Pattersons wanted customers to still purchase Winter Buggies even after owning an automobile. In fact, some of these ads came after the Pattersons introduced their own automobile.

Of their Patterson Perfect Winter Buggy, the ads stated that it "is not a luxury, it is a positive necessity, just as essential as a plow or binder. Once used no price would induce you to be without one." They further went on to state that the buggy could be used long into the year as the windows opened and it could still be used for summer use. During many months, the buggy "will prove a blessing to school children" as it would keep them out of all sorts of weather.[66] The Patterson Winter Buggy became an essential piece of equipment to those who owned one. One can only imagine the difference in comfort between riding in an open buggy exposed to the weather compared to riding in an enclosed cab. It is no wonder that this vehicle became so popular.

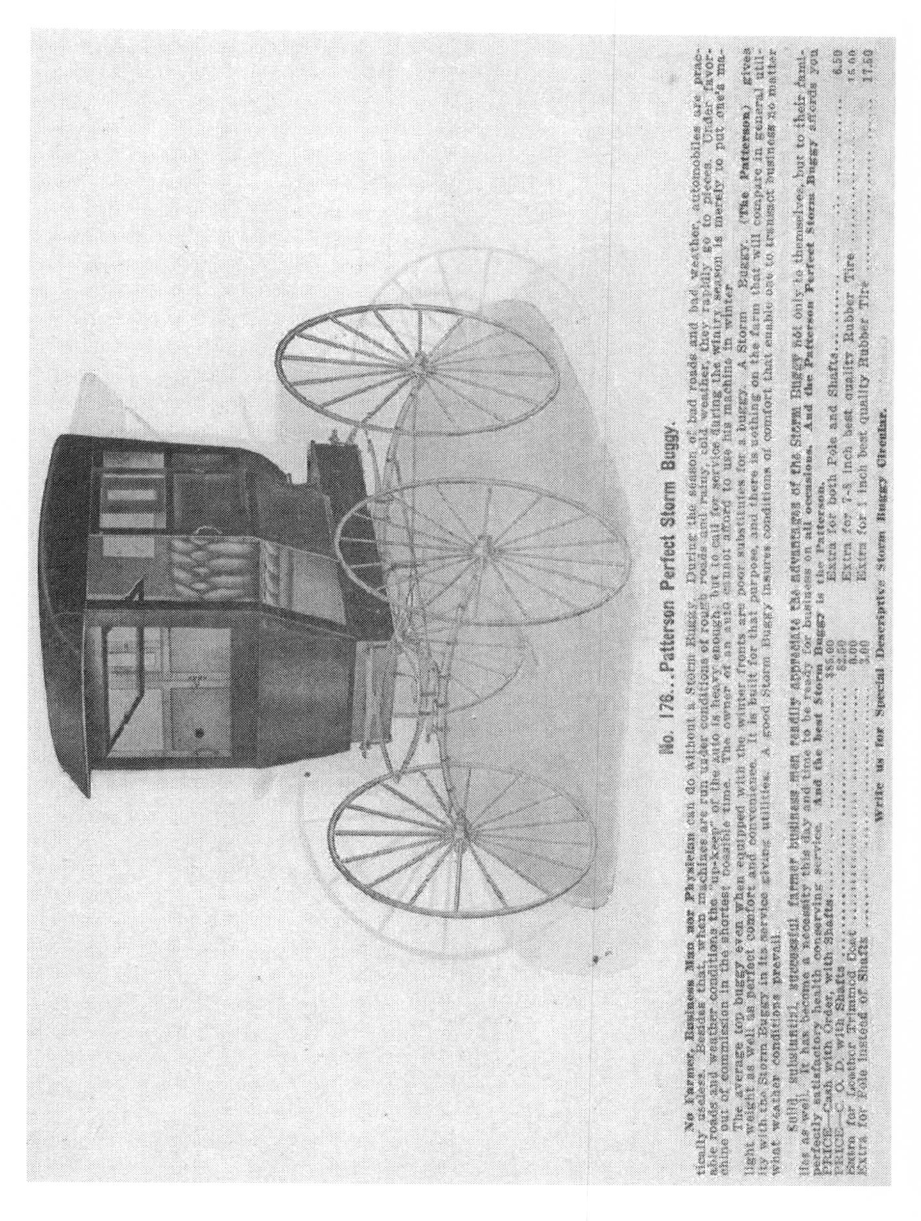

No. 176...Patterson Perfect Storm Buggy.

No Farmer, Business Man nor Physician can do without a Storm Buggy. During the season of bad roads and bad weather, automobiles are practically useless. Besides that, when machines are run under conditions of rough roads and rainy, cold weather, they rapidly go to pieces. Under favorable roads and weather conditions the "up-keep" of the auto is heavy enough; but to call for service during the wintry season is merely to put one's machine out of commission in the shortest possible time. The owner of an auto cannot afford to use his machine in winter.

The average top buggy even when equipped with the winter fronts are poor substitutes for a buggy. A Storm Buggy. (**The Patterson**) gives light weight as well as perfect comfort and convenience. It is built for that purpose, and there is nothing on the farm that will compare in general utility with the Storm Buggy in its service giving utilities. A good Storm Buggy insures conditions of comfort that enable one to transact business no matter what weather conditions prevail.

Solid, substantial, successful farmer business men readily appreciate the advantages of the Storm Buggy not only to themselves, but to their families as well. It has become a necessity this day and time to be ready for business on **all occasions. And the Patterson Perfect Storm Buggy** affords you perfectly satisfactory health conserving service. **And the best Storm Buggy** is the Patterson.

PRICE—Cash with Order, with Shafts	$85.00
PRICE—C. O. D. with Shafts	82.50
Extra for Leather Trimmed Coat	8.00
Extra for Pole instead of Shafts	3.00
Extra for both Pole and Shafts	6.50
Extra for 7-8 inch best quality Rubber Tire	15.00
Extra for 1 inch best quality Rubber Tire	17.50

Write us for Special Descriptive Storm Buggy Circular.

(from 1913 Patterson catalog, courtesy of R.P.)

Patterson advertisement listing several of the carriage companies that were using the patent for the Patterson Perfect Winter Buggy on a royalty basis. (*Hillsboro Dispatch* October 17, 1911)

Mail Delivery Buggy

The Patterson Company built several special purpose vehicles for its customers, including hearses, wedding carriages, and delivery wagons for all types of goods or products. One of these special purpose vehicles, a mail delivery buggy, was donated to the Greenfield Historical Society. This vehicle was labeled as "R. F. D. (Rural Free Delivery) Route No. 1, U. S. Mail." It is unclear which Post Office this vehicle was built for, but many locals believe that it was built for the Greenfield Post Office. While most of these would have most likely been custom made for each individual customer, and therefore differing from vehicle to vehicle, photographs were taken to record some of the features of this vehicle.

This mail delivery buggy accommodated a single passenger and a cargo of mail. Along with storage under the driver's seat, mail slots were also located in a tray in front of the driver. This buggy was completely enclosed, similar to the design of a Winter Buggy. The side doors of the buggy were of a split-level design. The upper half consisted of a sliding window that slid horizontally along a track. The lower half of the door was split in half vertically and was two swinging doors that opened to the outside. These were located on each side of the vehicle.

The front window of the buggy could be swung inward and attached to the ceiling of the vehicle to provide ventilation during the warmer months. When it was closed, slots were provided in the front wall of the vehicle for the reins to pass through. Also on this front wall was the Patterson Automatic Whipper, also known as the Hancher Whipper. Highland County resident Leonard Hancher originally designed this device, but the Patterson Company perfected it. It became a standard piece of equipment on many of its enclosed vehicles. It kept the driver from having to reach outside to whip a horse. This device consisted of a 1-5/8" ball and socket that passed through the front wall of the vehicle. On this inside was a handle for controlling the whipper, while on the outside was a standard whip socket. A

Patterson ad stated "you can get at a horse when he needs it – best cure on earth for a lazy horse – a safety device always ready. Just think, you don't let all the heat out and all the cold in – no doors to open."[67] This device was surely considered a convenience at the time.

Another convenience to the mail carrier was the installation of a heater in the buggy. A small oil tank was mounted on the backside of the buggy. A series of piping brought the oil down the back and along the bottom of the buggy where it attached to the heater and vent that was mounted in the floor of the buggy near the driver's feet. This heater was made by the Ashman Heater Company of St. Paris, Ohio, and was patented in 1907. The heater vent measured 8" by 13" and was made of cast iron. The oil tank measure 7" tall, 7-3/4" wide, and was 2-1/4" thick. This was surely appreciated by the mail carrier during the colder months.

Patterson advertisement for the Patterson Whipper. (*Hillsboro Dispatch* October 22, 1912)

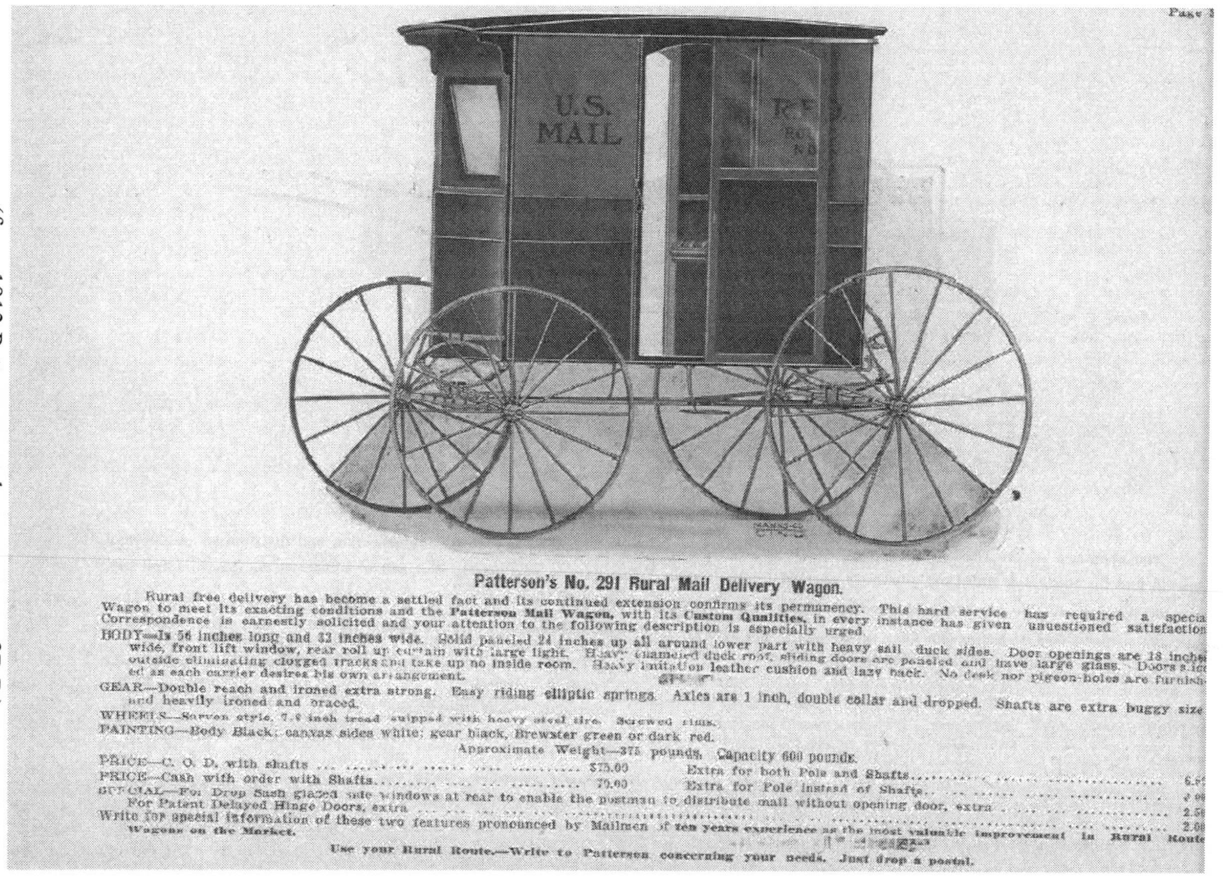

Page 3

Patterson's No. 291 Rural Mail Delivery Wagon.

Rural free delivery has become a settled fact and its continued extension confirms its permanency. This hard service has required a specia
Wagon to meet its exacting conditions and the **Patterson Mail Wagon,** with its **Custom Qualities,** in every instance has given unquestioned satisfaction
Correspondence is earnestly solicited and your attention to the following description is especially urged.

BODY—Is 56 inches long and 32 inches wide. Solid paneled 24 inches up all around lower part with heavy sail duck sides. Door openings are 18 inche
wide, front lift window, rear roll up curtain with large light. Heavy enameled duck roof, sliding doors are paneled and have large glass. Doors slid
outside eliminating clogged tracks and take up no inside room. Heavy imitation leather cushion and lazy back. No desk nor pigeon-holes are furnish
ed as each carrier desires his own arrangement.

GEAR—Double reach and ironed extra strong. Easy riding elliptic springs. Axles are 1 inch, double collar and dropped. Shafts are extra buggy size
and heavily ironed and braced.

WHEELS—Sarven style, 7-8 inch tread equipped with heavy steel tire. Screwed rims.

PAINTING—Body Black; canvas sides white; gear black, Brewster green or dark red.

Approximate Weight—375 pounds. Capacity 600 pounds.

PRICE—C. O. D. with shafts .... $75.00 Extra for both Pole and Shafts 6.5
PRICE—Cash with order with Shafts .... 73.00 Extra for Pole instead of Shafts 2.0
SPECIAL—For Drop Sash glazed side windows at rear to enable the postman to distribute mail without opening door, extra 2.50
For Patent Delayed Hinge Doors, extra 2.00
Write for special information of these two features pronounced by Mailmen of ten years experience as the most valuable improvement in Rural Route
Wagons on the Market.

Use your Rural Route.—Write to Patterson concerning your needs. Just drop a postal.

(from 1913 Patterson catalog, courtesy of R.P.)

Front oblique view of a Patterson-built mail buggy.

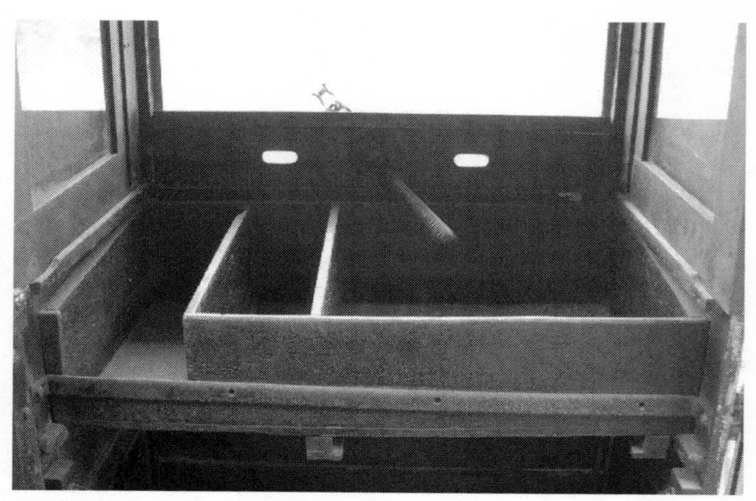

Mail tray in the front of the mail buggy.

The Patterson Automatic Whipper and rein slots. Notice the ball and socket mounted in the front wall along with the whip socket on the outside. Also in the photograph are the rein slots through the front wall.

Rear view of the mail buggy. The oil tank for the heater is mounted high up on the left.

School Wagons

In the early 20th century, as more rural schools became consolidated, the need for proper transportation conveyances grew. The Patterson Company was a pioneer in this work and began designing school wagons for the transportation of students. Early school wagons were often too heavy for the horses, exposed the children to the weather, or were just too dangerous to use. The Patterson Company tackled these problems and tried to build a better school wagon. By 1906, they had built a school wagon that was seen as being superior to others and it became a standard design.

After observing some of the Patterson school wagons, the Ohio State School Commissioner, E. A. Jones, stated "I am much pleased with the wagons furnished by you for the transportation of pupils. They seem to be constructed with reference to strength, durability, and safety. They are well arranged for convenience of pupils, well protected and well lighted. I have seen nothing better in this line." Another article attested that it was no surprise that the Patterson Company managed to become successful in this line of work. It went on to state "when you know, however, something of the spirited aggression of this firm and the style and merit of their work, it is not surprising that they are able to close contracts in the face of competition."[68] On several occasions, as any good business would do, the Patterson Company would try to step in and fill a void and provide a product that was in great demand.

None of the school wagons have survived, so little information is known about them except for details gleaned from newspaper articles and from the 1913 catalog. A 1906 article provided some information about the vehicles. It stated:

> The body is a swell side 12-foot arch with roll-up curtains all along the sides. The front door slides and this avoids interference with the horses. The driver sits inside and is not only protected from exposure but can

> exercise supervision over the children within. The wagon can easily accommodate from 20 to 28 pupils. The front truck shows full sills platform gear, with stiff tongue and heavy tongue straps. One of the most notable features of the wagon are the steps with rail to assist the very small children in entering.[69]

In Charles Napper's letter to the President of the Ohio State University, he describes how the school wagon improved children's studies at school. He mentioned that students previously would have to travel exposed to the weather and then be expected to sit down in a classroom and think with distracted minds. Now, "this vehicle affords the protection which often changes slow faculties benumbed with cold into apt minds alive to schoolwork."[70] They built many of these school wagons and sent them to several states. Its school wagons were seen as being superior to many of the other early pupil transports of the time. The experiences and contacts that the Patterson Company gained building these school wagons became important several years later when the company began building motorized school buses.

Although only a few examples of the Patterson Company products have been detailed here, the company built many types of vehicles. Although the company only advertised the main vehicles that it built on a regular basis, it also built many custom and special purpose vehicles at the customer's request. One article mentioned that "the extent of their business [was] determined only by the limitations of the use of vehicles and means of transportation...Patterson and Sons have not been content to merely follow or just keep up with the carriage industry but have led off in many ways."[71] Through the carriage-building phase of the company, they met success in many ways. They invented and patented several components to include a vehicle dash, a thill coupling, furniture casters, the automatic whipper, and several modifications to its Patterson Perfect Winter Buggy, while also pioneering the effort to provide vehicles for school transportation. They kept up with the needs of the consumers and made attempts to fill those needs. They offered a wide range of products in their carriage line but saw a decline of that business around 1910 owing to the rising automobile age. The Patterson Company began its shift to the automobile industry.

The Patterson Standard School Wagon

Your special attention is called to the specifications given below because there is a reason for every feature of the construction of these Wagons. **Long experience** in the manufacture and constant attention to requirements that give best service, enable us to produce the **Best Wagon** on the market today. The perfect satisfaction given by the Patterson School Wagons in every instance proves stronger than words the unequalled merit of their excellent quality.

KINDS—We make two sizes of School Wagons—differing only in the length of body. One twelve-foot, the other ten-foot.

BODY—Is straight sill swell side pattern, sills are best ash; panels best poplar screwed and plugged; Seats 1 foot. high, 1 foot. wide, 24 inches between seats, lazy back 16 in. high; Front is solid panel, two side lights, middle lift window with door on left side; Rear-two side lights, middle door, convenient steps. Basket Stalls Under Seat.

GEARING—Gear-platform; stiff tongue; Axles—1 1-2x8, both Full Coach; Springs—Full platform and ribbed; Wheels—Sarven Style, 2-in. tread, riveted rims, heavy steel tires; Doubletree, Singletree and Tongue—Best native timber well ironed and complete with chains and straps; Brake—Center lever, effective, convenient.

TRIMMING—Cushion and Rack heavily padded. Duck is preferable to carpet being cleaner and more sanitary; Curtains—Made of heavy oiled duck; Top—Covered with Rubberoid Roofing.

PAINTING—Lead and oil system, best color, durable varnishes; Body black, gear yellow; Lettering as desired and no extra charge.

PRICE—

12-ft. Wagon $192.50

10-ft. Wagon 187.50

F. O. B. Greenfield, Ohio.

You will find us $35.00 to $50.00 lower on the Same Grade Wagon.

TERMS—Cash or Note within 30 days after delivery.

We are the most interested of all parties in having our School Wagons possess every merit of service as well as price, and we will be glad to confer with prospective buyers where we can explain fully the essential features that experiences has proven to be necessary to Satisfactory Service in a School Wagon.

Write us on any point not covered and we will be glad to enter fully into every detail.

(from 1913 Patterson catalog, courtesy of R.P.)

(from 1913 Patterson catalog, courtesy of R.P.)

CHAPTER 2 NOTES

[1] Don H. Berkebile, *American Carriages, Sleighs, Sulkies, and Carts* (New York: Dover Publications, 1977), v.
[2] Ibid., v.
[3] David A. Hounshell, *From the American System to Mass Production, 1800-1932: The Development of Manufacturing Technology in the United States* (Baltimore: The Johns Hopkins University Press, 1984), 147-149; The Museums at Stony Brook, *19th Century American Carriages: Their Manufacture, Decoration, and Use* (Stony Brook, New York: The Museums at Stony Brook, 1987), 10, vi.
[4] Thomas Kinney, *The Carriage Trade: Making Horse-Drawn Vehicles in America* (Baltimore: The Johns Hopkins University Press, 2004), 5.
[5] The Museums at Stony Brook, *19th Century American Carriages*, 139, 143.
[6] Ibid., 139.
[7] Horace Greeley, *The Great Industries of the United States* (Hartford, Connecticut: J. B. Burr and Hyde, 1873), 805.
[8] John Howard Burrows, "The Necessity of Myth: A History of the National Negro Business League, 1900-1945" (PhD diss., Auburn University, 1977), 1-2, 5.
[9] D. J. Lake, *Atlas of Highland County, Ohio* (Philadelphia: C. O. Titus, 1871), 57.
[10] Bill Naughton, "Patterson-Greenfield Cars," *Old Cars Weekly* June 19, 1980. Perhaps Lowe and Patterson had bought out Dines and Simpson since the two businesses worked out of the same location, where the later Patterson shop would also be located. *Greenfield Republican,* "Mr. Ed Dines," December 1, 1921.
[11] Records of the National Negro Business League (microfilm), compiled by Kenneth M. Hamilton (Bethesda, Maryland: University Publications of America, 1994). 1st Annual Meeting 1900:66. Several sources claim that J. P. Lowe died in 1893 and

---

this left C.R. as sole owner, but Lowe moved on to other business ventures to include being President of the Home Building & Loan Company in Greenfield for many years. The H. B. & L. Company gave the Pattersons nearly half of the mortgages that they would obtain through their business, so it seems likely that C.R. and Lowe had separated on friendly terms.

[12] *The Highland Weekly News*, December 5, 1883.

[13] *The Cincinnati Daily Star*, April 26, 1880.

[14] *The Highland Weekly News*, "Burglars at Greenfield, Two Carriage Factories Robbed in One Night," December 23, 1880.

[15] *News-Herald (Hillsboro)*, November 14, 1889.

[16] Highland County Partnership Record, January 10, 1899. Available at Highland County, Ohio, courthouse.

[17] Booker T. Washington, *Booker T. Washington's Own Story of His Life and Work* (Naperville, Illinois: J. L. Nichols & Company, 1901), 323; John Howard Burrows, "The Necessity of Myth: A History of the National Negro Business League, 1900-1945" (PhD diss., Auburn University, 1977), 41.

[18] *Greenfield Times,* "Greenfield Nearly Made U. S. Auto Center," November 18, 1971; J. W. Klise, *The County of Highland: A History of Highland County, Ohio, From the Earliest Days* (Madison, Wisconsin: Northwestern Historical Association, 1902), 428; Thomas William Burton, *What Experience Has Taught Me* (Cincinnati, Ohio: Jennings and Graham, 1910), 105.

[19] National Negro Business League, 2nd Annual Meeting 1901:21; *JNMA,* "Books, Lay Press, etc.," Vol. 6, No. 4 (1914):268. The fire insurance maps of the known company buildings do not reflect a number remotely close to 50,000 square feet. The Patterson Company owned many properties, but it is not certain what the exact purpose of most of the properties. Perhaps they used some of these as storage repositories and have considered them in the 50,000 square foot total, or the number could be an embellishment.

[20] Ibid., 268; *GR,* "C.R. Patterson and Sons," Holiday Issue, December 1902.

---

[21] *New York Times,* "Future of the Negro: Nashville Convention Decides it is in His Own Hands," August 22, 1903; *Greenfield Journal,* "Stricken in Death," April 29, 1910; *GR,* "$5,000 Worth of Business," January 28, 1909.
[22] *JNMA,* "Books, Lay Press, etc.," Vol. 6, No. 4 (1914):268; *GR,* "C.R. Patterson and Sons," Holiday Issue, December 1902.
[23] *JNMA,* Patterson Ad, Vol. 1., No. 1 (1909): 268.
[24] *JNMA,* "Books, Lay Press, etc.," Vol. 6, No. 4 (1914): 268.
[25] *Hillsboro Dispatch* (hereafter shown as *HD*), Patterson Ad, April 19, 1910; March 19, 1912; April 12, 1912.
[26] *Courier (Cleveland),* "Negro Family Made 'First' Cars," December 10, 1965; *JNMA,* "Books, Lay Press, etc.," Vol. 6, No. 4 (1914):268.
[27] Records of the National Negro Business League (microfilm), compiled by Kenneth M. Hamilton (Bethesda, Maryland: University Publications of America, 1994). 8th Annual Meeting 1907:129-130.
[28] *The Crisis,* Patterson Ad, Vol. 19, No. 2 (1919); Vol. 19, No. 5 (1920); Vol. 20, No. 1 (1920); Records of the National Negro Business League (microfilm), compiled by Kenneth M. Hamilton (Bethesda, Maryland: University Publications of America, 1994). 8th Annual Meeting 1907:130; *Greenfield Times,* "Former Employee Recalls Buggy Company," February 14, 1992.
[29] Henry Howe, *Historical Collections of Ohio* (Cincinnati: C. J. Krehbiel & Company, 1904), 924; Records of the National Negro Business League (microfilm), compiled by Kenneth M. Hamilton (Bethesda, Maryland: University Publications of America, 1994). 1st Annual Meeting 1900:66; *GR,* Patterson Ad, June 28, 1910; Ohio Colored Men's Business Association, *The Ohio Business Directory and Information Guide of Ohio's Colored Business Men and Women* (Columbus: Ohio Colored Men's Business Association, 1913), 49; *JNMA,* "Books, Lay Press, etc.," Vol. 6, No. 4 (1914):268; The Industrial Commission of Ohio, *Directory of Ohio Manufacturers* (Columbus: F. J. Heer Printing Company, 1918), 30.

---

[30] Ibid., 29-31. The Patterson shop being in the top 28% of carriage companies in Ohio in regards to number of employees is pretty significant. Six States shared over one-half of the total carriage products of the United States – Ohio, Indiana, Illinois, Michigan, Wisconsin, and Missouri. Ohio led these States in production numbers; in fact, Cincinnati manufacturers alone produced twice as many carriages as the number 2 (St. Louis), and number 3 (South Bend, Indiana) cities in the United States. This indicates that being in the top 28% of Ohio carriage makers during that time was no small feat. Joseph Russell Smith, *Commerce and Industry* (Baltimore: Henry Holt and Company, 1916), 247-248.

[31] Greenfield Historical Society. *Greenfield, Ohio: 1799-1999* (Paducah, Kentucky: Turner Publishing Company, 2000), 131-132, 135, 138.

[32] *JNMA,* Patterson Ad, Vol. 1., No. 1 (1909); Frank Lincoln Mather, *Who's Who of the Colored Race: A General Biographical Dictionary of Men and Women of African Descent* (Chicago, 1915), 212; Records of the National Negro Business League (microfilm), compiled by Kenneth M. Hamilton (Bethesda, Maryland: University Publications of America, 1994). 1st Annual Meeting 1900:66.

[33] Jack Salzman et al., *Encyclopedia of African-American Culture and History* (New York: Macmillan Library Reference USA, 1996), 2110. In comparison to the production numbers of a large mechanized factory, the Studebaker Brothers Company of South Bend, Indiana, the Patterson numbers look small. While the Patterson Company was turning out 400-600 vehicles per year at $75,000 with 35-50 men employed, the Studebaker factory was producing 6,950 vehicles per year at $691,000 with a labor force of 325, and this was 28 years earlier in 1872, so by 1895, the numbers had increased exponentially to 75,000 vehicles and 1,900 employees, Hounshell, *From the American System to Mass Production*, 147-149.

[34] *The Ohio Farmer,* Patterson Ad, January 13, 1912.

---

[35] Kinney, *The Carriage Trade,* 52; Highland County Deed Record, Vol.82, Page 232. March 1890; Highland County Deed Record, Vol. 92, Page 483. December 1900. Available at Highland County, Ohio, courthouse; Records of the National Negro Business League (microfilm), compiled by Kenneth M. Hamilton (Bethesda, Maryland: University Publications of America, 1994). 1st Annual Meeting 1900:66; *GR,* "C.R. Patterson and Sons," December 1902; *HD,* Patterson Ad, May 28, 1912.
[36] *HD,* Patterson Ad, July 19, 1910; September 20, 1910; September 27, 1910; October 4, 1910.
[37] *GR,* Patterson Ad, February 27, 1908; November 14, 1907; *HD,* Patterson Ad, June 7, 1910; *JNMA,* "Books, Lay Press, etc.," Vol. 6, No. 4 (1914):268; *HD,* Patterson Ad, July 2, 1912; August 19, 1913; *Greenfield Times,* "Former Employee Recalls Buggy Company," February 14, 1992.
[38] Kinney, *The Carriage Trade,* 40, 41.
[39] Ibid., 47, 50-53, 59.
[40] Ibid., 47, 50-53, 59.
[41] Ibid., 36, 44, 47, 50-53.
[42] *HD,* Patterson Ad, June 7, 1910; *GR,* Patterson Ad, March 18, 1915; *HD,* Patterson Ads, March 31, 1914; April 21, 1914; March 11, 1913.
[43] *HD,* Patterson Ad, March 8, 1910.
[44] *GR,* Patterson Ad, March 12, 1908.
[45] Bill Naughton, "Former Slave, Descendants Were Early Vehicle Pioneers in Ohio," *Tri-State Trader,* July 5, 1980; *Greenfield Times,* "Patterson Family Earned Niche in History of Automaking," July 14, 1980.
[46] Henry W. Meyer, *Memories of the Buggy Days* (Cincinnati, Ohio: Brinker, Inc., 1965), 140.
[47] Henry W. Meyer, *Memories of the Buggy Days,* 140.
[48] *GR,* "C.R. Patterson and Sons," December 1902; *HD,* Patterson Ad, June 25, 1912.
[49] Records of the National Negro Business League (microfilm), compiled by Kenneth M. Hamilton (Bethesda, Maryland: Uni-

---

versity Publications of America, 1994). 1st Annual Meeting 1900:64.

[50] *The Ohio Farmer,* Patterson Ad, February 10, 1912; *JNMA,* Patterson Ad, Vol. 4, No. 2 (1912); *HD,* Patterson Ad, November 5, 1912.

[51] *GR,* Patterson Ad, February 16, 1905; *HD,* Patterson Ad, January 14, 1913; January 7, 1913.

[52] *The Ohio Farmer,* Patterson Ad, March 9, 1912.

[53] *HD,* Patterson Ad, May 6, 1913; August 13, 1912; April 14, 1914; *GR,* Patterson Ad, October 31, 1907.

[54] *HD,* Patterson Ad, July 30, 1912; August 6, 1912; April 1, 1913.

[55] *GR,* Patterson Ad, June 25, 1914; *HD,* Patterson Ad, January 1, 1910.

[56] *HD,* Patterson Ad, April 22, 1913; April 15, 1913.

[57] *HD,* Patterson Ad, May 21, 1912; August 12, 1913; July 1, 1913.

[58] *HD,* "Prestige for Greenfield is Won Through the Progressive Efforts of the C.R. Patterson and Sons Company," December 5, 1916.

[59] *GR,* Patterson Ad, December 3, 1908; *HD,* Patterson Ad, January 17, 1911.

[60] *HD,* Patterson Ad, December 6, 1910; October 31, 1911; October 7, 1913.

[61] *HD,* Patterson Ad, November 21, 1911.

[62] *HD,* Patterson Ad, December 28, 1909.

[63] *HD,* Patterson Ad, October 17, 1911; Letter from Charles W. Napper to Ohio State University President W. O. Thompson dated December 6, 1911. This letter is located at the Greenfield Public Library, Greenfield, Ohio, in the Patterson Company scrapbook; *HD,* Patterson Ad, October 17, 1911; *HD,* "The Hancher Whipper," October 22, 1912.

[64] Columbus Carriage and Harness Company, *Thirty-third Annual Catalogue, The Columbus Carriage and Harness Company* (Cincinnati, Ohio: C. J. Krehbiel, 1910), 54-55; Ohio Carriage

---

Manufacturing Company, *Split Hickory Vehicles & Ohio Oak Tanned Harness, Sixteenth Annual Catalogue* (Columbus, Ohio: Ohio Carriage Manufacturing Company, 1916), 86-87. Both available at the State Archives of Ohio, Ohio Historical Society, Columbus.

[65] *GR,* Patterson Ad, October 29, 1914; November 4, 1915; December 23, 1915; October 24, 1916.

[66] *HD,* Patterson Ad, December 12, 1911; January 28, 1913.

[67] *HD,* "The Hancher Whipper," October 22, 1912; *HD,* Patterson Ad, January 27, 1914.

[68] *GR,* "Centralized Schools Goal of Country Districts: Real Problem of Transportation is Solved by a Local Concern," March 1, 1906; *GR,* "School Wagons," August 29, 1907.

[69] *GR,* "Centralized Schools Goal of Country Districts: Real Problem of Transportation is Solved by a Local Concern," March 1, 1906.

[70] Letter from Charles W. Napper to Ohio State University President W. O. Thompson dated December 6, 1911. This letter is located at the Greenfield Public Library, Greenfield, Ohio, in the Patterson Company scrapbook.

[71] *JNMA,* "Books, Lay Press, etc.," Vol. 6, No. 4 (1914): 268.

3

THE SHIFT TO MANUFACTURING AUTOMOBILES

The Automobile Industry in the United States

Early in the 20th century, as more and more automobiles appeared on city streets, it became apparent that the time of the horse-drawn vehicle was coming to an end. Although automobiles first arrived in the United States in the mid-1890s, they did not become reliable enough and affordable enough to pose a threat to the carriage builder until after the turn of the century. During the 1890s several builders, and backyard tinkerers, experimented with building an automobile. Some of these people included names such as the Duryea brothers, Ransom E. Olds, Henry Ford, and Alexander Winton.[1]

Most of the car builders of the 1890s, and early 1900s for that matter, were assemblers and did not truly manufacture an entire automobile and all its parts. The parts of an early automobile already existed, but were in use for other purposes, such as "stationary and marine gasoline engines, carriage bodies, and wheels."[2] The gears, transmissions, chains, and other parts were already being used in industrial settings and could easily be added to the body. The biggest issue for the builder was creating a steering mechanism for the automobile since a tug on the reins to the left or right was no longer an option for steering as it had been with the horse. Most of the early builders used a tiller as a solution to this problem and it wasn't until later that a steering wheel was added.

In the early years, auto bodies looked much like a carriage. By 1908, the automobile body styles had added the hood and motor on the front of the body, which increased the wheelbase and created more room for passengers. Moving the engine out front, rather than its former position underneath the passengers, also allowed the car to set lower, which improved the handling. Prior to this, the carriage-style body was what people were used to seeing and anything different was considered "uncouth" by many of the early builders. The motor was tucked up under the seat and some type of drive mechanism, often a chain, ran to a sprocket on the rear axle. Since the automobile industry was so new, it had not yet become clear which type of motive power would be the best for use in the vehicle. Many forms of power production were used, including steam, electric, and gasoline, some with more success than others. Steam engines and electric motors both had some significant limitations at the time, so gasoline engines won out in the end and became the dominant and most accepted form of fuel for powering automobiles.[3] The gasoline engine had its own problems, but it seemed the best choice at the time.

By 1900, many people had tried their hand at building an automobile. According to an 1895 article in *The Horseless Age,* an estimated 300 Americans had attempted to construct automobiles. This number grew as time went on. An 1899 article in *Motor Age* estimated that perhaps "one thousand such shops exist in the United States today, and probably one hundred of them have been in operation for two years or longer without yet having advanced to the stage of manufacture, except in a very few instances." As of 1984, the Motor Vehicle Manufacturers Association stated that close to 1,900 automobile firms had operated in the United States since 1893 and there had probably been others that left no trace of their existence. In 1966, David Hebb stated that "the cars had to be bought as well as built, and for a while there were more builders than buyers."[4]

Early builders, prior to 1910 when the automotive field was well established, came from many different fields and backgrounds. "The logical moves were from wagon or carriage building, like the Studebakers; from bicycles, like Colonel Pope; from machine shop work, like Leland and the Dodge Brothers." Industries that already had shops, equipment, and skilled workers could enter the automobile field with little initial investment. In the early years, getting into the automobile field was somewhat easy. Since the cars were assembled utilizing mostly existing parts, anyone with a mechanical inclination could attempt to build one.[5]

Many parts manufacturers built specific assemblies for automobiles. They would try to attract many newcomers into the automobile manufacturing field to experience higher sales for their own products. One example of this was the Mack Brothers of Pennsylvania who built chassis for automobiles. They advertised that they wanted wagon builders to build commercial motor vehicles. They added "no need to enlarge your factory; no necessity to hire automobile experts. Simply build the bodies and mount them on." Others did the same. The Auburn Motor Buggy Chassis Company from Indiana advertised that "you can equip any buggy with our chassis and thus convert it into an automobile."[6] With this form of targeted advertising toward carriage manufacturers and the notion that the days of the horse-drawn vehicle were coming to an end, many carriage makers made the shift to manufacturing automobiles.

Before the turn of the century, poor roads made the use of an automobile often more trouble than it was worth, because of the constant broken parts and tire troubles caused by the rough terrain. As technological advancements, mechanical reliability, and road conditions improved, the automobile sales increased greatly. In 1902, there was one car for every 65,000 people, but by 1909, there was one car for every 800 people. A 1907 article in the *New York Times* emphasized the rapid development in the industry and stated that in the past 11 years there had been 1,170 automobile related patents granted and over 1,500 granted for tires alone. Automotive historian John B. Rae stated that "by the end of the first decade of the twentieth century the automobile in the United States could no longer be regarded either as a novelty or as a rich man's plaything; it was already potentially what it would become in fact – an item of incredible mass consumption."[7]

The early automobile industry was dominated by White manufacturers. Blacks finally entered the automobile industry in large numbers as laborers during the period 1916-1918, as a result of labor shortages because of World War I and restrictions on European immigration. In 1913, Henry Ford had switched his operations at his Highland Park plant to using the automated assembly line, which, by 1914, required 14,000 workers to mass-produce automobiles. At the onset of World War I, many of his workers were called to service for the war effort, so Ford opened his factory to Black workers. Not only did Ford start providing jobs for Blacks, but he also paid them the same high wages that the White workers received, $5 a day. This created a mass migration of Blacks to the Detroit area and aided tens of thousands of Blacks to enter the middle class.[8] According to Charles James Price, who eventually became the first Black worker at Ford to receive a salaried position, only he and one other Black worker were employed at Ford in 1915. To fill the shortage of employees caused by the war, Ford began hiring large numbers of Black workers and between 1917 and 1925 he had hired nearly 10,000 Blacks. Many of these were placed in positions that were considered the auto industry's "job ghettoes." These included jobs in the noxious paint spraying and wet-sanding departments, and the equally uncomfortable foundries, where the heat was intense, and the air was full of thick clouds of soot. More than half of Ford's Black workers were employed in the foundry, which was considered one of the plants "Negro jobs." Nonetheless, the automobile industry in the north provided Blacks with a much better economic opportunity than existed for them in the south.[9]

Blacks also became involved in the age of the automobile because of Jim Crow laws and other segregation biases. During the early 20th Century, several Blacks began operating jitney bus services, because White bus lines would not carry them. Although the Black automobile economy was on the rise by the late 1920s, segregation kept Blacks from experiencing the "open road" in the same way that the White automobile owners did. When Blacks traveled, "they encountered not a feeling of national camaraderie, but a contested terrain that was neither officially segregated nor completely welcoming. The growing numbers of White-owned hotels, resorts, and auto camps that grew up alongside the road often did not admit Blacks." Many service stations and repair shops refused to serve them. This prompted Black businessmen to build their own separate system of automobile related services. These were set up to allow Blacks to travel more freely without fear of not finding services that they needed during their trips. In 1936, at the height of racial segregation, Victor and Alma Green published their travel guide for Black Americans: "The Green Book." In the introductory section of the first publication, Victor Green poignantly wrote: "There will be a day sometime in the near future when this guide will not have to be published. That is when we as a race will have equal rights and privileges in the United States. It will be a great day for us to suspend this publication for then we can go wherever we please, and without embarrassment. But until that time comes, we shall continue to publish this information for your convenience each year." The Green Book filled a critical need for Black Americans which was the key to its success.[10] Many Blacks purchased automobiles for the same reasons that Whites did – the automobile offered escape and were status markers symbolizing a move into the middle class. Many early Blues songs included the automobile as a symbol of freedom.[11]

A Black man who had been trained as a Master Mechanic complained of not gaining employment in his field during the

1912 National Negro Business League meeting. During his confrontation with the Union, he simply said "in spite of all this preparation, you refuse me the privilege of even joining your Union, and yet you refuse me work merely because I don't belong to the Union."[12] This was an obstacle that many Blacks dealt with in the White dominated industry. Besides these few situations, there was not much other Black business activity in the automobile field.

The automobile quickly changed the face of America. Many supporting industries arose because of the needs of automobile manufacturers and automobile owners alike. In 1914 George Perris summed up the effect of the automobile quite nicely by stating "the automobile is, perhaps, the most characteristic embodiment of the luxury and ingenuity of our time. Roads are built for it; hotels and stores and repair shops spring up in its train; the earth is ransacked for supplies of oil and rubber which, when found, become matters for riotous speculation."[13] One of these repair shops sprang up on North Washington Street in Greenfield, Ohio, and became a part of the evolving Patterson business.

The Patterson Company Enters the Automobile Age

Sixty-five thousand autos ran on the streets and roads of Ohio by 1912 and this number grew to 168,000 by 1915, just three years later. Of these 168,000 automobiles, 716 were in Highland County.[14] This meant that by 1915, the Pattersons no doubt saw many of their neighbors, and former customers, driving automobiles daily. It had become very apparent that the popularity of the automobile was growing rapidly in their local area, so they added mechanics to their employment rolls and automobile repair work to their services. They did repair work for several years before ever building their first car.

It was quite common at the time for car owners to have their automobiles "maintained by one of the independent auto repair businesses that were sprouting up in blacksmith's shops, bicycle shops, and livery stables." The mechanical skills of these newly formed shops varied widely, from well-trained mechanics "down to the untutored, trial-and-error blacksmith mechanic trying to shift from the age of the horse to the age of the auto."[15] It is safe to assume that the Patterson shop, at least in the beginning, began as one of the trial-and-error shops.

By the second decade of the century, as carriage makers and blacksmiths, Patterson employees initially would not have gained the knowledge of motors and other automobile components needed for performing auto repairs; they had to learn new skills as they went along. According to Kevin L. Borg, this was part of the Patterson Company moving into the "middle ground." Borg described this as "middle not in the sense of being between a higher or lower level of technology or status, but as occupying an ambiguous space between production and consumption in which workers maintain and repair artifacts that they neither create nor own." Before this, the Pattersons, and other blacksmiths, had been both producer and repairer of metal objects, but because of mechanization "they shifted more firmly into the middle."[16] By adding auto repair to its services, the Patterson Company began its transition into the auto age and into the middle ground of technology.

The first advertisement that the Patterson Company placed in the *Hillsboro Dispatch* specifically promoting automobile repair at the shop appeared in February of 1913. This same advertisement was placed in the *Ohio State University Monthly* at the same time and became the very first advertisement ever to be placed in that publication. The shop had apparently been conducting automobile repair before this, although it was unadvertised, as evidenced by an article in *The Horseless Age* in September of 1909 that stated, "C.R. Patterson & Sons of Greenfield, Ohio, will soon begin the erection of a modern garage at that place. A representative of the concern has been sent to Columbus and Cincinnati to secure ideas for the new structure."[17]

A 1916 Patterson advertisement in the *Journal of the National Medical Association* stated that they had specialized in

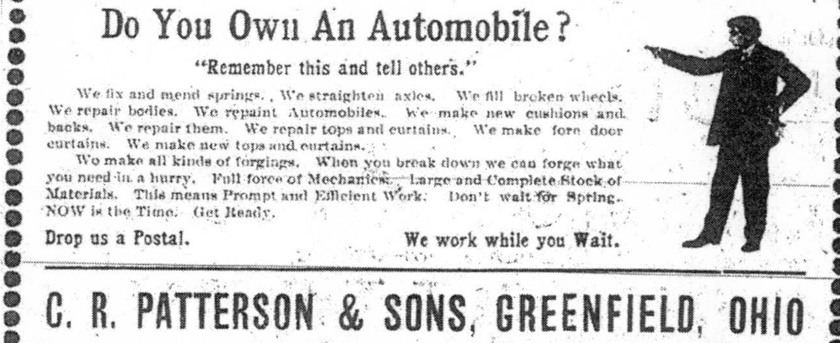

First Patterson advertisement mentioning automobile repair.
In *Hillsboro Dispatch*, February 11, 1913.

automobile work for 15 years.[18] This would have placed them conducting automobile repair as early as 1901, but with the low number of automobiles in the area at that time, the number of repairs would have been minimal. Although it is unlikely that they would have been doing repairs to motors or electrical systems during this time, it is quite possible that they were making repairs to parts such as tires, wheels, springs, and other iron parts that may have broken. Any repair that could have been done by a blacksmith is probably the type of repair that was made by the Patterson Company in the early days.

The advertisements placed for automobile repair beginning in 1913 included items such as fixing and mending springs, straightening axles, wheel repair, body repair and repainting, forgings of all kinds, and upholstery work to include seat cushions and backs, and new tops and curtains. The painting of automobiles seemed to be a popular service offered by the Pattersons, just as buggy painting had been. Automobile painting first showed in the Patterson advertisements in 1913 and was priced between $25 and $60 depending on the size of car. The Pattersons liked to make cars look like new. They had even developed their own line of products that were first advertised in 1915 such as Patterson's Auto Top Dressing to "make tops look new without injuring the fabric," Patterson's Lining Dye to "make that soiled, stained lining uniformly black," and Patterson's Body Polish that "brightened the dead varnish and give the paint a permanent luster."[19]

By 1916, the Patterson shop also offered general motor overhauling for any make of automobile. An advertisement claimed that when completed, an overhaul "will make your car as good as new – often far better." One testimonial included in an advertisement referred to a complete car overhaul completed by the Pattersons for Sheriff Ben South of neighboring Clinton County. "Everyone knew the condition this old machine (Overland) was in and it looked nearly new when it came from your shops." Many of the advertised parts and services included special repairs or upgrades for the Ford Model T, which was by far the most popular automobile at the time. The Pattersons offered services to re-magnetize the magneto magnets for Fords rather than paying the expense to have the parts replaced. They also offered an aftermarket lighting system that provided a steadier beam of light than the Ford system. One advertisement even introduced a coil protector for Fords that had been created by the Pattersons, for which they had applied for a patent.[20] The Patterson shop also carried general parts for automobiles. Some of these included items such as light bulbs, radiator and hood covers, windshield curtains, tires and tire chains, lap robes, driving gloves, and heaters.

The old carriage maker was well suited to enter automobile repair work. The basic structures of carriages and the early automobiles were very similar, and this allowed the carriage maker to have some familiarity with the hardware. A 1903 book on automobile theory and construction stated:

> A motor vehicle should be constructed in all of its iron work, its running gear and axles, the method of putting on its springs, etc., as nearly as possible after the methods now in existence in the carriage world, using, as far as practicable throughout the vehicle, standard carriage

> hardware. In this way the purchaser of an automobile has a resource at his own door for such repairs as he may need from year to year in addition to his regular painting, varnishing, and trimming repairs.[21]

This made the addition of automobile repair a good supplement to the Patterson Company at this time. The shop was basically already set up to perform the job. A major problem with early repair shops was that of financing. Those who tried to start a shop were warned "it is a big mistake to start a garage on the assumption that it will pay a profit from the start," as it takes a considerable length of time to accumulate the tools and the skills necessary in the business.[22] This was not an issue for the Pattersons since automobile repair was only a supplemental addition to their carriage business, at first, and they likely already had many of the necessary tools and equipment.

It is not known exactly what the inside of the Patterson shop looked like during this period, but there are several basic tools that were used for automobile repair that would have been present, such as small hand tools like hammers, chisels, files, scrapers, punches, reamers, taps and dies, measuring instruments, screwdrivers, saws (for wood and metal), wrenches, jacks, breast drills, metal snips, and pliers. Some larger tools and pieces of machinery that would have been needed included general blacksmithing tools, vises, forges, a chain hoist, a lathe and its' accessories, a drill press, an emory wheel or grinder, a vulcanizer, and possibly a milling machine. The shop would also be equipped with an air supply system, workbenches, shelves and racks for tools and stock, and sturdy cabinets with drawers for bolts, nuts, and other small hardware. According to Andrew Lee Dyke, an early automobile pioneer who later turned to writing manuals for auto repair, "about one-half of the work of the automobile repairman is for making adjustments and fitting parts," therefore, "a full machine shop is not necessary."[23] Many small shops only installed those machines that were necessary to the level of repair they were willing to conduct.

During the transition between automobiles and carriages, Patterson wanted to protect as much of the old trade while moving into the new. Since the company would have its foot planted in both markets at the same time, the old horse and the new car, Patterson preserved as much of his established carriage trade as possible while moving on to adding automobiles to the company's products. In advertisements in the month leading up to the introduction of their automobile, the Pattersons frequently used words like loyalty, faithfulness, and allegiance. One ad stated, "We still declare our allegiance to old Dobbin," the carriage horse. To follow this, three weeks later, in the same edition of the *Greenfield Republican* that they announced their automobile to the world, their advertisement stated, "No – the automobile has not displaced the horse and buggy. It never will. There is ample and imperative need for both."[24] The Pattersons left room in their shop to construct, service, and repair automobiles and carriages alike.

Frederick Patterson Becomes the First and Only Black Auto Manufacturer

Automobile repair was not enough for Frederick Patterson. He wanted to build automobiles. On September 23, 1915, an advertisement in the *Greenfield Republican* introduced the Patterson-Greenfield automobile, the first automobile to be built in Greenfield or anywhere in Highland County. Perhaps a much bigger first (which was not publicized at the time) was the fact that Frederick Patterson had just become the first Black manufacturer of automobiles. One photograph has surfaced of Frederick in an earlier style automobile. According to some sources, this may have been the first that he built, but by all indications this photograph is of a 1910 or 1911 Maxwell car. Regarding the Patterson-Greenfield, the editors of the *Greenfield Republican*, to ensure accuracy of information, allowed Frederick Patterson to describe the new automobile in his own words. Frederick wrote that "in my judgment there isn't a machine on the market that sells at or less than $1,000 that will equal the Patterson-Greenfield."[25] Frederick went on to say:

> Our car is made with three distinct purposes in mind. First – It is not intended for a large car. It is designed in size to take the place originally held by the family surrey. It is not intended as an omnibus or carry-all. It is a 5-passenger vehicle, ample and luxurious. Second – It is intended to meet the requirements of that class of users, who, though perfectly able to spend twice the amount, yet feel that a machine should not engross a disproportionate share of expenditure, and especially it should not do so to the exclusion of proper provisions for home and home comfort, and the travel and varied other pleasurable and beneficial entertainment. It is a sensibly priced car. Third – It is intended to carry with it (and it does so to perfection) every convenience and every luxury known to car manufacture. There is absolutely nothing cheap about it; nothing shoddy. Nothing skimp and stingy. Delicate and finished in its appointments and when compared to cars costing twice the price the chief difference lies in the size, and yet, to the average man, the smaller size is to be preferred.[26]

The Patterson-Greenfield Roadster as it was introduced in 1915. On the running board of the automobile is Frederick's son Postell, who was 9 years old at the time of the photograph. (public domain)

The Patterson-Greenfield 4-25 Touring Car as it was introduced in 1915. (public domain)

Photograph of Frederick with an early style car. Although sometimes identified as the first Patterson-Greenfield automobile, by all indications this is a 1910 or 1911 Maxwell car. It was likely in for repair, or perhaps Frederick's first personal car. (public domain)

Frederick also stated in the article that many of the parts were standard and could be purchased at any garage if there were any problems with the car. This suggests that much of the car (excepting the body and some of the other parts) had been assembled just like many of the early automobile builders had done. The first Patterson-Greenfield automobiles cost $685, fully equipped. Word of the new automobile spread across the county, and a follow up announcement made in the *Leesburg Citizen* stated that the car "bids fair to be a dandy." The following week the *Hillsboro Dispatch* stated, "The Patterson-Greenfield is a beauty and the sale has already started off in a brisk fashion." In 1916, the *Negro Review* mentioned that Patterson displayed his car at that year's Ohio State Colored Fair.

One person at the fair "at once canceled his order for a Dodge when he saw this car and put in his order for one."[27]

The week after the initial announcement of the Patterson-Greenfield in the *Greenfield Republican*, a Patterson advertisement stated, "it is not true that we have discontinued any department of our business, but we have added to our facilities and products so that we are now a complete carriage concern fully equipped to serve you best in any of your requirements."[28] Public concerns over discontinuing the carriage business are probably the reason that they had led up to the introduction of the Patterson-Greenfield with several carriage friendly advertisements. Beginning the following week, the Patterson local advertisements returned to a carriage focus with only a small mention of the Patterson-Greenfield automobile. The advertisements remained this way until March through July of 1917 when there was a large advertising push for the Patterson-Greenfield automobile in an apparent attempt to gain enough sales to keep the car in production.

Some of Patterson's customers, who had once bought his carriages, now bought his autos. A 1916 advertisement in the *Journal of the National Medical Association* included a testimonial from a Dr. John E. Hunter out of Lexington, Kentucky, about the new Patterson-Greenfield automobile that he had bought. In October of 1911, almost five years prior, a testimonial had been included in an ad by this same doctor about a Patterson Storm Buggy he had purchased.[29] This repeat business was good for the Patterson Company and showed that its switch to automobiles was not going to affect its entire clientele in a negative manner.

Patterson automobile production, though, did not last long – only until 1917. The last advertisement for the Patterson-Greenfield ran on July 12, 1917. Most accounts say that between 1915 and 1917, Patterson built only 30 to 150 cars.[30] In a 1976 *Dayton Daily News* interview with Postell Patterson, who was nine years old at the time his father built the first Patterson-Greenfield, he recalled that about 150 of the cars were built. In mentioning the quality of the cars, Patterson said, “there wasn’t anything like planned obsolescence back then. You didn’t dare even to throw anything away, let alone plan for it to break down.”[31]

Given the low number of Patterson-Greenfield automobiles produced, it is no surprise that none are known to have survived. In 1980, author Bill Naughton and the directors of Carillon Historical Park in Dayton, Ohio, conducted a nationwide search for any existing cars, trucks, or buses built by the Patterson Company. Carillon Park was hoping to locate one of the vehicles to place in an exhibit at their park. This search consisted of running articles in Highland County and surrounding counties in hopes that one of the locally produced vehicles might still be in the area. A film production company conducted a larger, yet just as fruitless, nationwide search in 1982. The D. P. N Company based in Washington D. C. looked for a vehicle and any other information pertaining to the Patterson Company, because it wanted to produce a short film to be run during Black History Month as a salute to the company’s accomplishments.[32] No vehicle showed up.

While Frederick Patterson became the first and only Black to manufacture an automobile, he was not the only Black to design one. In 1926, William Hale of Litwar, West Virginia, filed for a patent for a motor vehicle that could be driven from either end (U.S. Patent No. 1,672,212 – 5 June 1928). This concept was mainly applicable for motor buses and trucks so they would not have to turn around to make a return trip over a rough road. There were six wheels on this vehicle. The two placed in the middle were the drive wheels that had a drive shaft coming into a differential from both ends. The wheels on both ends could be steered (one end or the other always had to be locked into place) and there was a motor and driver’s compartment placed at each end. This concept didn’t catch on and none of these vehicles were ever constructed.

The Patterson-Greenfield Automobile Has Made Good!

Read **Consider** **Think**

Lexington, Ky.

C. R. Patterson & Sons,
Greenfield, Ohio.

Gentlemen:

I am glad to say that my Patterson-Greenfield Car is giving the very best of service. Last Friday I made a run of fourteen miles in the deep snow—running thirty miles an hour easily. It has an abundance of power and I am sure it is the best hill climber in this section. In my mind it is the best car under Fifteen Hundred Dollars and is better than many of the very fine cars which cost three times as much as yours. I do not see why this car should not take the lead with the public—as it certainly is up to date and delivers the goods.

Yours most cordially,
J. E. HUNTER, M. D.

Backed by Fifty Years Experience in the vehicle industry and Fifteen Years specialized on automobile work, the Patterson-Greenfield automobile brings to you Merits and Qualities impossible to obtain in any other car. We prove our claims in your service.

New 1916 Model Now Ready $775.00.

Fully equipped with every feature for dependable service and comfortable riding.

Write for complete description. We want to tell you all about this wonderful car. It will pay you to address us.

Responsible Physicians can do well with our Agency. No interference with other work. Write for territory desired. Contract early for Spring deliveries. Send to—

C. R. PATTERSON & SONS
Greenfield, Ohio

The only Negro Automobile Manufacturing Concern in the country.

A 1916 Patterson-Greenfield advertisement with testimonial. This testimonial came from repeat customer Dr. John Hunter who had also purchased carriages from the Patterson Company in the past. Note the included sentence at the bottom of the advertisement. From *Journal of the National Medical Association* Vol. 8, No. 1 (1916).

While it was common for carriage builders to make the transition to manufacturing automobiles, making a success of this was uncommon. At first, they resisted making a transition to the auto industry and "looked with disdain on the automobile, regarding it as a passing fad." This may have led to their own undoing. According to Mr. Connolly, President of the Carriage Builder's National Association, in a 1909 article, "the carriage builders certainly made a big mistake when they allowed the manufacture of automobile bodies to slip out of their hands."[33] While he felt that the salvation of the carriage industry was to enter the automobile field, he, among others, warned that it should only be for the construction of automobile bodies, a task that could easily be accomplished utilizing the skills of the carriage builder.

This resistance of the transition occurred even within the Patterson shop. Apparently not all Patterson Company employees felt that making a shift to automobiles in the preceding months showed promise. Two long-time Patterson employees, Clay Gordon and Frank Roberts, started their own blacksmith shop in town in July of 1915, just two months before the Patterson-Greenfield car would be introduced. Clay Gordon had been a long-time employee of the company and held at least one patent together with C.R. Patterson.[34]

While it is true that many early automakers were "assemblers," rather than manufacturers, by the time that Frederick entered the automobile field, the manufactured car led the industry. Assemblers entering the industry were said to "underrate the difficulties of building a commercial car from standard parts. They do not realize that fully as good and experienced an automobile engineer is required as though an entirely new car was to be designed from the ground up." It still took the skills of an engineer to put the proper parts together to create the assembly. Benjamin Briscoe, an early pioneer in the auto industry, labeled such assemblers as "manufacturing gamblers" and used terms such as "piratical skimmers" to describe this group. The mechanism of the automobile did not fit the skill set of the average carriage maker. According to Connolly, "the carriage builder naturally desires to become an automobile manufacturer, to enable him to maintain his relations with his agents; but in our opinion his proper place in the auto industry is that of a parts maker, furnishing bodies to the automobile manufacturer, as the assembling of automobile machinery is rather too much out of his line."[35] Despite these warnings, many carriage builders, including the Patterson Company, attempted the full transition into the automobile manufacturing field, rather than accepting the more obvious business of supplying automobile bodies. Most of these failed.

The Patterson-Greenfield automobile seemed to have been a very efficient car and had all the features of some of the top automobiles of the day, but Frederick Patterson just entered the field at the wrong time; he was too late. The same year that Patterson introduced his automobile, 1915, Henry Ford sold his one-millionth Model T. Some of the large automakers, such as Ford and General Motors, were already in place and were swallowing up the smaller car manufacturers and pushing them out of business. These large production automakers could produce cars much quicker and cheaper than the small companies, thus making their products more economical for the public to buy. A 1915 article in *The Automobile* stated that as of October of that year, 16 Detroit factories were producing 1,800 gasoline cars per day.[36] If Ford's plants in other states were added to this total, it would amount to 3,300 cars per day. The total number of Patterson-Greenfield cars produced from 1915 to 1917 was only between 30 and 150, which was a mere fraction of the daily production in Detroit. It is easy to see why the Patterson Company failed as an auto maker.

Frederick Patterson was not alone in his quest to enter the automobile manufacturing field during this time. Many had attempted to produce automobiles through the years and many more would try even after Frederick failed. In Larry Freeman's 1949 book, *The Merry Old Mobiles,* he gave a roll call of 1,523 known automakers between 1895 and 1950.[37] Of these, 827 had produced automobiles from the period 1895 to 1909, which equaled 54.3% of the total number of known manufacturers. The failure rate during this period came to 82.3%, and another 17% of these failed during the 1910 to 1919 period, leaving less than 1% of the original manufacturers still in existence after 1920. As time went on and people began to realize the difficulty of competing in the field, especially after leaders such as Ford and General Motors moved into prominent positions at the head of the industry, the number of those attempting to enter the field began to decrease. The period 1910 to 1919 saw 472 people enter the field, equaling 30.99% of the total number of known manufacturers. The failure rate for this group during this period equaled 78.4%, which included the Patterson Company, and another 18.4% failed during the 1920s. Another 1.3% of this group made it to the 1930s before they failed, which left only 1.9% of this group still active in the 1940s. The last group, from 1920 to 1950, saw only 224 enter the field, which equaled 14.71% of the total number of known manufacturers. Of the 1,523 listed on the roll call, only 19 automakers were still active for the 1950 model year, which equaled only 1.25% of all that had entered the field since 1895. These numbers show that it was not easy to enter the automobile manufacturing field and become competitive. Only those with the strongest will, and, more importantly, sufficient capital survived. The Patterson Company may have had the drive, but it lacked the capital.

Another reason for the failure of the Patterson-Greenfield to thrive was that the prices of parts kept increasing. The Patterson Company did not make its own parts, so it ordered them from several different parts suppliers. Once World War I started, many of these suppliers turned their attention toward war orders. This left fewer manufacturers of auto parts, so demand caused the prices to rise. Also, materials were scarcer owing to the war effort, which also resulted in higher prices. Since car "assemblers" bought the parts on credit, the suppliers withheld parts from many of the smaller companies "because of a lack of faith in their financial credit and because of current predictions to the effect that the smaller manufacturers will be unable to weather the materials, labor, and other war conditions now existing and promising to become even more perplexing in the near future."[38] The combination of plummeting sales and a lack of credit made for a deathblow to many small automobile manufactures to include the Patterson Company.

It is very apparent that the company was struggling during this period. A 1916 letter from Frederick to Emmett J. Scott, who became President of the National Negro Business League after Booker T. Washington's death, made this fact very clear. Mr. Scott had inquired as to whether Frederick was going to make it to that year's meeting. Frederick responded that he most likely would not make it owing to the state of his business. Frederick stated:

> You know for the past several years we have been always busy during that period making school wagons and while I am frank to tell you at the present time the prospects for this business do not appear very flattering, at the same time if it did come in it would be imperative to devote all our time and attention to that one thing. I am frank also to tell you that we have on our shoulders this automobile business and it is pretty much of an undertaking and one that is taxing every ability we have and every direction.[39]

This not only explains that the Patterson automobile business was not faring so well at that time, but also that the established carriage business was declining rapidly as well.

Although in 1916, the average cost of a car was $1,600, a person could purchase a Ford Model T for only $360.[40] Ford's "car for the masses," and the other cheaper cars from the larger manufacturers were just too much for the Patterson Company to compete with, so production of the Patterson-Greenfield stopped after just a few years. Fortunately, during this period, the company did not have to depend on sales of the Patterson-Greenfield automobile alone for income. The auto-building period coincided with the late carriage-building period, although the carriage business was slowing dramatically as well. The Patterson Company had also repaired autos and other vehicles and equipment during this time. This service allowed the company to keep busy and maintain some income while Frederick tried to figure out a way to overcome the decline of the business and get things moving forward again.

In an interview with Richard Patterson, grandson of Frederick and Estelline, the author was provided with portions of the following information, which provided new leads to follow regarding the topic. Late in 1917, sometime after September, with the company floundering from a lack of sufficient orders for automobiles and the rapid decline of the carriage industry, there were a lot of struggles at the company, as well as at home. Frederick and Estelline were struggling in their marriage, which was caused by Frederick's frantic working to try to pull the company back into a steady business as well as the financial woes of both the company and family during this time. Estelline, along with her mother (Pauline Postell) who was still living in Greenfield at this time, took the two children (Frederick Postell and Postell) and moved to Philadelphia to live in a house that Pauline had purchased there. Sometime in the next year, Frederick and Estelline filed for a legal separation in Philadelphia. During the court hearing, the children were asked to testify in court. This led to the oldest son, Frederick Postell, developing a stuttering issue that remained with him the remainder of his life. Younger brother, Postell, attended middle school and started high school in Philadelphia, while Frederick Postell completed high school there. Interestingly, in the 1920 census, both Frederick and Estelline reported themselves as "widowed." This form of census documentation for marital status was a common occurrence during the period when divorce or legal separation had a stigma surrounding it.

It is interesting to note that for nearly two decades the company ran a different weekly advertisement in local newspapers. However, in late 1917, coinciding with Estelline leaving Frederick and moving to Philadelphia, the pattern of advertisements changed. The exact same ad ran for nearly two years from August 1917 to July 1919. It is likely that the struggles that Frederick were facing in keeping the company afloat took his priorities away from certain activities that he had always kept on top of, such as variety in the company advertising.

An additional member of company primary leadership gave in to the strained conditions at the company at this time, generally leaving Frederick alone to find a way to turn the company around. Frederick's nephew, Charlie Napper, left the company during this time and focused on his Geology career.

Technical Details of the Patterson-Greenfield Automobile

Not all technical details concerning the Patterson-Greenfield automobile are known. Advertisements described it as having "easy riding cantilever springs – ample, dependable, and powerful motor – full floating rear axle – three speeds forward and one reverse – demountable rims – worm drive – electrically lighted – electrically started – electric horn – 108" wheelbase – pockets on all the doors – elegant one-man pantasote top – tank in cowl – stream line body – curtains, bell hideaway style – in fact a perfect car of ample proportions at a very reasonable price."[41]

A 1916 advertisement in the *New Atlas of Highland County, Ohio* added that it was left-hand drive, center controlled, and had a ventilating windshield. The ad also stated that the car came as a roadster or a touring car. The ad mentioned that the Patterson-Greenfield was a strictly custom-made car, and it possessed "every feature and convenience demanded by modern motoring. It is a wonderful car, appealing for handsome appearance and enduring qualities." A Patterson ad in 1917 supported the custom-made car idea by stating "we incorporate your ideas in its construction."[42] Apparently, the Pattersons were open to certain requests from buyers pertaining to the car that was being built for them, perhaps making custom upholstery or paint color, among other things.

A 1915 advertisement in the *Journal of the National Medical Association* described the Patterson-Greenfield as having rich upholstery, a speedometer and ammeter, an extra rim, and crowned fenders. A 1917 ad touted that the car came with an "ammeter for the current, sight feed for the oil, speedometer for the mileage. Day and night you can instantly tell whether things are right or wrong. It pays to know." The following week, an ad talked about tire removal being made easier using demountable rims on the Patterson-Greenfield and how there was only one ring to remove, which allowed the tire to be slid off easily.[43] Another advertisement in the *Journal of the National Medical Association* stated, "the Patterson-Greenfield automobile is not a makeshift attracting attention because of the oddness or freakishness. Parked anywhere among any number it will hold its own. The Patterson-Greenfield automobile has a mechanism that is simple and readily understood. Easy to guide, sensitive on the accelerator as a spirited horse, with that surplus power instantly had – makes its driving a pleasure and not a strain." The article also included the fact that the Patterson-Greenfield weighed less than 2,000 pounds, which helped to save on tire expense and achieve higher gas mileage.[44]

A Patterson advertisement in the 1916 *New Atlas of Highland County, Ohio,* included the words "our special motor has that surplus power and greatest pull. Try it on your test hill."[45] Many advertisements noted that a demonstration could be arranged at any time and included statements such as "what hill do you want to climb?" This special motor is often misidentified as being a Continental motor. An advertisement that the Pattersons placed in the *Journal of the National Medical Association* in April of 1917 stated that they used "the G. B. & S. [Golden, Belknap, & Swartz] Motor in Patterson-Greenfield automobiles" and, in over 1,200 Patterson advertisements analyzed for this study, as well as contemporaneous newspaper and journal articles, there was never a single mention of a Continental motor. The Pattersons described some of the details of the motor and explained that the combination of the superior motor and 65 years of vehicle

Advertisement for the Patterson-Greenfield automobile. This advertisement appeared on the inside of the front cover of the Highland County Atlas in 1916. From H. W. Hunter, *New Atlas of Highland County, Ohio* (Hillsboro, Ohio: H. W. Hunter, 1916).

construction experience is what you have in a Patterson-Greenfield. According to Henry May, Frederick examined some of his competitor's cars to get ideas on how to build a car.[46] One of his nearest competitors was in Columbus, Ohio, the Cummins-Monitor Motor Company. It utilized this same G. B. & S.

motor in its 4-cylinder cars, so it is possible that it is where the idea to use this specific motor for the Patterson-Greenfield came from.

Ellsworth Belknap and John Swartz patented the Golden, Belknap, & Swartz motor (G. B. & S. as it was known) in 1911 (U. S. Patent No. 1,007,842 – filed Sep 15, 1910), and they manufactured it at their factory in Detroit, Michigan. An article in *The Automobile* in 1914 highlighted the G. B. & S. motor. The motor produced 22.5 horsepower, but at 2800 rpm the motor produced a full 36.9 horsepower. The reason for this additional horsepower is the fact that the stroke of 4.25 inches was considerably greater than the bore of only 3.75 inches.[47] This bore and stroke made the piston displacement equal to 187.9 cubic inches in the 4-cylinder motor.

The crankcase of the motor was constructed of lightweight aluminum and the pistons and cylinders were made of gray iron. The motor had an L-type removable head and the cylinders were cast in pairs. The camshaft, magneto, and generator were all driven by ½" pitch silent chains to cut down on noise. Also cutting down on vibration, wear, and the resulting noise were the balanced crankshaft and flywheel. The makers of this motor claimed that it was balanced well enough to be free of vibration at speeds up to 40 mph.[48]

The motor used a combination force-feed and splash oiling system. An oil pump was driven by the No. 4 exhaust cam lobe, which forced pressurized oil first through the sight feed gauge mounted on the driver's dash and then continued to the main crankshaft bearings and the chain drives. The splash system took care of the rod bearings and pistons that were splashed with oil by the rotating crankshaft. The cooling system for the motor was a thermo-syphon system that utilized large water intake ports and large water jackets that surrounded the cylinders. This system was backed up by a fan that was driven by a belt attached to the front of the crankshaft.[49] The outside of the motor was very neat and tidy to keep things simple.

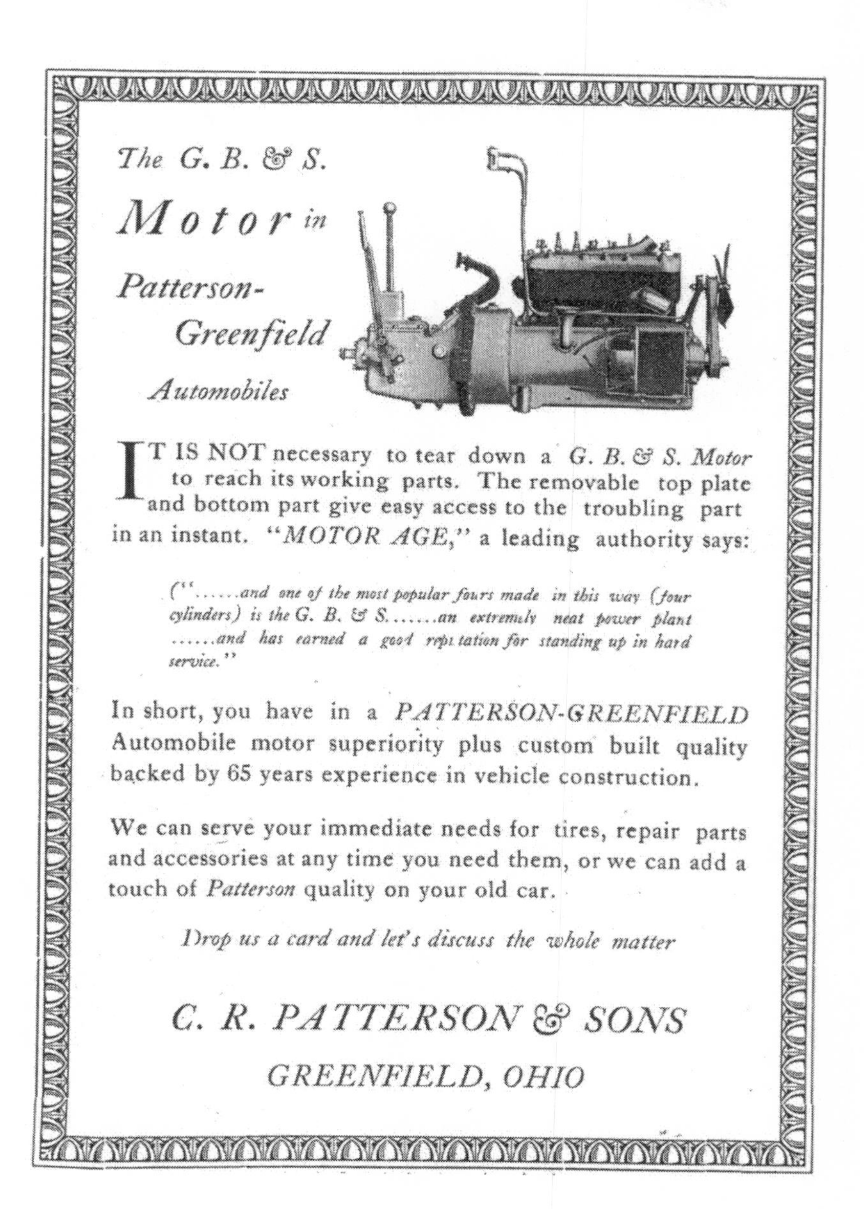

Advertisement placed by the Patterson Company to highlight the G. B. & S. Motor that was used in the Patterson-Greenfield automobile. From *Journal of the National Medical Association,* Vol. 9, No. 2 (1917).

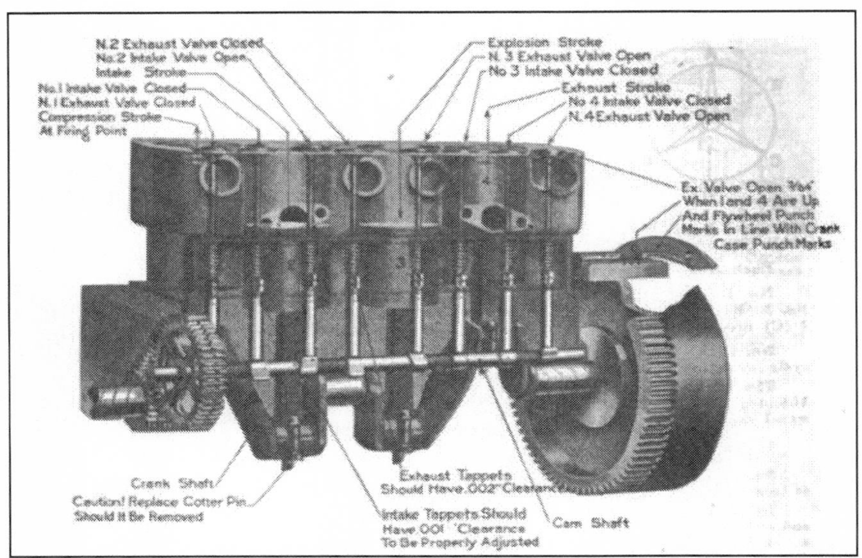

Illustration providing an exploded view of the location of the valves, camshaft, crankshaft, and the silent chain drive for the camshaft in the G. B. & S. 4-cylinder motor. From A. L. Dyke, *Dyke's Automobile and Gasoline Engine Encyclopedia* (St. Louis: A. L. Dyke, 1920), 120.

Diagram of the G. B. & S. motor showing an inside view of the internal cooling system. The outer water jacket has been removed the show the circulation pathways around the cylinders. From Dyke, *Dyke's Encyclopedia*, 188.

Illustration showing the right side and the front views of the G. B. & S. motor. Note the large removable plates bolted to the right side of the motor for easy access to bearings. From *The Automobile*, "Simplicity Feature of G. B. & S. Motor," February 19 (1914): 465.

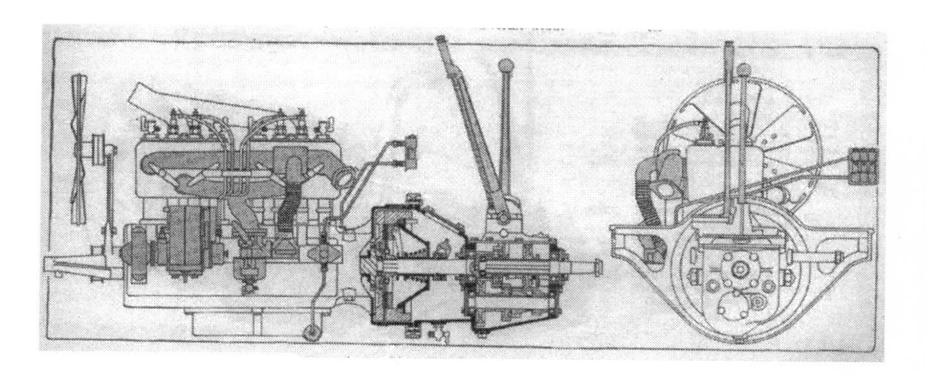

Illustration showing the left and right sides of the G. B. & S. motor with a sectional view of the clutch and gearbox. From *The Automobile*, "Simplicity Feature of G. B. & S. Motor," February 19 (1914): 465.

The G. B. & S. Company produced only this one 4-cylinder motor. They felt that by concentrating on one motor, they could make it the absolute best. With no other motor to fall back on, they had to make sure their product was top of the line, or the company wouldn't survive. According to an advertisement placed by the company, "you want flexibility, balance, power, economy, quietness, and endurance in a motor – and that is exactly what you get in this G. B. & S. motor."[50]

The frame is one of the automobile's most important parts. Even the best motor can't function properly if it is placed in a frame that can't handle the load properly. Although some early automobile frames were made of wood, which would have been easily within the capabilities of the Patterson Company, it seems highly likely that the company purchased frames for its Patterson-Greenfield automobiles. Several reasons have led to this conclusion. First, since the Pattersons came into the automobile building field late, most frames were made of steel by that time, which means that the Pattersons most likely would not have wasted their time trying to use wood if they hoped to be competitive with other manufacturers. Secondly, the Patterson shop would not have had the equipment needed to build a proper automobile frame. It took highly specialized and expensive presses and dies to create the frame rails. Since the Pattersons specialized in woodwork and regular blacksmithing operations, rather than being a fully functional machine shop, it seems unlikely that they would have owned this type of equipment. Even the larger manufacturers of automobiles, such as Henry Ford in his early days of auto making, contracted with the Dodge Brothers of Detroit or the A. O. Smith Company of Milwaukee to build his frames.[51]

The following photograph of a man standing by a Patterson-Greenfield chassis clearly shows that the frame rails are made of pressed steel that was probably shaped by a press and die. Companies right in Ohio did this type of work, such as the Parish and Bingham Company of Cleveland. They specialized in pressed steel frame rails and offered them in 12 different standard styles,

Photograph of a possible Patterson family member or employee standing beside a bare Patterson-Greenfield automobile chassis. From the radiator and firewall styling, it appears to be a chassis for the larger touring car body. At one time, this photograph was misidentified as being Frederick. (public domain)

or they made custom frames for a specific need.[52] The frame in the photograph is also riveted together in a manner that suggests that a regular frame builder had assembled it, rather than a smaller shop that was not equipped for this type of operation. Of course, it is also entirely possible that the Pattersons had purchased entire rolling chassis from a company and added their own body to it. This was also common in the era since many makers, such as Ford, Chevy, and Buick, offered chassis for sale.

The bodies of most early cars consisted of a framework of wood that was then covered with a skin of either sheet steel or aluminum. The body-building process fell more solidly into the realm of the Patterson Company's existing skills. It had built wooden carriages bodies for several decades and, combining this with their blacksmithing skills and experience, building auto bodies was a naturally suited task for them. First, the wooden framework was constructed using hand cut wood that could be steamed and shaped into the desired design. This framework would be fastened together with glue and screws, with wrought iron braces placed at certain stress points to add strength. Once this was finished, sheet steel or aluminum would be stretched over the wooden framework and shaped using hammers and special metal blocks, called dollies, to get the metal to conform to the proper shape.[53] From here, the body was ready for paint preparation.

Several sources have repeated Postell Patterson's words in a 1976 *Dayton Daily News* interview that "several features made the auto [Patterson-Greenfield] superior to the Model T Ford." One intended element of this current study was to make a comparison between the Patterson-Greenfield and the Ford Model T. Unfortunately, not enough specific information was located about the Patterson car to make a full comparison possible. A few generalizations can be made, but no solid opinions can be formed because of the lack of details available. One plus for the Patterson-Greenfield would be the sleeker body style as compared to the 1915 and 1916 Model T (of course, this is in the eye of the beholder). The Patterson-Greenfield had a slightly more powerful motor than the Model T (22.5 hp for the Patterson and 20 hp for the Ford). This slight difference in horsepower was likely due to a slightly longer stroke of the Patterson motor. Both had 4-cylinder motors, but the Model T had cylinders cast en bloc, while the Patterson cylinders were cast in pairs. The lubricating systems were slightly different as the Model T used a combination gravity and splash system, while the Patterson used a combination force feed and splash system. The cooling systems were both the thermo-syphon style. Both were shaft driven and, while the Model T only provided two forward gears and a reverse, the Patterson offered three forward gears and a reverse. The 108" wheelbase for the Patterson was slightly longer than that of the 100" wheelbase of the Model T. Both had wooden artillery style wheels, and both were left hand drive. Perhaps the biggest advantage that the Patterson-Greenfield had over the Model T was a matter of convenience; it came with an electric starter. The Model T did not add the "option" of an electric starter until 1919.[54]

Perhaps the best way to judge the Patterson-Greenfield is to compare it to other cars of the period, besides the Model T. A table showing the "comparison of features of the average American car for six years" appeared in a 1915 issue of *The Automobile*.[55] Some of the data show averages, while other data give a percentage of specific types of parts used in different categories. Utilizing the data for the year 1916, the Patterson car fell below the average of 28.66 horsepower (Patterson = 22.5). Of course, one reason for this is that 45.8% of 1916 automobile models used a more powerful 6-cylinder motor and the Patterson had a 4-cylinder, which only represented 39.2% of automobile models that year. This throws some of the other figures off as well; to include other motor related categories such as bore size and length of stroke.

Other categories that can be compared include the shape of the cylinders and their arrangement in the motors. The Patterson fell into the group of 73.3% that used the L-head type block but fell well below the group of 78.4% of motors whose cylinders were cast in block (Patterson cylinders were cast in pairs = 10.2%). The Patterson fell into the largest groups that used an electric starter (98.8%), had left hand steering with center control (87.1%), had a full floating rear axle (51.8%), and used wooden wheels (82.5%). The Patterson fell short of the 61.2% of motors that used the pump circulating type of cooling system (Patterson used thermo-syphon style cooling system = 38.2%) and the 52.7% that used the splash type oiling system (Patterson used combination splash and pressure oiling = 23.35%). Not enough details are known to compare additional categories. From the data that is available, the Patterson-Greenfield automobile seemed to have fallen somewhere in the middle of the group but was in no way to be considered an inferior piece of machinery.

The price of the original Patterson-Greenfield Touring Car was $685 in 1915. The price given in the *Journal of the National*

Medical Association in January of 1916 was $775 for the touring car body style. In October of 1916, there were two models of the Patterson-Greenfield roadster priced at $650 and $750 depending on the model. The 1917 model touring car was advertised at the price of $810.[56] Although this was a mid-priced car for the time, many larger manufacturers offered their automobiles at less cost. Not only was the cost a determining factor in a purchase, but customers also bought from the established names that they knew. If Frederick had built his car five to 10 years earlier, he might have had a better chance at success, but he entered the field too late to compete and was forced to stop production of the Patterson-Greenfield automobile.

Patterson Automotive Repair Continues

Although the production of the Patterson-Greenfield automobile did not work out as planned for Frederick Patterson, he still stayed in the automobile business by conducting automobile repairs at the shop. He knew that although he couldn't make money building cars, there was ample opportunity for business in the auto repair field. Automobiles were machines, and machines break down, so an auto repair shop would always be needed. Borg stated that "the repair shop is where the weaknesses of technology is laid bare; where progress is stalled, repaired, and sent back on the road; where technological failure is the stock-in-trade and the ideal of the well-oiled machine meets the reality of our entropic world."[57] Frederick shifted his focus from producing an automobile to expert repair of automobiles.

The Patterson Company prided itself in providing quick service to its customers. For most repairs it offered "one day service." Even on more difficult tasks, such as making new forgings and other iron work, the Pattersons claimed "we'll make that new part at once." They could offer this quick service because they provided a full work force of 30 mechanics and a large stock of supplies that allowed repair work to get done quickly. They also claimed to have a complete machine shop by 1919.[58]

Additions to the Patterson services were also highlighted in advertisements. One of these additions included radiator repair and the opening of a complete battery repair shop with a qualified expert on hand. Patterson also offered car washing and polishing; they first advocated these services in 1917. A later advertisement added that cars were washed on Saturday nights and up until noon on Sundays.[59] An illustration of the following page shows a Patterson Repair Service advertisement that appeared in several local newspapers. Glancing through this advertisement, it is easy to see that a variety of services were offered at the Patterson shop during this period. The business was a complete "job shop" and would perform almost any service or repair that was needed by the public, ranging from furniture and stove repair to servicing a carriage or automobile. The Patterson Company showed flexibility during this period and adapted to the needs of its customers, while at the same time, multiplying its sources of income.

Perhaps one of the most popular products offered by the Patterson Company during this time was the closed top that it offered for cars. A 1916 advertisement offered closed tops built for any car from $62.50 to $125 that could be removed for summer use. The Ford Sedan top was offered for $90 to $115 and could be built with only five days notice. Beginning in 1918, the Pattersons offered special "De Luxe" bodies for the Ford. A 1919 advertisement said that this De Luxe body would take your Ford "out of the common car class and place it in the Boulevard class." Of the Patterson De Luxe equipment, this description is given: "Streamlined body, hood and radiator cover; sloping, ventilating windshield; one-man top; crown fenders and front splash pans; extra long running boards; gasoline tank in rear with vacuum feed system. Fits any Ford pleasure car chassis."[60] In a 1920 advertisement, they even resorted to the "sex sells" style of

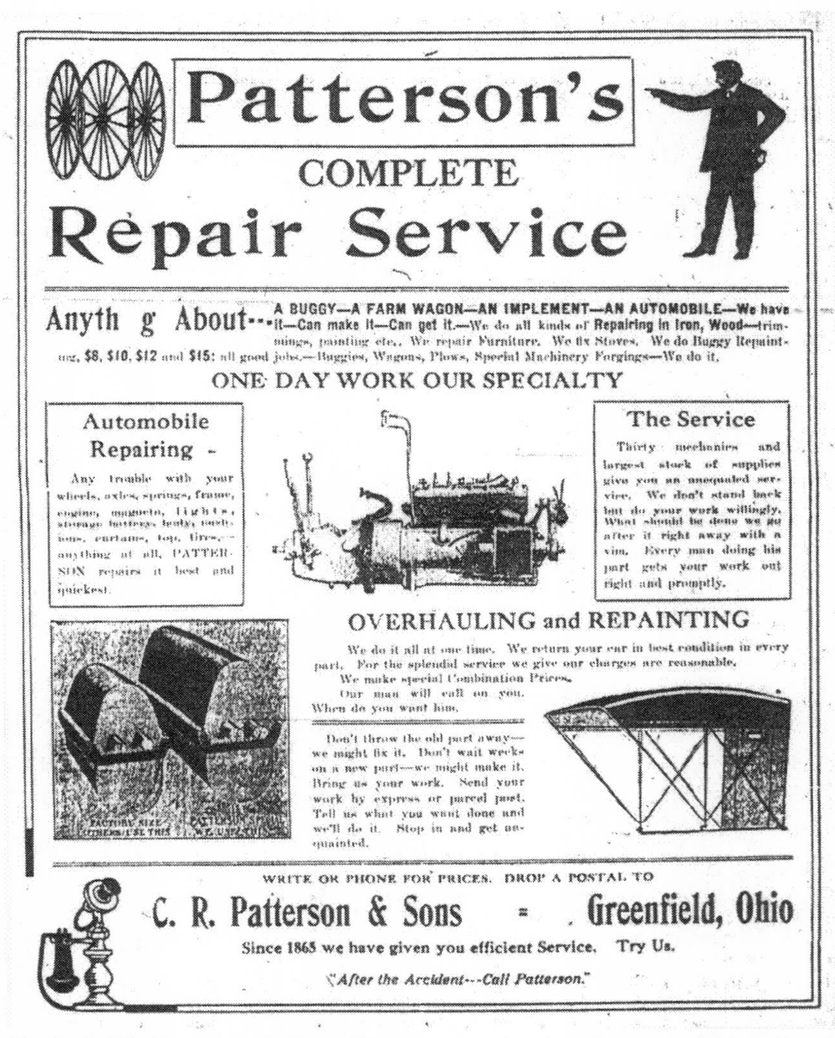

Typical C.R. Patterson & Sons Company repair advertisement. They offered a wide range of repairs and services as seen in this 1917 advertisement. From *Greenfield Republican,* Patterson Ad, March 15, 1917.

advertising through a cartoon added to the advertisement. Its' underlying theme suggested that, with the Patterson De Luxe body on your car or a fresh coat of paint, you could feel young again, or rejuvenated, and gain the attention of the younger, fairer members of the opposite sex. They had even named the speedster style body the "Red Devil," which may have planted the images of youthful rebellion in the minds of the readers. This advertisement just may have caught the attention of more than one middle-aged man in the local area.

Many auto repair businesses felt that they were not sufficiently prepared to handle the complexity of electrical repairs. Several specialty shops that handled automotive electrical repair came on the scene by the 1920s. The Patterson Company filled this niche in its shop. It had several specialists on hand including Mr. Harry McCarter, an ignition expert and master mechanic who had been trained at Ohio University. According to a 1920 *Greenfield Republican* article, this position was extremely important because 90% of all car trouble at the time was in the ignition system. McCarter took over as the new manager of the automobile repair department of the Patterson Company and apparently the business improved greatly as evidenced by its increasing patronage that was discussed in a 1921 article of the *Greenfield Republican.*[61]

The Patterson automobile repair department received recognition in an article in the *Hillsboro Dispatch* that highlighted major businesses in Highland County. The article said that at the Patterson repair department, "here service of the most magnanimous nature is rendered competently and efficiently by master mechanics. These men know the automobile from A to Z and can execute any work from radiator to rear axle in the most expert manner. The repair department is by no means confined to auto work for Pattersons is prepared to do all kinds of repairing, especially farm work."[62]

The Patterson Company became involved in the automobile business in other ways as well. In 1917, used automobiles appeared in its advertisements, many times with a newly rebuilt motor or new paint. Presumably the company was taking these vehicles in on trade, or purchasing them at a discount, and doing necessary repairs to resell them. The Patterson Company annou-

Patterson ad for the De Luxe Body for Fords. At this time, 1920, they have resorted to using different advertising tactics to draw attention to their products, including the addition of cartoons and this example of a "sex sells" style ad. From *Greenfield Republican*, Patterson Ad, March 25, 1920.

nced in 1919 that it had become a dealer for Briscoe cars.[63] This venture did not appear to last long since only a few advertisements were placed for this portion of the Patterson business.

In at least one instance, the purchase, repair, and reselling of used cars got Frederick into some legal trouble. In 1919, the Sheriff of neighboring Ross County served him with a warrant for receiving stolen goods. A Ford car had been offered to him at a bargain that supposedly belonged to a soldier deployed overseas at the time (World War I). Frederick had purchased several soldiers' cars in this manner, providing needed money for their families, so he didn't think much about it. He bought the car for $350. Later that same day, a woman called the company asking about a car that had been stolen from her and asked Frederick to watch out for it. Well, it turned out to be the same car. Frederick gave the car back to the woman. She thanked him and even gave him a small reward of $25. Of course, this was not enough to repay him for what he had lost in the sale.[64]

Several suspicious things had happened during the sale and during the exchange with the woman to whom he had returned the car. According to the *Greenfield Republican,* it turned out at the court case that, for some undisclosed reason, this had been an attempt by the plaintiffs to discredit Mr. Patterson and hurt his business and reputation in the community. Immediately after the judge ordered all charges dropped against Frederick Patterson, the Sheriff served the plaintiffs with papers for a damage suit for $25,000 that had been filed by Mr. Patterson.[65] This case did not go through and apparently was an action taken more to prove Mr. Patterson's point than being a serious suit.

For many years, the Patterson Company had placed a different advertisement every week in local newspapers, such as the *Greenfield Republican* and the *Hillsboro Dispatch*, but by late 1917, the advertisements slowed. Between August 1917 and July 1919, a period of almost two years, the Pattersons ran the exact same advertisement each week in these newspapers. It covered repair work of all kinds of vehicles from horse-drawn to automobiles. The normal advertising pattern picked up again for a while until October of 1920 through June of 1921 when an advertisement for their battery shop was all that ran during that time.[66] Sporadic advertisements were placed after this until February of 1922. Between February 1922 and October 1929, the company placed no advertisements in any of the local newspapers, a period of 7-1/2 years. This erratic pattern coincided with the busy period of bus body building and the shift away from conducting repair services.

Automobile repair was a major factor in the Patterson Company business during this period. It covered a longer span of time than the building of the Patterson-Greenfield automobile. This meant that the repair service had carried the company through a difficult period, but was not enough, by itself, to carry the entire load. Just like many other early auto repair shops, Patterson added several other sidelines to the business.[67] He sold accessories for automobiles such as parts, vehicle enhancements, and comfort items. He also sold used cars, the Briscoe, and, ironically, the Patterson-Greenfield can almost be considered a sideline item of this period owing to its lack of profitable income and low sales. It also offered special sidelines including vulcanizing, upholstering, painting, and the addition of building aftermarket bodies for automobiles would have been considered a special sideline as well. During all of this, the Pattersons still offered the services related to the carriage business, general blacksmithing, and furniture repair. There was a lot going on at the Patterson shop while they were in this transition phase that prepared them for their next venture.

CHAPTER 3 NOTES

[1] Stephen W. Sears, *The American Heritage History of the Automobile in America* (New York: American Heritage Publishing Company, 1977), 26.
[2] James J. Flink, *The Car Culture* (Cambridge, Massachusetts: The MIT Press, 1975), 43.
[3] Ibid., 50; Robert Casey, *The Model T: A Centennial History* (Baltimore: The Johns Hopkins University Press, 2008), 3; Sears, *American Heritage History,* 26; Flink, *The Car Culture,* 45.
[4] Sears, *American Heritage History,* 20; *Motor Age* Vol. 1, No. 4 (1899), quoted in James J. Flink, *America Adopts the Automobile, 1895-1910* (Cambridge, Massachusetts: The MIT Press, 1970), 31; John B. Rae, *The American Automobile Industry* (Boston: Twayne Publishers, 1984), 17; David Hebb, *Wheels on the Road: A History of the Automobile from the Steam Engine to the Car of Tomorrow* (New York: Crowell-Collier Press, 1966), 74.
[5] Ibid., 81; John B. Rae, *American Automobile Manufacturers: The First Forty Years* (Philadelphia: Chilton Company, 1959), 46.
[6] Hebb, *Wheels on the Road,* 104, 106.
[7] William Reed, "100 Years of Car Making: Blacks Then and Now," *New Pittsburgh Courier*, July 31, 1996; *New York Times*, "Auto Patents Reveal Trend of Industry," February 3, 1907; Rae, *American Automobile Manufacturers,* 103.
[8] William Reed, "100 Years of Car Making: Blacks Then and Now," *New Pittsburgh Courier,* July 31, 1996; Russ Banham, *The Ford Century: Ford Motor Company and the Innovations that Shaped the World* (New York: Artisan Books, 2002), 242; Gerald D. Jaynes, *Encyclopedia of African American Society* (Thousand Oaks, California: Sage Publications, 2005), 67.

---

[9] Detroit News, "Mr. Ford, Blacks, and the UAW," January 31, 2000; Steve Babson, *Working Detroit: The Making of a Union Town* (Detroit: Wayne State University Press, 1986), 41-42.
[10] Excerpt from the biography of Victor and Alma Green into the Automotive Hall of Fame as part of the Class of 2022. Dearborn, Michigan.
https://www.automotivehalloffame.org/honoree/victor-green/
[11] Juliet E. K. Walker, *Encyclopedia of African American Business History* (Westport, Connecticut: Greenwood Press, 1999), 50-51, 566; Kevin L. Borg, *Auto Mechanics: Technology and Expertise in Twentieth-Century America* (Baltimore: The Johns Hopkins University Press, 2007), 59,95; Bruce Sinclair, *Technology and the African-American Experience: Needs and Opportunities for Study* (MIT Press, Cambridge, 2004), 136; John Alfred Heitmann, *The Automobile and American Life* (McFarland, Jefferson, North Carolina, 2009), 115.
[12] Records of the National Negro Business League [microfilm]: 13th Annual Meeting, 1912: 57.
[13] George Herbert Perris, *The Industrial History of Modern England* (New York: Henry Holt and Company, 1914), 478.
[14] *HD*, "65,000 Autos Will be Licensed in Ohio in 1912 According to State Registrar Shearer's Prediction," April 30, 1912; *HD*, "Automobiles in Ohio Number over 168,000," September 14, 1915; In 1915, this number of registered automobiles placed Ohio as having the second most automobiles in the United States behind New York, which had 212,000; *HD*, "716 Autos in Highland County up to July 1st," October 8, 1915.
[15] Borg, *Auto Mechanics,* 16; Sears, *American Heritage History,* 193.
[16] Borg, *Auto Mechanics,* 2, 3.
[17] *HD,* Patterson Ad, February 11, 1913; *Ohio State University Monthly,* Patterson Ad, February (1913): 44; *The Horseless Age,* "Garage Notes," Vol. 24, No. 11 (1909): 306.
[18] *JNMA,* Patterson Ad, Vol. 8, No. 1 (1916).

---

[19] *HD,* Patterson Ad, February 11, 1913; February 25, 1913; *GR,* Patterson Ad, April 15, 1915.
[20] *HD,* Patterson Ad, February 8, 1916; October 6, 1916; August 29, 1916; *GR,* Patterson Ad, February 15, 1917; November 27, 1919.
[21] James E. Homans, *Self-Propelled Vehicles: A Practical Treatise on the Theory, Construction, Operation, Care, and Management of all Forms of Automobiles* (New York: Theo. Audel and Company, 1903), 63.
[22] American Technical Society, *Automobile Engineering* (Chicago: American Technical Society, 1917), 381.
[23] Andrew Lee Dyke, *Dyke's Automobile Encyclopedia* (St. Louis, Missouri: A. L. Dyke, 1911), 406-413; American Technical Society, *Automobile Engineering,* 442-454.
[24] *GR,* Patterson Ad, September 2, September 9, and September 16, 1915; September 23, 1915.
[25] *GR,* "The Patterson-Greenfield 4-25 Touring Car," September 23, 1915.
[26] Ibid.
[27] Ibid; *Leesburg Citizen,* "Patterson Automobile," September 30, 1915; *HD,* "Are Manufacturing Automobiles," October 5, 1915; *Negro Review (Cleveland),* "The Patterson-Greenfield Automobile in Evidence at the State Colored Fair," November 10, 1916.
[28] *GR,* Patterson Ad, September 30, 1915.
[29] *JNMA,* Patterson Ad, Vol. 8, no. 1 (1916); Vol. 3, no. 4 (1911).
[30] *GR,* Patterson Ad, July 12, 1917; Beverly Rae Kimes, *Standard Catalog of American Cars, 1805-1942* (Iola, Wisconsin: Krause Publications, 1989), 1117; Bill Naughton, "Former Slave, Descendants Were Early Vehicle Pioneers in Ohio," *Tri-State Trader*, July 5, 1980:1; Production numbers are uncertain and could not be determined through research of the Ohio Bureau of Motor Vehicle records of the period. The registration process was still new at that time and the frequent changes in

---

record keeping styles (not to mention missing records) made the tracking of production numbers impossible.

[31] Steve Konicki, "150 Built: Better Than Model T," *Dayton Daily News,* March 21, 1976; There are some accounts that state that Frederick Patterson built his first car as early as 1902, but did not officially produce them until 1915. This does not seem likely because that is something that would have been remembered in Greenfield, so there wouldn't have been so much excitement in the town in 1915 when the Patterson-Greenfield was introduced as being the first car built in the county. Curiously, though, one photo of a supposed Patterson car is of a much earlier style body. By all indications, this is a 1910 or 1911 Maxwell car. This may have been a car that Frederick Patterson had experimented with or perhaps one that was in for repair that he posed with. It looks nothing like the Patterson-Greenfield that was introduced in September of 1915. Everything about it is different including the body style, frame design, headlights, and it is right hand drive. The Patterson car of 1915 had a very sleek body style and there was nothing crude about it at all, unlike the car in the other photo. The familiarity of the Pattersons in building vehicle bodies would have surely let them produce a better body than the crude one in the photograph. It is fairly safe to stick with the 1915 date for the first car produced by Frederick Patterson.

[32] *Leesburg Citizen,* "Locally-Made Vehicles Are Subject of Search," n.d. (1980); *Greenfield Times,* "Salute for C.R. Patterson Planned," January 23, 1982.

[33] *The Horseless Age,* "The Plight of the Carriage Industry," Vol. 24, No. 17 (1909): 455.

[34] *GR*, "New Blacksmith Shop," July 8, 1915.

[35] *The Horseless Age,* "Building a Car from Standard Parts," Vol. 24, No. 23 (1909): 643; David Hochfelder and Susan Helper, "Suppliers and Product Development in the Early American Automobile Industry," *Business and Economic History,* Vol. 25,

---

No. 2 (1996): 42; *The Horseless Age,* "The Plight of the Carriage Industry," Vol. 24, No. 17 (1909): 455.

[36] Hebb, *Wheels on the Road*, 113; *The Automobile,* "Build 1800 Gasoline Cars Daily," October 7, 1915: 669.

[37] Larry Freeman, *The Merry Old Mobiles* (Watkins Glen, New York: Century House, 1949), 226-233. All numbers extracted from the roll call of auto manufacturers. Percentages were figured based on 3 time periods of entering the field and when the failure occurred. Additional percentages were figured for the period 1915-1919, when the Patterson-Greenfield was made. During this short period, 116 automakers started producing automobiles and failed by 1919. This equaled 24.6% of the starts and failures of the 1910 to 1919 period, suggesting that the first 5 years of the period saw an extremely heavy failure rate, which surely discouraged others after this to enter the field. One has to wonder why Frederick felt that he could compete against these odds.

[38] *The Automobile,* "Small Car Makers Having Difficulties," September 27, 1917: 558.

[39] Records of the National Negro Business League [microfilm]: Letter dated July 25, 1916, from Frederick D. Patterson to Emmett J. Scott.

[40] *The Automobile,* "Comparison of Features of the Average American Car for 6 Years," December 30 (1915): 1172; Lindsay Brooke, *Ford Model T: The Car that Put the World on Wheels* (Minneapolis: Motorbooks, 2008), 90.

[41] *GR,* "The Patterson-Greenfield 4-25 Touring Car," September 23, 1915.

[42] H. W. Hunter, *New Atlas of Highland County, Ohio* (Hillsboro, Ohio: H. W. Hunter, 1916); *GR,* Patterson Ad, July 5, 1917.

[43] *JNMA,* Patterson Ad, Vol. 7, no. 4 (1915); *GR,* Patterson Ad, April 19, 1917; April 26, 1917.

[44] *JNMA,* Patterson Ad, Vol. 8, No. 2 (1916).

[45] Hunter, *New Atlas of Highland County*; This special motor has been misidentified for several years as being a Continental mo-

---

tor. Out of over 1,200 Patterson advertisements that were analyzed during this study, not a single one mentioned anything about a Continental motor being used in the Patterson-Greenfield automobile. The first mention that was located of a Continental motor appears in a 1980 article by Bill Naughton in the *Tri-State Trader* that was entitled "Former Slave, Descendants were Early Vehicle Pioneers in Ohio." Since that time, several, if not most, articles that have appeared about the Patterson Company states that they used a Continental motor. These have used the Naughton article as a source of information, which is how they arrived at the errant information.

[46] *HD,* Patterson Ad, June 20, 1916; *JNMA,* Patterson Ad, Vol. 9, No. 2 (1917); Henry A. May, *First Black Autos: The Charles Richard "C.R." Patterson & Sons Company: African-American Automobile Manufacturer of Patterson-Greenfield Motorcars, Buses, and Trucks* (Mount Vernon, New York: Stalwart Publications, 2006), 34.

[47] *The Automobile,* "Simplicity Feature of G. B. & S. Motor," February 19 (1914): 464-465.

[48] *The Automobile,* "Gearless Differential in Mecca 30," December 2 (1915): 1016; *Motor Record,* "Specifications for Pleasure Cars 1917," March (1917): 9-10; *The Automobile,* "Simplicity Feature of G. B. & S. Motor," February 19 (1914): 464-465.

[49] *The Automobile,* "Simplicity Feature of G. B. & S. Motor," February 19 (1914): 464-465.

[50] *The Automobile,* G. B. & S. Ad, August 17 (1916): 88; *The Automobile,* G. B. & S. Ad, July 13 (1916): 71.

[51] Charles K. Hyde, *The Dodge Brothers: The Men, the Motor Cars, and the Legacy* (Detroit: Wayne State University Press, 2005), 42.

[52] *The Horseless Age,* Parish and Bingham Company (Cleveland) Ad, Vol. 14, No. 1 (1904): 31.

[53] Hugo Pfau, *The Custom Body Era* (New York: A. S. Barnes and Company, 1970), 187-189.

[54] Konicki, "Better Than Model T"; Brooke, *Ford Model T*, 93.

---

[55] *The Automobile,* "Comparison of Features of the Average American Car for 6 Years," December 30 (1915): 1172. All averages in next two paragraphs come from this source.
[56] *GR,* "The Patterson-Greenfield 4-25 Touring Car," September 23, 1915; *JNMA,* Patterson Ad, Vol. 8, No. 1 (1916); Vol. 8, No. 4 (1916); Vol. 8, No. 2 (1916).
[57] Borg, *Auto Mechanics,* 4.
[58] *GR,* Patterson Ad, June 3, 1915; July 22, 1915; March 15, 1917; June 19, 1919.
[59] *HD,* Patterson Ad, June 13, 1916; *GR,* Patterson Ad, November 20, 1919; March 29, 1917; May 27, 1920.
[60] *HD,* Patterson Ad, November 3, 1916; December 29, 1916; *GR,* Patterson Ad, August 22, 1918; March 20, 1919.
[61] Borg, *Auto Mechanics,* 9; *GR,* "Patterson Service," December 9, 1920; *GR,* "Patterson Garage Service: Greatly Improved with Enlarged Facilities for Handling Business and New Management," April 7, 1921.
[62] *HD,* "Prestige for Greenfield is Won through the Progressive Efforts of the C.R. Patterson and Sons Company Which is Known Throughout this Section of the Middlewest – Makers of the Famous Patterson-Greenfield Automobile," December 5, 1916.
[63] *GR,* Patterson Ad, February 8, 1917; April 10, 1919.
[64] *GR,* "Patterson Vindicated: Dismissed by Mayor Story of Chillicothe After Hearing Side of Prosecution," October 9, 1919.
[65] Ibid.
[66] *GR,* Patterson Ad, August 30, 1917 to July 24, 1919; October 14, 1920 to June 16, 1921.
[67] American Technical Society, *Automobile Engineering* (Chicago: American Technical Society, 1917), 380-381.

4

PATTERSONS ENTER THE COMMERCIAL BODY BUSINESS

The Motor Bus and Truck Industry in the United States

By 1918, it had become obvious to Frederick Patterson and his family that the Patterson-Greenfield automobile had not been a successful business venture. They had entered the field too late to stand up against the already established automobile giants such as Ford and General Motors. Their attempt was not in vain though, because it introduced them to the ins and outs of the automotive field and prepared them for their next venture. The Patterson Company went into the production of commercial vehicles, a newly developing branch of the automotive field that included motorized trucks and the buses used for both public transit and school transportation.

The Rise of the Transit Bus Industry

Streetcars had been the dominant form of public mass transit in cities since the late 1880s. In the early 1900s, the rapid spread and sprawl of cities made it difficult for the streetcar industry to keep up with the demands of the public to provide transit routes to all areas. The streetcar's dependency on running along tracks laid in the street limited it to certain areas. The lack of flexibility and the expense of laying new track to every new neighborhood became quite costly to the streetcar companies. This led to the development of alternative forms of public transportation, mainly the motor bus.[1]

When the motor bus first made its appearance between 1900 and 1910, it was very primitive and not easy to maintain. Streetcar service remained the better option at the time. Heading into the 1920s, the bus closed the gap between it and the streetcar. By this time, many flaws of motor vehicles had been corrected, and this made the bus a more viable form of transportation. The bus offered more flexibility owing to its capability to alter routes easily to avoid obstacles or detours as it did not ride on rails. While the streetcar lines tended to dictate where the new settlement of a city developed, the flexibility of the motor bus allowed it to go wherever the new settlement developed.[2]

The bus industry quickly caught up with the streetcar industry throughout the 1920s. In 1923, there were 83,066 streetcars and 27,250 buses; in 1925, totals were 81,635 streetcars and 42,425 buses; and in 1929, the numbers reached 71,219 streetcars and 50,500 buses.[3] The streetcar industry remained relatively stagnant and did not grow (in fact, experienced a slight decline), while the bus industry grew in exponential numbers. In 1923, the number of streetcars to buses was approximately three to one, but by 1929 it was closer to 1.5 to one. Buses were not only ideal for urban settings, but they were suited for rural and isolated areas as well. A rural bus company could be as small as a single bus. A bus could be purchased and immediately available for service, while a new streetcar required a complex infrastructure and a large ridership.

This rapid growth of the bus industry appealed to several manufacturers and bus building plants opened across the country. In the early 1920s, no chassis were designed specifically for bus manufacture and companies used those developed for trucks.

This left the bus bodies high up off the ground and top heavy. By the mid-1920s a small number of manufacturers built buses from the ground up; most companies built either the body or the chassis. For the integrated companies, the trade magazines covered their sales and profits well in their columns, but they "did not accord bus body builders the same attention."[4] This has left a void in the ability to conduct a thorough study of bus body company production numbers, including those of the Patterson Company.

The Rise of the School Bus Industry

The main stimulus to the rise of the school bus industry was the movement to consolidate schools, which became a widely debated issue across the country at the turn of the century. There had been a noticeable difference in the quality of education being received in the larger city schools and that received in the small one-room schools that dotted the countryside. The consolidation of several of these smaller schools into one centralized location allowed the students to get "the benefits that the city schools get and at a cost but little, if any more, than under the old district system."[5] Transportation became the main obstacle to the creation of larger districts covering wider geographical areas.

Much discussion occurred amongst parents about the topic of consolidation and the matter of transportation. According to research conducted by Albert Callon in 1930, "in most instances, patrons would not object to having their children attend consolidated schools if problems of transportation were not involved." Although some school districts in Ohio consolidated on their own earlier in the century, it wasn't until 1914 that the Ohio School Laws authorized and recommended the consolidation of school districts. The General Code of Ohio (section 7731) stated that any pupil who lived more than two miles away from school must be provided transportation to and from school at the expense of the local Board of Education.[6] This ended the debate of whether to consolidate, but intensified concern over how to transport students.

As discussed in an earlier chapter, the Pattersons were pioneers in the development of the school wagon, the first publicly provided mode of school transportation. Entering the 1920s, most school districts still utilized the horse-drawn school wagon to transport pupils, but the motorized school buses quickly became a more popular choice. As road-building projects became more common across the nation, the improved surfaces made the use of motorized school buses more practical. Callon summed up the thinking of the period by saying "the development of the consolidated school has made the school bus a necessity."[7]

The school bus had several important advantages over the horse-drawn school wagon. Some of these, as listed by Joseph Copeland in 1931, included "equal [educational] opportunities for each child, greater regularity in attendance at school, the removal of distance as a factor, proper protection and supervision while going to and from school from both harm and inclement weather, and less time spent traveling and more time to help with the duties of the home."[8]

The development of school bus building followed the same general timeline as the transit style bus. By the year 1926, a total of 26,685 school buses operated in the United States and by 1931, only five years later, the number had reached 48,775. Since early school and transit buses often shared the same type of chassis, school buses, too, encountered height and balance issues. By the late 1920s, however, several manufacturers produced bus specific chassis that allowed the body to sit much lower and provided a more passenger-friendly suspension system.[9] The biggest differences between the school bus and transit style bodies were those specifics required by the local laws of the state the buses were going to, or those features requested by the purchasing school board. For the most part, until actual standards for building school bus bodies were set in place, the bodies were entirely custom built for each customer.

The Rise of the Motor Truck Industry

The first commercial vehicles quickly followed the appearance of the automobile. As early as 1895, a carriage builder in Chicago, Charles Woods, built a delivery wagon propelled by electricity. The Winton Company built the first delivery wagon powered by a gasoline motor in 1898. The single cylinder motor was mounted under the seat and had a planetary transmission and chain drive.[10]

Several early automobile manufacturers built commercial vehicles. At the very first automobile show held at Madison Square Garden in 1900, several delivery wagons were present, including vehicles ranging from those that could handle light loads all the way up to a 4-ton electric truck. Passenger automobiles became so popular during this time that the large demand for them made most manufacturers abandon their commercial vehicle development to increase production and remain competitive in the automobile field.[11]

The production of commercial vehicles remained low in comparison with passenger vehicles throughout the first decade of the century. The number of truck registrations in the United States was only 1,000 in 1905 and gradually increased to 10,000 by 1910. Beginning in 1915, and including the years during World War I, the numbers rose much more quickly. In 1915, the number of registrations finally broke the 100,000-vehicle mark and reached 159,000, and finally broke the million-vehicle mark in 1920 when it reached 1,108,000 vehicles.[12]

A major reason for the rapid rise during this period was the high production of trucks to support World War I, both on home soil and on the battlefront. The war provided the testing and proving grounds that demonstrated what a truck was capable of handling. In S. V. Norton's 1918 book *The Motor Truck as an Aid to Business Profits,* he stated, "the use of the motor truck in the Great War has demonstrated its merits to the public more than many thousands of full-page advertisements. The businessman finds more convincing proof of the utility of the truck when he realizes that the conditions met with at the front are often much worse than those encountered in his own business."[13] The businessman could now see the motor truck as a durable vehicle that could be depended on as a tool to aid in his business.

Trucks had many advantages over horse-drawn delivery wagons, such as increased range, greater speed, dependable service, greater carrying capacity, and access to distant markets.[14] Other benefits existed as well, including publicity value. With the greater range of a motor truck, the business name painted on the truck advertised the business over a wider area than would normally be reached within the range of horse-drawn vehicles.

The care and maintenance of a motor truck was cheaper than providing for harness animals as well. A horse could only be pushed as hard and as quickly as its body allowed. It needed rest, food, and medical care to keep its usefulness to a maximum. The animal's workday was limited, unlike a motor truck that could be driven during continuous shifts throughout the entire day. With the significant increase in the amount of work that could be extracted from the motor truck, no real advantage existed to the continued use of horse-drawn vehicles by the 1920s.

The Greenfield Bus Body Company – 1920s

According to Kate's handwritten history of the company, she spoke of her brother, Frederick, and stated, "After trying it out, building the Patterson-Greenfield, he changed his mind. Learning first-hand the expense entailed, and realizing our already limited banking and credit facilities, he promptly turned his attention to the building of the school bus." The experience gained by building the Patterson-Greenfield automobile, by repairing automobiles, and by building carriages, played an important role in the next phase of the C.R. Patterson and Sons Company. By taking advantage of this experience, the Patterson Company felt it could succeed by starting to build motorized school buses.

The Patterson Company expanded to add this new trade and became also known as the Greenfield Bus Body Company.

Although the sign on the building stated that they were a bus body company, they built bodies for many types of vehicles. By the time they began building these specialized bodies, most work occurred at the east building on the SE corner of North Washington and Lafayette Streets. According to a 1928 article in the *Greenfield Republican*, the east building had approximately 12,000 feet of floor space. Kathleen Patterson mentioned in her autobiography that the bus bodies were built on the first floor and there was a room in the back for glass cutting and another room for the blacksmith shop. The second floor housed the offices and behind these were areas for building bus seats.[15] The third floor could be used for storage.

The Patterson Company still built carriages, wagons, sleighs, and other horse-drawn vehicles during this period of transition to bus bodies. It continued to do so well into the 1920s and possibly into the early 1930s whenever a request was made. During the first few years of the 1920s, the Pattersons still offered auto repair and ran a battery shop on the premises to provide supplemental income while the bus business grew. With their experience of building horse-drawn school wagons, they were quite familiar with the needs and wants of school boards regarding student transportation. This also gave the Patterson Company a distinct advantage in attracting business, as many school boards were already familiar with the company and its vehicles. The Pattersons were able to enter a new field with an already established clientele. For the Patterson Company, this was a continuation of what it had done in the past, only, this time, these school wagon bodies were mounted on motorized chassis. The frame construction of early bus bodies were all wooden structures, so the shop came equipped to take on this new product.

W.E.B. DuBois had asked Frederick for some information about the company in relation to an article or study being completed by the Crisis, magazine of the NAACP. In November of

1920, Fred wrote a letter to Mr. DuBois indicating that they were currently too busy with a possible reorganization of the company that would take place at the beginning of the year to submit the materials at that time, but he will get them submitted as soon as possible. It is unclear exactly when the Patterson Company produced its first motor bus, but according to information located in Ohio Secretary of State reports for 1921, the Pattersons filed as a new corporation, the "Greenfield Bus Body Company," with a capital stock of $50,000 dollars. The listed members of this corporation included Fred Patterson, John R. Rudd (Fred's brother-in-law), Dollie Rudd (Fred's sister), John Simms (a long-time employee of the company), Alvah E. Bell, and Robert Nichells.[16] This became the official start of the Greenfield Bus Body Company, although the Patterson Company had perhaps been building bus bodies as much as three years prior to this date. Based on the timeline of events, this reorganization of the company was likely what Frederick was referencing in his letter discussed above.

Funds were low for the company, and they had struggled to obtain mortgages over the past few years. In 1920 and 1921, small mortgages were received, but with Estelline still gone in Philadelphia, Frederick and his mother Josephine had to rely on bringing in additional signatories on the mortgages. Some that were included on the smaller mortgages during this period were Frederick's sister, Dollie, along with Dollie's husband John R. Rudd, as well as long-time employee John Simms and his wife Kate.

An article in the *Greenfield Republican* on June 23, 1921, highlighted some of the company's bus building work. The article stated, "the Pattersons are pioneers in this work, and they have given much study to the matter of getting out and presenting to the public a school bus that will meet every need while providing a comfortable conveyance for the children from and to their homes." The article also described shipments of buses that had been sent to Michigan, Iowa, and Mississippi, and noted that there had been many follow up orders "showing the satisfaction

Early style bus body being built in the Patterson factory. This is a powerful photograph as it shows an integrated workforce working side by side in the shop. (public domain)

that goes with their work."[17] This business from distant locales suggests that their buses had been around for a long enough period for the word to spread about their products. This makes it very likely that bus building started around late 1917, overlapping the period when they ceased production of the Patterson-Greenfield automobile. An April 1922 article in *Commercial Car Journal* stated that the Greenfield Bus Body Company's buses were in large demand due to the various types and styles of bodies that the company offered for any make of chassis. An official of the company stated, "We foresaw the coming demand for bus bodies several years ago and prepared for it and we believe that the bus business is still in its infancy. We are constantly making careful surveys and investigations to

A view into the inside of a mid to late 1920s Patterson bus with center row and side row seating. Image from Greenfield Bus Body Company catalog. (R.P.)

determine the best types of bodies to meet particular transportation problems and thus we are able to offer several different types of bodies, each one meeting a particular requirement."[18]

While the company was struggling to keep up with the large demand for their buses and trying to retool the Patterson shop for this relatively new endeavor, an unforeseen circumstance occurred that helped turn the tide in the company's favor. Pauline Postell, Frederick's estranged wife's mother, died in 1922 in Philadelphia. Her estate amounted to $52,656.02, which was split between her four living children, including her daughter Estelline ($13,164 to each of her living children). Estelline returned home to Greenfield with sons Frederick Postell and Postell and reconciled with her husband Frederick. The company had transitioned to building bus and truck bodies by this time but had not had major success until the year prior. As the business picked up, the factory had to upgrade, retool, and build up a stock of supplies. For the past several years, dating back to 1917, the number of mortgages that the Pattersons were able to obtain had dwindled. Suddenly, in 1922, upon Estelline's return to Greenfield, rather large mortgages were able to be obtained amounting to $13,000 in October 1922 alone. Perhaps with Estelline's financial backing, the company was able to borrow more freely once again. Upon Estelline's return to Greenfield, she took a more active role in the finances of the company, likely as a condition that she set when using her own money as collateral or as a direct contribution to the company. This introduces a whole new element into the Patterson story as we now not only have the Patterson Company being the largest manufacturing company owned by Blacks in the United States, but now we have a Black woman in charge of the financial success of the company. It is unclear when Estelline started dumping her own money into the company, but apparently, she did throughout the next few years as the family and company were very limited on funds by the time the Great Depression hit in the late 1920s, indicating that Estelline's funds had also dried up.

By October of 1922, the Patterson Company claimed that their products had become the standard for other manufacturers. As reported in Bus Transportation in October 1922, it states, "The Greenfield line of bus bodies is always known as the most popular and desired motor bus body on the market. Motor truck manufacturers have standardized with Greenfield products. Truck dealers and bus line operators, experience has shown that Greenfield Bus Bodies are the best choice."

A 1921 article in the *Greenfield Republican* gave some details of the buses. It mentioned that the bodies could be made to fit any type of chassis and came in 12-, 14-, or 16-feet lengths. These buses had a capacity of 24 to 36 pupils. The body styles were "coach-shaped, of excellent proportion, not top heavy, and are easy of entrance and egress." It went on to say, "they are made so as to be ventilated without endangering the health of the children and are equipped with heaters for the severe weather of winter." Another source mentioned a later safety feature built in by the Pattersons and highlighted the fact that "the bus side panels extended down to cover the running board, providing extra strength and safety."[19] Although no set standards existed during this early time of bus building, the Pattersons still built their bodies with the children's health and safety in mind.

In a 1980 article in the *Tri-State Trader*, Kathleen Patterson described how the interiors of the buses had been covered with plywood and the seats also had plywood backing. The seats were then covered with imitation leather. She also mentioned that the metal bows for the seats and the roof supports were bent in the blacksmith shop as part of the bus assembly. Depending on the option chosen by the purchaser, the bus seats were either mounted lengthwise, across the middle, or a combination of both. Any of these arrangements were common for the time. An article in *School Bus Fleet* described a group of 1930 Patterson buses at a school in Ohio that had "a row of bench seats completely around the perimeter of the bus, plus one row of seats up the center." The article also stated that of the 5 buses owned by the school, one carried 48 passengers and the other four carried 54 passengers.[20]

Like Patterson-Greenfield automobiles, Patterson bus bodies have vanished, even though they were built in large numbers for 20 years. The bus bodies would have made great storage sheds or chicken coops for farmers several decades ago, and it is quite possible that is where many of the bodies ended up, or at least in similar situations. Fortunately, photographs document the two decades of bus building at the Patterson Company. Some information is known from newspaper articles, a detailed study for a 1930 University of Cincinnati Master's thesis by Albert Callon, as well as two late 1920s and early 1930s Greenfield Bus Body Company catalogs that have come into the author's possession.

In his thesis, Callon wanted to evaluate bus bodies currently in use in the Midwest United States and set standards for school bus bodies according to what both manufacturers and school boards felt were important safety features.[21] Information collected by Callon about the Patterson bodies provided some technical details about their work and craftsmanship. Callon's work included the types of materials used for the different body components. To summarize, the Patterson Company built the floors and roof using tongue and groove pine, while the rest of the body, such as the sills, body posts, door posts, rails, and ribs were a mixture of ash and oak, all secured with steel braces. This wooden frame was then covered with a 20-gauge sheet metal skin and covered with Duco style paint. The bus passenger compartment lengths varied from 12' to 16' with 33.33% of the body weight extending beyond the rear axle. The upholstery of the bus consisted of spring seats and thick felt covered with imitation leather.

The Greenfield Bus Body Company only constructed bodies for the buses and trucks. Most buses during the 1920s and early 1930s utilized chassis and bodies from different makers assembled to form the complete bus. Many of the larger bus makers began to build the entire bus from the ground up during the 1930s.[22] The Patterson Company quoted its prices to prospective clients in two different ways. The client could provide the chassis for the bus and the Pattersons would build the body and mount it, or the Pattersons could provide the chassis and body for the client.

In the earlier years, many clients provided the chassis themselves, since they were sometimes modifying an older truck for carrying passengers. Until strict standards were set for school bus bodies in the mid-1930s, limiting the years of service for both bodies and chassis, several schools would reuse the same chassis with a new body mounted on it.[23] Sometimes the Patterson Company would have to modify and lengthen the chassis to equip it with a bus body.

If the Pattersons were to provide the chassis, they did not seem to discriminate and utilize a certain make. Several chassis makers that the Patterson Company used are listed in articles. These include Ford, Chevrolet, Dodge, Reo, International, Commerce, Gramm-Bernstein, Biederman, and Graham.[24] This seems to suggest that, although the Pattersons were to provide the chassis for certain orders, the clients may have had preferences for certain makes of chassis. This also could indicate that certain size buses (passenger capacity) may have worked better with certain chassis.

The Patterson Company also shipped completed bodies to clients by railroad without mounting them on a chassis first. For this type of order, the company utilized a type of construction referred to as "knock-down construction." According to a 1923 article in *Motor Coach Transportation,* "this type of body is easily crated, and the maker claims this method saves one-third on the freight charges. Easily handled it can also be reshipped with the maximum of convenience. Assembling is very simple, and the company claims this can be accomplished without possibility of mistake. These bodies are built in different sizes to suit requirements."[25]

The Pattersons built the main fleet of school buses for the Greenfield school. According to an interview with Kathleen Patterson, "E. L. McClain wanted the school buses painted the color green for Greenfield and with gold leaf lettering."[26] This may have taken place prior to 1921, since an article in the *Greenfield Republican* in that year noted that two additional school buses were to be built since some of the existing ones were already too overcrowded. By January of 1922, all of the school's buses were housed in a building rented from the Pattersons at their business location.[27]

While the Pattersons periodically placed advertisements in trade journals, usually Frederick promoted his vehicles while sitting in front of school boards and at large conventions or shows. A 1923 article in the *Greenfield Republican* recounts Fred's trip to the National School Supply Association's annual convention in Chicago that year. It stated that the "product of the Greenfield Bus Body Company is a center of attraction at the Chicago meeting." According to the article, which likely derived information from Frederick himself, after reviewing some samples of the work that the Patterson Company had built,

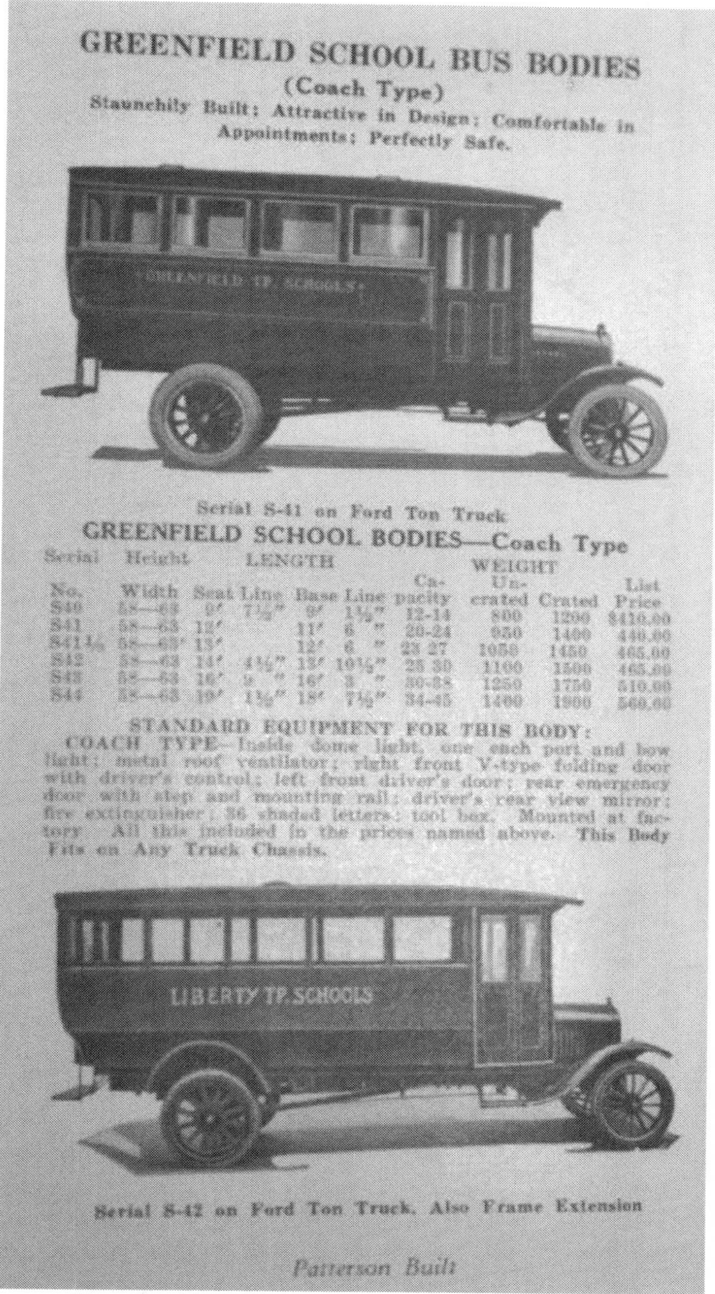

GREENFIELD SCHOOL BUS BODIES

(Coach Type)

Staunchily Built: Attractive in Design; Comfortable in Appointments: Perfectly Safe.

Serial S-41 on Ford Ton Truck

GREENFIELD SCHOOL BODIES—Coach Type

Serial No.	Height Width	LENGTH Seat	Line	Base	Line	Capacity	WEIGHT Uncrated	Crated	List Price
S40	58—63	9'	7½"	9'	1½"	12-14	800	1200	$410.00
S41	58—63	12'		11'	6 "	20-24	950	1400	440.00
S41½	58—63'	13'		12'	6 "	23 27	1050	1450	465.00
S42	58—63	14'	4½"	13'	10½"	25 30	1100	1500	465.00
S43	58—63	16'	9 "	16'	3 "	30-38	1250	1750	510.00
S44	58—63	19'	1½"	18'	7½"	34-45	1400	1900	560.00

STANDARD EQUIPMENT FOR THIS BODY:

COACH TYPE—Inside dome light, one each port and bow light; metal roof ventilator; right front V-type folding door with driver's control; left front driver's door; rear emergency door with step and mounting rail; driver's rear view mirror; fire extinguisher; 36 shaded letters; tool box. Mounted at factory. All this included in the prices named above. **This Body Fits on Any Truck Chassis.**

Serial S-42 on Ford Ton Truck. Also Frame Extension

Patterson Built

(from late 1920s Patterson bus catalog, courtesy of R.P.)

GREENFIELD SCHOOL BUS BODIES

Series 45 and 45A. For Every Chassis

Series 45 and 45A. For Every Chassis

Serial 48 on International

GREENFIELD SCHOOL BODIES—Commercial Type

DESCRIPTION

Serial No.	Height Width	LENGTH Seat Line		LENGTH Base Line		CAPACITY	Side seats	Center	List Price
		9'	7½"	9'	1½"	20-26	2	1	$490.00
46	60—80	12'		11'	6 "	26-32	2	1	543.75
46½	60—80	13'		12'	6 "	30-36	2	1	586.75
47	60—80	14'	4½"	13'	10½"	36-40	2	1	586.75
48	60—80	16'	9 "	16'	3 "	42-48	2	1	639.75
49	60—80	19'	1½"	18'	7½"	48-56	2	1	697.75

STANDARD EQUIPMENT FOR THIS BODY:

Inside dome light, one each port and bow light; metal roof ventilator; right front V-type folding door with driver control; left front driver's door to open; rear emergency door with step and mounting rail; driver's rear view mirror; fire extinguisher; 36 shaded letters; tool box. Mounted at factory. All this included in the prices named.

This Body Fits Any Truck Chassis. This body carries three (3) seats and especially adapted for short frame lengths and yet affording large seating capacity.

DESCRIPTION

Serial No.	Height Width	LENGTH Seat Line		LENGTH Base Line		CAPACITY			List Price
45	60—82	9'	7½"	9'	1½"	26-32—	Four	Seats	$500.00
46-A	60-82	12'		11'	6 "	32-38	"	"	563.75
46½-A	60-82	13'		12'	6 "	36-42	"	"	616.75
47-A	60-82	14'	4½"	13'	10½"	44-50	"	"	616.75
48-A	60-82	16'	9 "	16'	3 "	52-58	"	"	669.75
49-A	60-82	19'	1½"	18'	7½"	60-66	"	"	732.75

Same equipment as three-seated job. The seating arrangement is four rows of seats, seats being back to back.

Patterson Built

(from late 1920s Patterson bus catalog, courtesy of R.P.)

GREENFIELD SCHOOL BUS BODIES

Utility Type. For Every Chassis.

Serials U120—U121—U122—U123—U124

A Body of Strength and Durability.
A Rare Combination of Price and Quality.
12 to 45 Pupils.

Serial U-122 on Ford with Olson Extension.

GREENFIELD SCHOOL BODIES—Utility Type

DESCRIPTION

Serial No.	Height Width	LENGTH Seat Line		LENGTH Base Line		Capacity	WEIGHT Uncrated	WEIGHT Crated	List Price
U-120	58-63	9′	7½″	9′	7½″	12-14	800	1200	$300.00
U-121	58-63	12′		12′		20-24	875	1400	320.00
U-121½	58-63	13′		13′					340.00
U-122	58-63	14′	4½″	14′	4½″	25-30	990	1500	340.00
U-123	58-63	16′	9 ″	16′	9 ″	30-38	1095	1700	370.00
U-124	58-63	19′	1½″	19′	1½″	38-45	1300	1900	400.00

STANDARD EQUIPMENT FOR THIS BODY:

Right front driver's door; rear emergency door with mounting rail; adjustable steel ventilator; rear view mirror; tool box. **This Body Fits Any Truck Chassis.**

Patterson Built

(from late 1920s Patterson bus catalog, courtesy of R.P.)

Bus built for the Greenfield School District by the Patterson Company.
(Greenfield Public Library)

several members of the association "who have had large experience with school buses called it one of the best bus bodies that they had seen and paid Mr. Patterson high compliments for the goods his company is turning out."[28]

The same article also described a similar reaction to the Patterson products at the New York Auto Show in January of that year. "Experts of large auto companies, and men who have had wide experience with handling interurban buses, all agreed that the Greenfield Bus Body Company bus bodies stood at the very front, the leading product of their line." As a result of the Chicago convention, an order was received for 30 bus bodies to go to Virginia, and it spurred several other inquiries as to "how soon bodies can be turned out." According to Fred, they were getting so many orders that "we cannot handle all the orders that we might get in." Fred stated that production capabilities at the time amounted to nine school bus bodies and two interurban bus bodies per week, but he had "hope in the near future to be able to increase this output somewhat." Four months later, the Patterson Company ran a large advertisement on the front page of the *Greenfield Republican* asking for several additional men with skills as cabinet makers and woodworkers to help the company get out more work.[29]

In addition to regular school and transit bus bodies, the Patterson Company also designed special sedan vehicles. A 1925 advertisement in *Motor Truck* magazine stated, "Greenfield Sedan – Built in any capacity for any chassis. One of the best jobs we have ever offered. Blueprints furnished for any special body. We are the largest exclusive builders of bus bodies and are equipped to handle any request. Send for our illustrated catalog. Greenfield Bus Body Co., Greenfield, Ohio."

Row of early style Patterson buses lined up along the street.
(Greenfield Public Library)

The Greenfield De Luxe Sedan, which was fully customizable and built to fit any chassis and in any passenger capacity. (R.P.)

One piece of information that we know little about was the cost of the Patterson buses. Since the company didn't really advertise buses to the public, costs are something that didn't come up in a review of literature. Many times, the costs could vary as the company was bidding against other bus manufacturers. Additionally, the company was always willing to customize the designs based on the customer needs. However, a 1927 list was recently provided to the author that indicates the cost of different standard bus bodies to the dealer, as well as a manufacturer's suggested retail price. This list aids in visualizing the Patterson Company income, especially for those years when they were producing upwards of 500 bus bodies. The 1927 dealer's confidential price list is shown on the following pages.

With the significant increase in the number of trucks being purchased during and after World War I, it seems perfectly logical that the Patterson Company would add the construction of these types of vehicles to their products. Just prior to the start of the war, several major automobile manufacturers had announced their intentions of entering truck manufacturing. According to an October 1918 article in *Automotive Industries* these manufacturers were "obliged to change their programs owing to the government's ruling relative to starting of new projects."[30] Those that had already started preparing their assembly plants had to set aside their plans to focus on the war support work required of them by the government. Existing truck building companies, such as Federal, still built vehicles, but a majority of those were going to support the war effort. This left a huge opening in the industry and created an opportunity for smaller companies, such as the Patterson Company, to get their foot in the door while the larger companies were locked into conducting work for the government.

Bus advertisement for the Greenfield Bus Body Company.
The Webster Avenue address is a shipping address for the company.
From *Bus Transportation,* Vol. 1, No. 12 (1922): 51.

The Most Complete Line of School Bus Bodies Manufactured

THE GREENFIELD LINE

Coach Type - Parlor Car Type - Multiple Seat Type - Utility Type
The Advanced Line

THE GREENFIELD BUS BODY COMPANY

GREENFIELD, OHIO, U. S. A.

Builders of Motor Bodies and School Wagons

Longest in Service - Unequalled in Quality
Lowest in Price - Unapproached in Design

Meeting the Legal Requirements of
Safety - - Service - - Comfort - - Convenience

These Bodies Designed for Every Chassis on the Market

Dealer's Confidential Price List

EFFECTIVE MARCH 1, 1927

SUPPLY STATIONS
Cincinnati, Ohio
Columbus, Ohio
Clarksburg, W. Va.
Greenfield, Ind.

Page from Patterson Company 1927 Dealer's Confidential Price List. (R.P.)

Greenfield School Bodies—Coach Type—*The Best School Bus Made*

DESCRIPTION

Serial No.	Height Width	LENGTH Seat Line	LENGTH Base Line	Capacity	WEIGHT Uncrated	WEIGHT Crated	List Price	Dealer's Price
S40	54—63	9′ 7½″	9′ 1½″	12—14	800	1200	$410.00	$307.50
S41	54—63	12′	11′ 6″	20—24	950	1400	440.00	330.00
S41½	54—63	13′	12′ 6″	23—27	1050	1450	465.00	348.75
S42	54—63	14′ 4½″	13′ 10½″	25—30	1100	1500	465.00	348.75
S43	54—63	16′ 9″	16′ 3″	30—38	1250	1750	510.00	382.50
S44	54—63	19′ 1½″	18′ 7½″	34—45	1400	1900	560.00	420.00

STANDARD EQUIPMENT FOR THIS BODY:

COACH TYPE—Inside dome light, one each port and bow light; metal roof ventilator, bus type; right front V type folding door with driver's control; left front driver's door; rear emergency door with step and mounting rail; driver's rear view mirror; fire extinguisher; 36 shaded letters; tool box. Mounted at factory. All this included in the prices named above. **This Body Fits on Any Truck Chassis.**

GREENFIELD SCHOOL BODIES—Multiple Seat Coach Type

THREE LENGTHWISE SEATS — DESCRIPTION — THREE LENGTHWISE SEATS

Serial No.	Height Width	LENGTH Seat Line	LENGTH Base Line	CAPACITY	List Price	Dealer's Price
45	54—80	9′ 7½″	9′ 1½″	20-26—2 side seats, 1 center	$490.00	$367.50
46	54—80	12′	11′ 6″	26-32— " "	543.75	407.82
46½	54—80	13′	12′ 6″	30-36— " "	586.75	440.07
47	54—80	14′ 4½″	13′ 10½″	36-40— " "	586.75	440.07
48	54—80	16′ 9″	16′ 3″	42-48— " "	639.75	479.81
49	54—80	19′ 1½″	18′ 7½″	48-56— " "	697.75	523.31

STANDARD EQUIPMENT FOR THIS BODY:

Inside dome light, one each port and bow light; metal roof ventilator, bus type; right front V type folding door with driver control; left front driver's door to open; rear emergency door with step and mounting rail; driver's rear view mirror; fire extinguisher; 36 shaded letters; tool box. Mounted at factory. All this included in the prices named.

This Body Fits Any Truck Chassis. This body carries three (3) seats and especially adapted for short frame lengths and yet affording large seating capacity.

FOUR LENGTHWISE SEATS — DESCRIPTION — FOUR LENGTHWISE SEATS

Serial No.	Height Width	LENGTH Seat Line	LENGTH Base Line	CAPACITY	List Price	Dealer's Price
45A	54—82	9′ 7½″	9′ 1½″	26-32—Four seats	$500.00	$375.00
46A	54—82	12′	11′ 6″	32-38— " "	563.75	422.82
46½A	54—82	13′	12′ 6″	36-42— " "	616.75	462.57
47A	54—82	14′ 4½″	13′ 10½″	44-50— " "	616.75	462.57
48A	54—82	16′ 9″	16′ 3″	52-58— " "	669.75	502.31
49A	54—82	19′ 1½″	18′ 7½″	60-66— " "	732.75	549.57

Same equipment as three-seated job. The seating arrangement is four rows of seats, the middle seats being back to back.

Page from Patterson Company 1927 Dealer's Confidential Price List. (R.P.)

GREENFIELD SCHOOL BODIES—Utility Type

DESCRIPTION

Serial No.	Height Width	LENGTH Seat Line	LENGTH Base Line	Capacity	WEIGHT Uncrated	WEIGHT Crated	List Price	Dealer's Price
U120	54—63	9' 7½"	9' 7½"	12-14	800	1200	$330.00	$264.00
U121	54—63	12'	12'	20-24	875	1400	352.00	281.60
U121½	54—63	13'	13'				374.00	299.20
U122	54—63	14' 4½"	14' 4½"	25-30	990	1500	374.00	299.20
U123	54—63	16' 9"	16' 9"	30-38	1095	1700	407.00	325.60
U124	54—63	19' 1½"	19' 1½"	38-45	* 1300	1900	440.00	352.00

STANDARD EQUIPMENT FOR THIS BODY:

Right front driver's door; rear emergency door with mounting rail; adjustable steel ventilator; rear view mirror; tool box. **This Body Fits Any Truck Chassis.**

FACTORY SERVICE You may ship chassis to factory. We unload, mount body, and reship at additional charge.......................... $10.00

ACCESSORIES

DESCRIPTION	List Price	Dealer's Price
Dome lights, ea., where not specified	$ 4.00	$ 3.00
Letters, each, in excess of 36......	.25	.18
Hanger straps, 2 rails..........	8.00	6.00
Fenders attached	6.66	5.00
Heavy duty type for wheels 33-38.	12.00	9.00
Buzzer system with buttons......	16.00	12.00
Tire carrier, single............	6.00	4.50
Roof type baggage rack.........	20.00	16.00
Baggage rack ladder............	5.00	3.75
Left front driver's door.........	16.00	12.00
Rear door control..............	16.00	12.00
Rear door and folding step control	21.00	15.75
Exhaust heater (floor register type)	20.00	15.00
Exhaust heater (steel pipe with heater guards) 12-14 ft........	32.00	24.00
Exhaust heater (steel pipe with heater guards) 16-18 ft........	36.00	27.00
Roof ventilator	8.00	6.00
Pyrene fire extinguisher.........	10.00	8.00
Front door driver's control.......	6.00	4.50
Port and bow lights............	3.20	2.40
Folding seat at right front door...	6.00	4.50
Crating KD	15.00	15.00
Crating set up................	10.00	10.00
All metal windshield............	15.00	11.25

DESCRIPTION	List Price	Dealer's Price
Driver's seat assembly..........	20.00	15.00
Long spring cushion (40")......	12.00	9.00
Standard spring cushion (27")...	10.00	7.50
Pointed cushion	12.00	9.00
Driver's seat cushion...........	8.00	6.00
Short spring cushion...........	8.00	6.00

FOR FORD TON TRUCK CHASSIS ONLY

DESCRIPTION	List Price	Dealer's Price
Steel frame extension........... Lengthens and widens frame.	$30.00	$22.50
Side spring frame extension......	100.00	75.00
Installation	10.00	10.00
Tire carrier, 32x4½, single......	4.40	3.30
Tire carrier, 30x3½, single......	3.20	2.40

We do not lengthen chassis wheelbase, but can furnish Rowe or any extension desired.

We are not responsible for the installation of an accessory when the chassis is not fitted for it.

SPECIAL SERVICE A large number of school routes are taken care of by drivers who furnish their own equipment and who consequently desire special body arrangement of varied nature to enable them to use their equipment during the vacation months. Greenfield makes such changes at the very least cost. Ask about them.

Page from Patterson Company 1927 Dealer's Confidential Price List. (R.P.)

GREENFIELD SCHOOL BODIES—Parlor Car Type—Series 70A

DESCRIPTION

Serial No.	Length	Adults	Children	Aisle	Sash	List Price	Dealer's Price
Width, 70"—Two Lengthwise Seats							Height, 62"
71A	13′	15-17	23-27	24"	3 32x18	$ 830.00	$664.00
72A	16′	20-22	30-38	24"	4 32x18	970.00	776.00
73A	19′	25-27	34-45	24"	5 32x18	1,170.00	936.00
Width, 84"—Three Lengthwise Seats							Height, 62"
74A	13′	21-23	30-36	40"	3 32x18	$1,016.00	$812.80
75A	16′	26-28	42-48	40"	4 32x18	1,069.33	855.47
76A	19′	31-33	48-56	40"	5 32x18	1,127.00	901.60
Width, 84"—Four Lengthwise Seats							Height, 62"
77A	13′	27-29	36-42	40"	3 32x18	$1,037.00	$829.60
78A	16′	31-33	52-58	40"	4 32x18	1,077.00	861.60
79A	19′	32-34	60-66	40"	5 32x18	1,128.00	902.48

Description: T steel roof, our own construction, with Haskelite roof panels; wheelhouse; sedan type front doors, right manually controlled from drivers seat, left driver's emergency door; rear center emergency door; cowl; brass sash windows; all-metal windshield, ventilating type, spring cushions and backs. Detailed specifications on request.

Series 70B—IN LENGTHS AND CAPACITIES SAME AS ABOVE

Serial No.	List Price	Dealer's Price	Serial No.	List Price	Dealer's Price	Serial No.	List Price	Dealer's Price
71B	$710.00	$568.00	74B	$869.00	$695.20	77B	$914.66	$731.73
72B	816.00	668.80	75B	889.00	711.20	78B	943.66	754.93
73B	953.56	762.85	76B	973.00	778.40	79B	984.66	787.73

Description: Regulation coach type roof (wood cross ribs with beaded tongue and groove ceiling); wheelhouse, right and left front doors; rear center emergency door; wood sash windows; built-in windshield; spring cushions and padded lazy backs.

Net extra over above prices for forward-looking cross seats $10.00 per seat.

GREENFIELD SCHOOL BODIES—Metropolitan Type

These types cover the field of special construction extra in all of their appointments and designed to meet the requirements of large city boards of education, colleges and institutions of various kinds, where the very finest is demanded. They are built to special order and to specifications. Our Engineer is open to consultation at all times.

CAPACITY	List Price	Dealer's Price
22 Passengers	$2,200.00	$1,650.00
26 Passengers	2,400.00	1,800.00
30 Passengers	2,600.00	1,950.00

PRICES All prices are F. O. B. Factory. No charge for mounting School Bus Bodies on chassis brought to our factory. Whatever the Government tax may be 3% effective March 28, 1926, will be added to the invoice price. No tax on horse-drawn vehicles.

DELIVERY Motor bodies are carried in stock for immediately delivery at all times.

NOTICE—Mounting Service includes the removal of temporary buck seats only, and not the dismantling of cab or body structure.

Special gas tank carriers entailing special manufacture and attachment carry an additional necessary cost to cover this service.

When additional materials in excess of $1.00 are needed, such as lighting wire, gas line, etc., etc., these materials will be added to the price of body.

We are not responsible for the installation of accessory when the chassis is not fitted for it.

Page from Patterson Company 1927 Dealer's Confidential Price List. (R.P.)

Little is known about the commercial vehicle bodies produced by the Pattersons as they were generally absent from the company advertisements. The Patterson Company had a standard line of products, but each vehicle ordered was customizable based on the needs of the customer. As with the bus bodies, truck customers could provide a chassis and the Pattersons would build the body for it, or the customer could purchase a package deal with the Pattersons providing the body and chassis. Luckily, a late 1920s catalog of the Patterson bus and truck products was provided to the author by descendants of the Patterson family. Several examples of the Patterson standard truck line are provided on the following pages. Since each was fully customizable, few specific details were provided in the catalog.

According to *Motor Trucks of America,* by 1917, many companies offered commercial attachments for Ford passenger car chassis and specialized bodies were becoming more popular. Advertisements placed by the Pattersons in April of 1915 showed that they were already constructing this type of body even before the introduction of their first Patterson-Greenfield automobile later that year in September. The ads stated, "Don't junk that old car – Bring it to Patterson who will make it over into a serviceable, profitable truck." The advertisements appealed to grocery, hardware, livery, undertakers, and any other businessmen who were "anxious to improve their service at least cost." The Patterson Company completely worked over any make of vehicle, lengthened the chassis if necessary, and built a body to a customer's specifications and made "it into an efficient motor truck." A 1916 advertisement stated that it could complete a car to truck conversion in only 10 days.[31]

Although the Patterson Company continued to build truck bodies throughout the next few years while it focused on the Patterson-Greenfield automobile, it wasn't until around 1919, after the automobile venture had been abandoned, that it concentrated more heavily on building truck bodies. The company created specialized bodies to haul items such as ice, furniture, milk, baked goods, as well as building moving vans and hearses.[32]

(images from late 1920s catalog, courtesy of R.P.)

(images from late 1920s catalog, courtesy of R.P.)

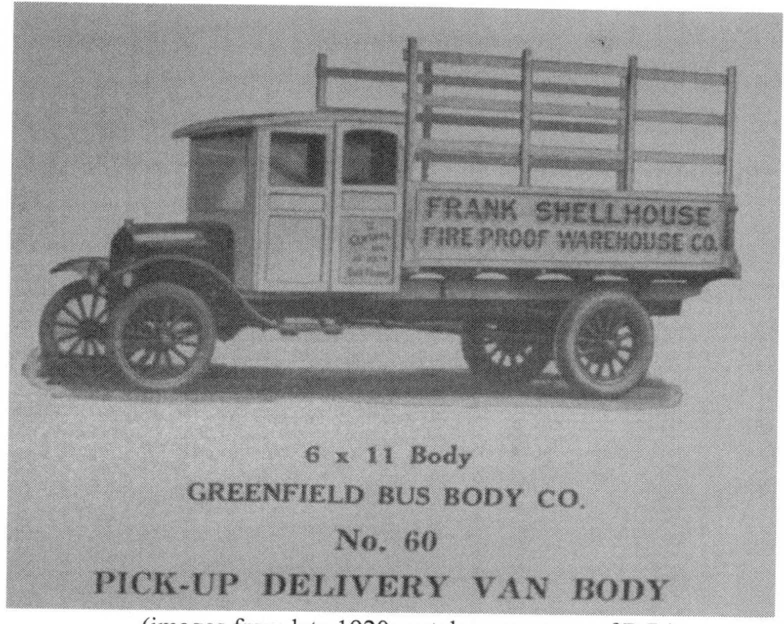

6 x 11 Body
GREENFIELD BUS BODY CO.
No. 60
PICK-UP DELIVERY VAN BODY

(images from late 1920s catalog, courtesy of R.P.)

9 ft. Body
GREENFIELD BUS BODY CO.
No. 20
VESTIBULE FULL PANELLED BODY

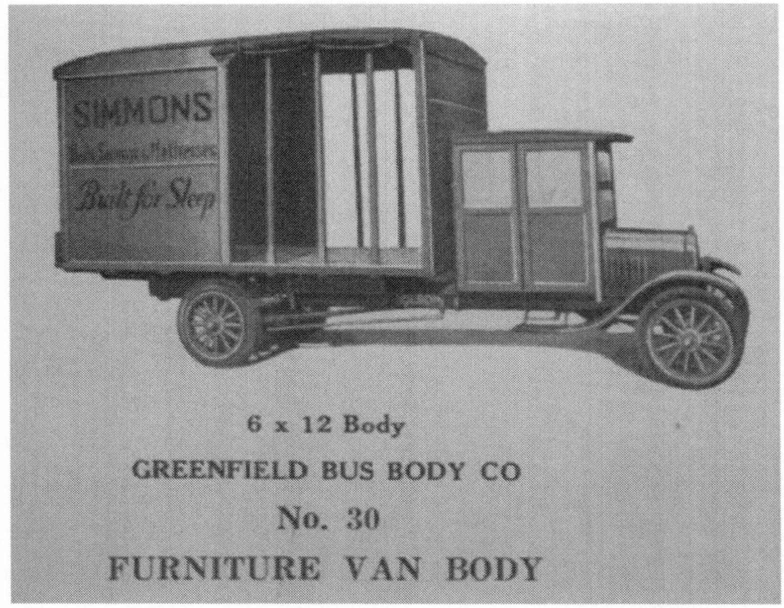

(images from late 1920s catalog, courtesy of R.P.)

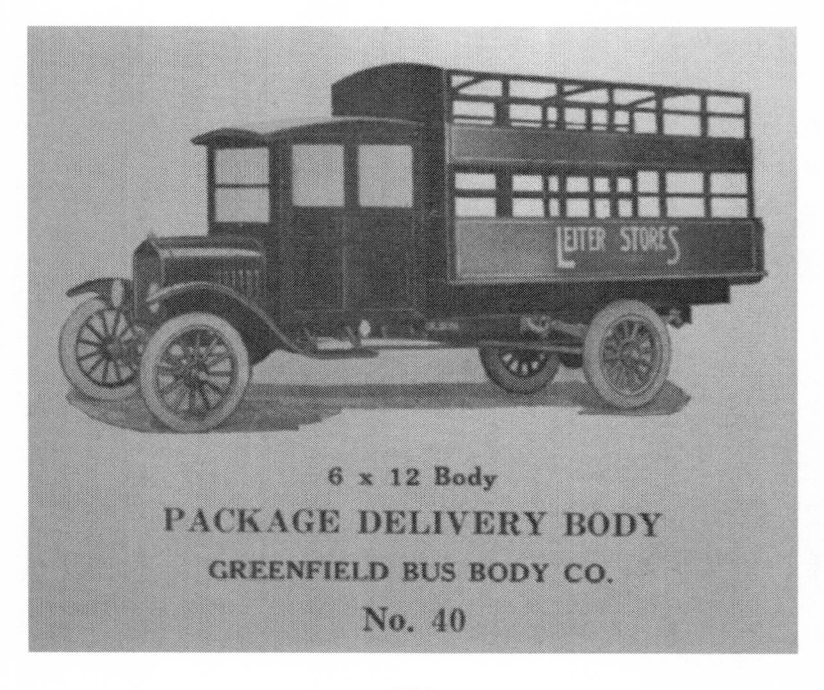

LEITER STORES
6 x 12 Body
PACKAGE DELIVERY BODY
GREENFIELD BUS BODY CO.
No. 40

Throughout the 1920s, the workload of the Greenfield Bus Body Company continued to grow. According to a 1929 article in *The Journal of Negro History*, J. H. Harmon stated that the company had an output of 500 bodies per year and an annual income of just over $150,000, with a maximum payroll of approximately $5,000 per month. Harmon also noted that the company "enjoys the reputation of building the best bus body in the State of Ohio and with a sale of the largest number of bodies of any corporation of its character in Ohio." The article mentioned that the Patterson Company held contracts with school boards across Ohio, West Virginia, and Kentucky, and had commercial body contracts in several states. One of these contracts was with International Harvester Trucks. The contract was to build some of their special purpose bodies. The Patterson bodies were placed on an International Harvester chassis and tagged with the International nameplate. Additional research indicated that International Harvester focused mainly on their pickup trucks during this time, and it was common for them to contract out special purpose bodies to be constructed under the International Harvester name. We know that the Pattersons still had this contract in 1929 (the date of the article), however, we do not know when the contract ended, but was likely by 1932 at the time of Frederick's death. While the payroll numbers do not match those mentioned by Harmon above, according to an article in the *Greenfield Republican*, "in 1929 as high as 70 men were employed and the payroll amounted to as much as $1,900 per week."[33] This discrepancy could be the result of seasonal fluctuations of the business during the time each author acquired his data.

School bus bodies were built more frequently than both transit style bus bodies and truck bodies. Although no known figures exist of how many school buses were built and sent to other states during the 1920s, some numbers are available regarding the Ohio schools. In a 1930 Master's thesis by M. D. Hartsook entitled *A Survey of Bus Transportation to and from Ohio*

Row of 15 International trucks with Patterson special purpose bodies that had just been delivered from Greenfield. (R.P.)

Schools, 1929-1930, percentages of buses by make of chassis and body were laid out to compare transportation vehicles in school districts in Ohio. These were separated by city schools and county schools to determine if a preference existed for certain makes of buses by district type. Buses made by the Greenfield Bus Body Company equaled 15.5% of all school buses in Ohio used by county schools. More importantly, the percentage of the company's buses used in the city schools in Ohio totaled 35.8%.[34] These totals are quite impressive; especially the percentage for city schools, considering this survey occurred in 1930, well after some of the larger body companies were overtaking many of the smaller builders.

The Pattersons had several competitors in the bus business by the end of the 1920s. One large competitor from nearby Dublin, Indiana, was the Wayne Body Corporation. Apart from not being a Black-owned company, this corporation was virtually a twin of the Patterson Company. It had started out in the mid-1800s (1868) building carriages, and like the Pattersons, they later constructed horse-drawn school wagons. From 1901 to 1917, the company constructed the Richmond automobile, also known as the Wayne. In 1914, the Wayne Corporation began placing school wagon bodies on extended Model T chassis as a forerunner to the motorized school buses they produced just after World War I, also in alignment with the Patterson Company. Unlike the Pattersons, Wayne had enough capital to afford the transitions to keep up with the competition and remained in business until 1956 when it merged with other companies. The Wayne name passed through the vehicle manufacturing business until 1992 when it finally disappeared.[35] Along with the Wayne Body Corporation, several other competing Ohio manufacturers supplied buses and bus bodies.

By 1929, ninety-nine Ohio manufacturers, mostly located in the northern part of the state, built buses, bus bodies, or bus parts. While many of these competed directly with the Patterson Company and focused mainly on school buses, others dealt only with transit style buses. Those companies that built bodies only were in more direct competition with the Greenfield firm. Some of the manufacturers of complete buses included Twin Coach, White, Flxible, and Beck. Those larger companies that built bus bodies in Ohio during this time included Bender, Anderson, Superior, Lang, and Fremont Body Companies.[36] These companies lasted into the 1930s or longer.

The Patterson buses would have been priced close to the average range, which is how they had managed to stay competitive for so long. In the 1920s, no set standards existed for school bus bodies and amateur constructed bodies were perfectly acceptable. School districts tended to buy the cheapest bus possible as noted in a report by Johns in 1928. In the report, Johns recorded that 189 out of 320 school buses were locally built bodies riding on truck chassis. Barden also supported this and stated that "the home-made body was used to a much larger extent than any other make" with 45.1% of county schools and 26.8% of city schools utilizing amateur constructed bodies in his study area. Kellmer found that, for the most part, the amateur constructed buses were poorly built. According to Barden "commercial or factory-made bodies are usually better than home made bodies in respect to convenience, comfort, and safety...the home-made bodies are usually much cheaper, which would account for the larger number of them being used." Regarding factory-built bodies Kellmer declared "as a rule we find substantial bodies and well powered chassis."[37]

In 1921, the front page of the *Greenfield Republican* highlighted an early large order of bus bodies. The Gramm-Bernstein Motor Car Company, based in Lima, Ohio, had brought a Model 10 Speedster truck chassis to Greenfield to be fitted for a bus body. The company placed an order for 22 bus bodies, but apparently, they would have to wait their turn since they were behind an order, from an unspecified customer, for 210 bodies.[38] According to the article, the company turned out one complete body per day at this time, so these two orders alone would keep the company busy for nearly a full year.

Several documents (including published interviews with the Pattersons) indicate that the Greenfield Bus Body Company had placed the first buses on the streets of cities such as Cincinnati and Cleveland.[39] In a 1976 interview for the *Dayton Daily News,* Postell Patterson stated that "Patterson-built buses were the first to ever carry passengers through the streets of Cincinnati."[40] Although it could not be determined if they were the first buses in Cincinnati, a photograph at the Greenfield Historical Society shows some early Patterson buses marked "Cincinnati Motor Bus Company" along their sides. This company, established in 1918 or 1919, was an early bus company in that city. This photograph does not support that the Pattersons placed the "first" buses on the streets of Cincinnati but does show that they provided some of the first.

Articles in the *Greenfield Republican* document some of the buses sent to different cities. According to a 1921 article, the

Patterson Company sent two large passenger buses to Cleveland. The article also referred to three passenger buses sent to Dayton and two shipped to Toledo that week. Mr. Dreese of Cincinnati ordered a passenger bus in October of 1922. This bus was placed on a Biederman chassis and the materials for the body had been purchased from The American Wagon Works. The article described the bus as "probably one of the best and most efficient of its kind, and the Greenfield Bus Body Company can be commended for their ability to turn out a serviceable and classy looking bus."[41]

A new transit service started in December 1921 in neighboring Sabina, Ohio, approximately 20 miles away from Greenfield, which used Patterson buses. Homer and Will Ray of Sabina started this small company. Originally part of the Red Star Bus Line, it later became part of the White Star Bus Line.[42] Several small companies such as this purchased Patterson buses.

Although most of the public attention focused on the school bus bodies that the company built, some trucks occasionally gained attention in the local newspapers. Reports of truck sales appeared sporadically in the *Greenfield Republican* to inform the public of some of the important sales, many times local, that the Patterson Company had made and to remind the public that the company made more than just bus bodies. In the summer of 1921, local reporters noted that the Pattersons had just sent two trucks to Greenfield, Indiana, and a large shipment of truck bodies to Hennessy, Oklahoma.[43] On a national scale, an article appeared in *The Automotive Manufacturer* in 1926 listing bids that were made by several bus body companies for the construction of 35 parcel delivery bodies for the United States Postal Service. While the Patterson's bid was only $400 per body, there were a couple of companies that came in with a lower bid and most likely won the contract.[44]

Some other noted sales were publicized in 1927. One involved a large moving van made for Bert Rich of Covington, Kentucky. This body was built on a Biederman Six truck chassis that produced 110 horsepower. This truck had a total length

Early transit style body, The Greenfield Inter-City Coach,
that was designed and built in Greenfield. (R.P.)

of 25 feet and provided 18 feet of loading space. The cargo box measured seven feet in width and height and had a padded interior along the walls. The same article highlighted the construction of a small delivery truck built for the Morris & Company meat packers in Columbus, Ohio, which it built on an International chassis and had a strong body for heavy work. The article stated, "The Bus Body Company has had a number of high-class jobs this year and their fame is spreading throughout the country."[45]

Another 1927 article in the *Greenfield Republican* described a new truck built for the Greenfield Furniture Company. This truck included a special design for handling furniture and storing the padding and other packing materials used in delivering furniture. The truck was built on a Graham chassis purchased from the local Dodge dealer, E. C. Gehres. The manager of the company, Mr. Brizius, reported being very well pleased with the truck, which had been purchased strictly from Greenfield businesses, and ordered three more of them for his stores in other locations. In 1931, this truck received an overhaul, which added a steel roof and sides, and had been repainted by the Greenfield Bus Body Company. To start the New Year off right in 1934, the truck became "dolled up" once again and painted to look like new.[46]

While the bus business became busy and grew during the 1920s, Frederick Patterson still made room for other work on the side. In August of 1926, Frederick joined Harry B. Alexander and James A. Jackson to form a business to manufacture and sell church furniture. Each partner put $700 toward the new business dubbed the "Greenfield Seating Company." For unknown reasons, this partnership did not last long, only until the spring of the following year. During this time, the company manufactured and sold furniture to at least three different churches: Jerusalem Baptist Church of Duquesne, Pennsylvania; First Pentecostal Church of Dayton, Ohio; and the Wayman African Methodist Episcopal Church of Dayton, Ohio.[47] Presumably, this furniture had been manufactured at the Patterson shop, which had all the necessary equipment to construct the furniture.

The bus business really boomed in the 1920s, but all was not perfect for the Pattersons. In February of 1926, a fire occurred one night at the east building.[48] The fire had started on the third floor over the paint department and, according to the article, "for a short time it looked as though the entire plant would be destroyed...but the fire department soon had streams of water pouring into the windows." Luckily for the Pattersons, this building adjoined the Greenfield Fire Department, which allowed a quick response to the fire. Several motor trucks being fitted with Patterson bodies were removed from the first floor. The quick actions of the fire department limited the amount of damage to approximately $1,000.

Another problem that the Patterson Company encountered was actually a desirable one; too much business and not enough room. Frederick attempted to remedy this problem in 1928 when the old "South Side School Building" in Greenfield went up for auction by the Greenfield Board of Education. Frederick purchased the building and several adjoining lots in what the article referred to as "a deal of more than average importance to Greenfield." He planned to remodel the building to make it suitable for a bus building plant and to erect an additional structure on one of the open lots. According to the article, the move became necessary because of the increase in business for the company. In its present building, the company had been "employing 35 men with good salaries, but the proposed expansion will carry with it a larger payroll and increased production."[49]

At this time, the company had already received offers to move the shop to a larger city, but Frederick "has preferred to remain in Greenfield where he was born and grew up in the business, the buggy manufacturing business having been carried on here for many years under the name of C.R. Patterson & Sons, the Patterson buggies being known all over the country." The Patterson buses were by now known across several states and the demands for buses became greater all the time. The article stated, "Mr. Patterson has developed the most complete school bus body there is on the market and has been a leader in this line for many years."[50] Expansion at this point would only be an asset to the company. Business was good and the demand was high. Unfortunately, no information surfaced that detailed whether the "South Side School Building" was ever converted into a part of the business, although it is known that the property remained in the Patterson family into the mid-1930s.

The Greenfield Bus Body Company – 1930s

Throughout the 1920s, Frederick D. Patterson ran the business, but in 1932, upon his death, it passed to his sons, Frederick Postell Patterson and Postell Patterson, the third generation of Pattersons to run the company. Fred P. was born in 1903, Postell in 1906, and both finished high school and then attended college at the Ohio State University during the 1920s. Fred P. studied mechanical engineering and became the Greenfield Bus Body Company's designer during the late 1920s. According to

Kate's handwritten history, Frederick was aging, and the strain of the business was causing his health to fail by the late 1920s, so he called his sons in from their studies at Ohio State to help him with the business. It appears that neither of Frederick's sons received a degree from Ohio State.

A few of the company's bus designs are highlighted in a 1932 catalog that was provided to the author by Patterson family descendants. Some of the details of the buses of this period are highlighted on the following pages. These are pre-1935 designs as the bus styles were forced to change to all metal construction during that year, which will be discussed later.

The 1930s started out well for the Patterson Company. A 1930 article in the *Greenfield Republican* referred to the Greenfield Bus Body Company as "a distinct asset to the community." The article commended the company for providing jobs in the community for many people, who in turn spent their money in the local area, which further stimulated the local economy. The article described how the Patterson Company had been operating 24 hours a day for some length of time and "turning out school bus after school bus, and no sooner is each bus finished and checked than some driver takes it away to its destination."[51]

After the stock market crash in 1929 and the onset of the Great Depression in the early 1930s, many businesses slowed to the point of closing their doors. Through the first few years of the Great Depression, the Greenfield Bus Body Company's business did not drop off drastically. Schools were required by law to provide transportation for their students, so school bus sales did not come to a complete standstill. Bus sales slowly trickled in at this time, which allowed the Patterson Company to continue to provide some jobs in the Greenfield community.

The Highland County Bank closed its doors on July 2, 1930. As covered in the *Greenfield Republican* on July 3, 1930, the State Superintendent of Banks, O.D. Gray, came to Greenfield that same day to oversee the conditions of the closure and indicated that "the closing is the result of the constant withdrawals from the bank for the past several months, large sums having

THE question of School Pupil Transportation is a most important part of the development of the Centralized School. Pupils must be transported from their homes to School and this must be done safely, comfortably, healthfully and economically.

As Body Builders we have turned our years of experience and study toward the solution of this problem. Four generations in vehicle building represented in the conduct of the GREENFIELD BUS BODY COMPANY is assurance of merit and devotion rarely encountered in these days of many changes.

GREENFIELD SCHOOL BODIES ARE SAFE.
They have the proper number of entrances and exits.

GREENFIELD SCHOOL BODIES ARE COMFORTABLE.
They have ample seating room. Special type automobile cushions insure easy riding qualities.

GREENFIELD SCHOOL BODIES ARE HEALTHFUL.
Full size, clear vision windows and bus type roof ventilators afford ample ventilation.

GREENFIELD SCHOOL BODIES ARE ECONOMICAL.
They are priced right.

In these pages we shall not deal in glittering generalities or verbose essays upon the technicalities of construction. These things are of minor importance to the layman. We offer Greenfield School Bodies for your consideration because

They are Correct in Design

They are Honest in Construction

They represent unsurpassed Monetary Values to the careful purchaser

---

Greenfield Bus Body Company

Greenfield, Ohio

Greenfield Bodies are "Patterson Built"

Page from Patterson 1932 bus body catalog. (R.P.)

The Greenfield School Bus Body

For Every Chassis

THE SUPREME BODY FOR SCHOOL TRANSPORTATION

Not Lowest in Price But Highest in Quality

WITHOUT doubt Greenfield School Bodies are the most popular school bus bodies ever offered. They are absolutely the last word in perfect quality—superior design—unequalled standard equipment and superior finish. In design and construction as well as finish nothing has been overlooked, nothing left undone to make this body the best that can possibly be produced—the kind that will last and serve admirably well the entire life of the chassis.

The Greenfield School Bodies are outstanding in their superiority in service. They outlast the chassis. They are built for service. They embody the experience of more than sixty-five years of vehicle construction.

GENERAL SPECIFICATIONS

Construction—The Greenfield School Body is constructed from the combination of wood and metal which insures long life, low upkeep, freedom from squeaks and rattles, sturdy strength, neat appearance, light weight, all at a moderate cost. This combination of the rigidity of metal and the elasticity of wood produces the most satisfactory motor bodies.

The frame construction is of native hardwood, oak and ash, air seasoned, tough and durable. The frame parts are put together in lead with screws and bolts (not simply nailed). Brace and angle irons are bolted and screwed in place and there are no swaying and creaking joints. The cross bunks are oak placed in channel steel, the foundation is as rigid as the chassis frame.

The floor and roof are "B" and better yellow pine. The panels are of 20 and 22 gauge auto body steel treated to prevent rusting. The roof construction is almost indestructible.

There are no water lodgments the way this body is constructed, no checking panels, loose plugs, nor rotting recesses, no distressing rust-outs.

The aisle is wide, the body low, and the vision clear

THE safety and welfare of the children is given first attention. The right front door under driver's control, the left front driver's door of easy egress, the rear emergency door, the overhead roof ventilator that circulates fresh air without drafts, the accessibility of every inch of the body for cleaning, the unequaled vision, the comfort of the deeply upholstered seats and the full standard equipment are included in Greenfield School Bodies and at a price so reasonable as to defy comparison.

Sixty-seven Years Experience In Vehicle Building

Page from Patterson 1932 bus body catalog. (R.P.)

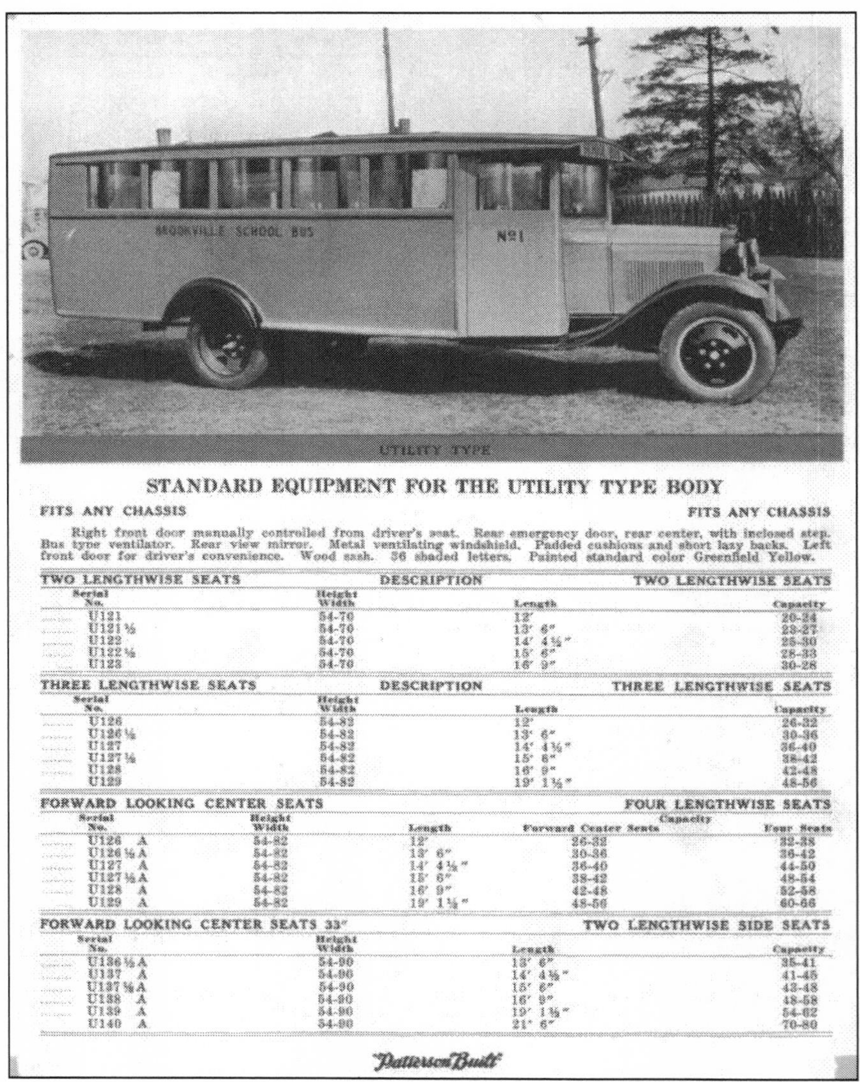

UTILITY TYPE

STANDARD EQUIPMENT FOR THE UTILITY TYPE BODY

FITS ANY CHASSIS — FITS ANY CHASSIS

Right front door manually controlled from driver's seat. Rear emergency door, rear center, with inclosed step. Bus type ventilator. Rear view mirror. Metal ventilating windshield. Padded cushions and short lazy backs. Left front door for driver's convenience. Wood sash. 36 shaded letters. Painted standard color Greenfield Yellow.

TWO LENGTHWISE SEATS — DESCRIPTION — TWO LENGTHWISE SEATS

Serial No.	Height Width	Length	Capacity
U121	54-70	12'	20-24
U121½	54-70	13' 6"	23-27
U122	54-70	14' 4½"	25-30
U122½	54-70	15' 6"	28-33
U123	54-70	16' 9"	30-28

THREE LENGTHWISE SEATS — DESCRIPTION — THREE LENGTHWISE SEATS

Serial No.	Height Width	Length	Capacity
U126	54-82	12'	26-32
U126½	54-82	13' 6"	30-36
U127	54-82	14' 4½"	36-40
U127½	54-82	15' 6"	38-42
U128	54-82	16' 9"	42-48
U129	54-82	19' 1½"	48-56

FORWARD LOOKING CENTER SEATS — FOUR LENGTHWISE SEATS

Serial No.	Height Width	Length	Capacity: Forward Center Seats	Capacity: Four Seats
U126 A	54-82	12'	26-32	32-38
U126½ A	54-82	13' 6"	30-36	36-42
U127 A	54-82	14' 4½"	36-40	44-50
U127½ A	54-82	15' 6"	38-42	48-54
U128 A	54-82	16' 9"	42-48	52-58
U129 A	54-82	19' 1½"	48-56	60-66

FORWARD LOOKING CENTER SEATS 33" — TWO LENGTHWISE SIDE SEATS

Serial No.	Height Width	Length	Capacity
U136½ A	54-90	13' 6"	35-41
U137 A	54-90	14' 4½"	41-45
U137½ A	54-90	15' 6"	43-48
U138 A	54-90	16' 9"	48-58
U139 A	54-90	19' 1½"	54-62
U140 A	54-90	21' 6"	70-80

"Patterson Built"

Page from Patterson 1932 bus body catalog. (R.P.)

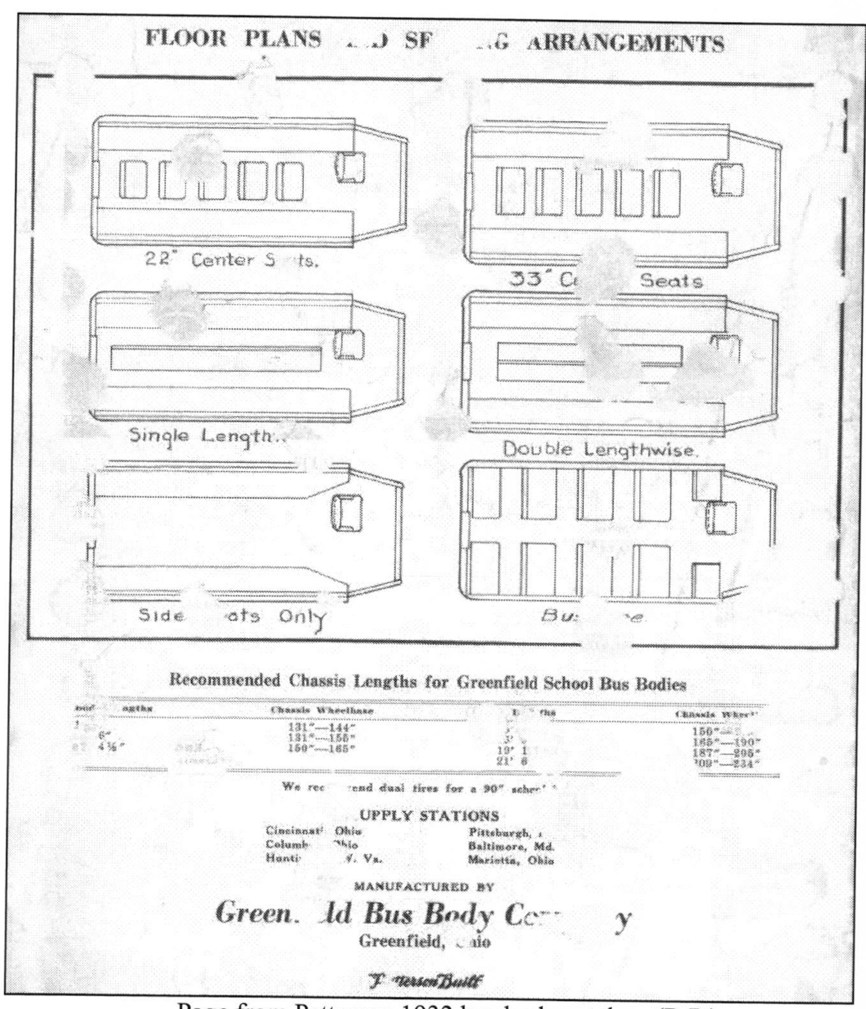

FLOOR PLANS [illegible] SE[illegible]G ARRANGEMENTS

Recommended Chassis Lengths for Greenfield School Bus Bodies

[illegible]ngths	Chassis Wheelbase	[illegible]	Chassis Whe[illegible]
[illegible]	131"—144"	[illegible]	150"—[illegible]
[illegible] 6"	131"—158"	[illegible]	165"—190"
[illegible] 4½"	159"—165"	19' 1[illegible]	187"—205"
		21' 8	[illegible]09"—234"

We rec[illegible]end dual tires for a 90" scho[illegible]

[illegible]UPPLY STATIONS

Cincinnat[illegible] Ohio
Columb[illegible] [illegible]hio
Hunti[illegible] [illegible]. Va.
Pittsburgh, [illegible]
Baltimore, Md.
Marietta, Ohio

MANUFACTURED BY

Green[illegible]ld Bus Body C[illegible]y
Greenfield, [illegible]hio

P[illegible]tersonBuilt

Page from Patterson 1932 bus body catalog. (R.P.)

been withdrawn at irregular intervals. The reserve fund required under the state laws has been depleted, which, together with frozen assets, is responsible for the action of the banking department." The article went on to state that "while the closing was quite a shock to the people of the community, there was very little excitement following the closing." Superintendent Gray also indicated that "there is no cause for alarm or panic. The Peoples National Bank is well able to take care of the public needs for the present until some plan of reorganization can be put into effect." Several months later after the closure, in December of 1930, a dividend of 10% was paid to each of the 3,700 depositors that had funds in the bank. Another 10% dividend was paid in April of 1931. In December of 1931, an additional 8% dividend was paid. The next dividend wasn't paid until September of 1932 when an additional 5% dividend was paid to depositors. Nearly 1.5 years later, in April 1934, an additional 5% dividend was paid. The final dividend came a few years later, in May of 1937, when an additional 7.93% dividend was paid. This came to a total of 45.93% of funds recovered for each depositor, so it wasn't a total loss for the depositors, but only recovering approximately half of the funds over a 7-year period did damage several businesses and private citizens alike.[52]

Fortunately for the Patterson Company, they did not bank with the Highland County Bank alone. Much of their banking was with the Home Building & Loan Company, which was under the presidency of C.R.'s former partner, J.P. Lowe. Lowe died in 1928, but there is little reason to believe that the Pattersons would stop banking with that organization. The Patterson Company also had funds in The Leesburg Bank. However, despite the losses that the company incurred through the closure of the Highland County Bank, only recovering 45.93% of their deposits, it is unlikely that it would have amounted to a substantial amount considering they were generally depleted of funds and had lost their creditworthiness by overextending mortgages by the time that the Great Depression started in October of 1929. By the time the depression set in, the company was generally using one order to pay for the next and there was little left over to continue growing the company.

During this time, to keep a decent revenue stream, the Pattersons offered some of their former services along with some new ones. A general advertisement for body and fender repair, upholstering, and Duco refinishing ran from October 3, 1929, to February 25, 1932, in the *Greenfield Republican*. The last advertisement placed by the Patterson Company, which ran from March 3, 1932, until November 24, 1932, in the *Greenfield Republican,* only advertised work for automobile and truck springs.

Things really began to deteriorate for the company in January of 1932, when the driving force behind the business, Frederick D. Patterson, suddenly died at the age of 61. He was inflicted with a sudden stroke of paralysis while at a meeting with a school board in Powell, Ohio, where he was attempting to close a deal on some school bus bodies. His chauffeur put Frederick in the car and drove him back to his home in Greenfield where he passed away four days later without ever regaining consciousness. The article further stated, "while Mr. Patterson had been in ill health for the past few months, he had recovered to the extent of being able to attend to business matters a portion of the time." The somewhat lengthy article had many nice things to say about Frederick, such as "Mr. Patterson was an outstanding member of his race and had an interesting career. He was the first colored graduate of the Greenfield High School, being a member of the class of 1888. He then entered Ohio State University where he was a prominent student of his class." The article's final words about Frederick were simply "Mr. Patterson was a well educated, refined gentleman, a capable business man, and added much to the industrial life of Greenfield, working energetically at all times for the good of the community and for the uplifting of his race."[53] Frederick had accomplished many things in his life.

After Frederick's death, sons Frederick P. and Postell took over the business with their mother, Estelline, in charge of the firm's clerical and financial duties. A combination of several factors diminished the business around this time. New management, the economic conditions during the Great Depression, rising competition, and new requirements and standards for bus bodies all played a part in the decline of the business. While Frederick D. Patterson had gained C.R.'s business savvy, it appears that Frederick's sons did not inherit enough of that trait from their father -- at least not enough to deal with all the major issues and changes that fell upon the company after his death.

Although struggling to keep the business afloat after Fred's death, the company still made sales. A July 20, 1933, article in the *Greenfield Republican* highlighted the fact that the Greenfield Bus Body Company had received orders for bus bodies from five different school districts during the preceding few weeks.[54] In the article, Postell told the newspaper that they expected a good amount of work for the coming fall.

Other factors from within the Patterson shop caused some complications with the business at this time. Just over a year following Frederick's death, prohibition ended in the United States in March of 1933. While Frederick Postell was a teetotaler and did not touch alcoholic beverages, his brother Postell was quite the opposite and enjoyed intaking these beverages. Frederick Postell served as the main designer and engineer for the Patterson shop. One reason for this is because of his stuttering issue that developed as a young child when forced to testify at his parent's legal separation hearing. Because of this speech impediment, he did not like to deal directly with potential customers. This was left to his brother, Postell, who would travel across the United States and "wine and dine" prospective clients. A struggle with alcohol became a life-long issue for Postell, as discussed in depth in his wife Kathleen's autobiography that she wrote late in her life.

Few details of the bus making process at the Patterson shop are known. In her 2001 autobiography, Kathleen Patterson described the process of making ball corners for the back of the bus roof on the later 1930s style buses. "This was done by having a round mold and taking a 16-inch square of steel sheeting and moving it back and forth over the mold and the steel hammer run by electricity would come down hard on the metal while you moved it back and forth over the mold." Once the switch was made to all-steel bodies, she stated that the ball corners could be ordered already made and could be welded on with the rest of the body. She also mentioned that their supply of steel came from the Ryerson Company in Cincinnati.[55]

The *Greenfield Republican* reported an order for two delivery trucks for a bakery in Manchester, Ohio, in 1934. This is quite possibly the same bakery in Manchester, "Baby Bear Bread," that Kathleen Patterson discussed in her autobiography. This bakery ordered 3 new truck bodies in 1937 at a time when money was desperately needed to keep the company afloat. The bakery provided the chassis for the trucks and delivered them to Greenfield when the bodies were ready to be mounted. Once completed, the trucks were delivered to Manchester by Kathleen, Postell, and employee John Humphreys and his wife.[56]

In 1935, the company made a very interesting sale, after representatives from the Republic of Haiti came to tour the Patterson shop and look at their buses.[57] The January 17 *Greenfield Republican* article stated that the island had no current bus or streetcar transportation and was looking to introduce a public bus system in the capitol city of Port-Au-Prince and eventually expand to other cities in the nation. The buses would be of special construction to insulate against the intense heat on the island. In April, the order was confirmed, and the contract stipulated that the Patterson Company would produce 3 buses as soon as possible with 3 additional buses to be built during the summer.[58] According to the article the buses were built on a V-8 Ford extended length chassis.

Although it may seem odd that the government of Haiti came all the way to a small company in Ohio to order buses, rather than one located more closely in the southeastern United States, there may have been an underlying reason. Haiti was a territory of France into the beginning of the 19th century and had imported between 500,000 and 700,000 African slaves to use as labor on plantations located there. When war between Britain and France broke out early in the century and Napoleon sold the Louisiana Territory to the United States to raise war funds, Haiti no longer held strategic importance to France, so it became an independent nation. In 1804, the Republic of Haiti became the first nation in the Americas ruled by those of African descent. In 1915, political turmoil and violent uprisings broke out in Haiti and U. S. troops entered and occupied Haiti to stop the bloodshed. From 1915 to 1934, American administrators controlled the Haitian government. During this time, a large amount of infrastructure had been created there, but no means of public transportation had been put in place. By 1935, when the representatives came to meet with the Pattersons, the government once again came under Black control.[59] It is possible that this nation of former slave descendants may have purposely sought out the Patterson Company, after learning of its existence as a Black-owned company that made the products that were needed.

By the mid-1930s the Patterson Company really felt the effects of financial burdens. By 1935, the company had relied heavily on remortgaging their properties to keep operations going.[60] A newspaper article in August of 1935 stated that the Patterson Company had been having a fine year – one of its best in several years – but, less than two months later, the Fidelity Building and Loan Company foreclosed on the property and a sheriff's sale was ordered. The article went on to say that, despite the current situation, Estelline Patterson had plans to "carry on and was trying to make arrangements to continue the business."[61]

Apparently, they successfully negotiated some type of arrangement. A December 1935 article in the *Greenfield Republican* announced that the business had been reopened on the first of December under the name of Patterson Body Company. The company opened under a new manager, Kathleen Patterson, the wife of Postell who had married into the family earlier that year. While living in New York City she had gained experience in management at the Alhambra Theater and at the

Brooklyn branch of the Amsterdam News.[62] It was hoped that the business could have a fresh start.

A problem of keeping the company moving forward was a lack of up-front finances. When a school board ordered buses, copies of the orders were taken to the bank and a loan would be obtained to cover the expenses of construction. The school board paid for the buses upon delivery and inspection, but by the time the bank loan had been repaid and the employees were paid, not much money remained for the family and company. This cycle continued for quite a while.[63]

Working on this type of system did not allow room for mistakes. In 1936, the company received an order for several school buses. These buses were to be built over the summer and were required to be delivered by the beginning of the school year. Labor problems arose and the employees went on strike, leaving only the family and one blacksmith, Arlis Bailey, to build the bus bodies. The buses were completed late because of this and the Pattersons had to reimburse the school for alternative transportation costs for the first part of the school year.[64] This left very little money for the company and family.

While the Great depression affected nearly everyone, it took a devastating toll on small businesses such as the Patterson Company. The company had enjoyed so much success in the early part of the 1920s, but by the end of that decade, business had slowed rapidly, and success was almost non-existent in the 1930s. Kathleen Patterson wrote in her autobiography that, at times, the company was doing barter work to survive. One instance that she described took place in the late 1930s when they traded some work for a 1928 Packard and a crate of chickens. It turns out that they were so broke at that time that they didn't even have enough money to feed the chickens. The chickens were dying faster than the family could eat them. She stated, "they were eating them as quickly as they could in the forms of fried chicken, roast chicken, stewed chicken, and noodles and chicken," but they still couldn't get rid of them fast enough.[65] Money was short, and the business was failing. Interviews with

Richard and Betty Patterson, children of Frederick Postell, resulted in a recollection of their mother, Bernice, speaking of having to pawn her wedding and engagement rings to bring in a small amount of money needed to survive. The rings were never recovered from the pawn broker.

While things were bad for the company during this time, the issues also probed some of the family members, as well. One example, which was not funny at the time, does add a bit of humor to their situation. Kathleen Wilson Patterson, Postell's new wife, had never learned to drive a car. While living in Greenfield as a child, she had moved to New York where she spent her teenage and early adult years. New York had such an extensive public transportation system that she never learned to drive a car. After moving back to Greenfield, Kathleen indicated that she wanted to learn to drive, and various people would let her drive vehicles to practice. In April of 1935, Kathleen was making a turn in town and lost control of the car. She skidded across the First Presbyterian church lawn and hit the iron banisters at the front of the church. Kathleen describes this event in her autobiography and talked about the embarrassment of newspaper headlines coming out saying "Patterson car runs wild over church lawn." When the story came out the headline just said, "Car runs into Church," but her description leaves more to the imagination.[66]

In 1937, Frederick Postell Patterson designed and built a large moving van for Allied Movers on East First Street in Dayton, Ohio. One of the truck bodies that he designed received recognition for its styling and streamlined design. An unusual vehicle that Frederick P. designed was an 8-foot-wide hearse that could hold a casket crosswise near the front of the bed and had room for 40 – 50 mourners.[67] The capacity to hold so many mourners helped to eliminate the long line of traffic that usually accompanied a funeral procession to the cemetery. Unfortunately, for the time, this vehicle was too wide to fit through many of the smaller lanes that led to rural cemeteries and, therefore, the hearse never saw production.

At the beginning of the camping craze, during the 1930s, Frederick P. designed one of the first two-wheeled travel trailers. According to Kathleen Patterson, the camper had all the luxuries of the time such as "a shower, toilet, cook stove, dinette, bedroom, and refrigerator." She also discussed that this took a lot of money to construct, perhaps too much, especially at a time when the company already had financial problems. The first ones were built with plywood sides that were covered with imitation leather, while the later ones had a thin metal skin, much more like they are today. While on sales trips, the Pattersons traveled with the camper and demonstrated it to promote sales. According to Naughton's article, "they said they went as far as Cape Cod and New York to sell their vehicles, but they said a lot of people couldn't believe that a Black manufacturing company would be building school buses, trucks, and house trailers."[68] Both the hearse and the two-wheeled travel trailers that were built are examples of the Pattersons trying to once again find a new niche in the vehicle market where they could succeed, since the bus and truck business had almost reached a standstill for them by the mid-1930s.

One of the company's biggest problems resulted from new standards for school buses in Ohio around the last half of 1935. As late as 1928, there had only been one standard for school bus bodies in Ohio, and that was that the driver had to have an unobstructed view of the road in front of him and to the left and right. An Ohio State Board of Education report went on to say, "the law does not prescribe further standards for the vehicle to meet." A study by Callon in 1930 stated that although every state required that transportation be provided for school children, "not one of the states provides standards or specifications for the construction of school bus bodies in which these children are transported." Around this time, some serious school bus accidents caught the attention of the public and drew some criticisms toward lawmakers for not setting proper safety standards.[69] After many studies and research projects, Ohio and several surrounding states set school bus standards around late 1935, which included a shift to an all-steel body.

As a result of these new body standards, the Pattersons had to retool their shop to do all metal work. This cost quite a bit of money and required the employees to learn new skills that many did not take to easily. Postell went to the Hobart Manufacturing Company in Troy, Ohio, and took classes on how to weld. He purchased a welder from them and brought it to the shop where he began to teach the employees how to weld. This became quite a shift in processes for those who had been in the woodworking business for so long. A *Greenfield Republican* article stated that the firm was "installing electric welding equipment and will engage in the manufacture and repair of auto bus bodies, with special attention being given to the manufacture of metal bodies."[70] This transition became quite costly and came at a time when finances were already short. The company just could not afford this type of expense.

Competition from larger bus body companies, such as Wayne Body Corporation and the Superior Body Company, intensified in the late 1930s. The large factories were much better equipped to meet the new standards, and some had already been building all-steel bodies, so the change did not affect them as much. Many small shops like the Patterson Company could not afford the expensive dies required for mass-producing steel body parts.[71] The Greenfield Bus Body Company had held its own before this, because it had the woodworking equipment and know-how required to build the wood-framed bus bodies. Building wooden frames and bodies for vehicles had been its expertise for 65 years, extending back into its early carriage building days. Until now, each build had primarily been a custom build, so production lines were not an issue as they had now become.

Other advances caused the Patterson Company problems as well. During the 1930s, advanced body design became a key to success in the industry. Streetcars disappeared at this time and buses became the main public transportation. This increased competition between bus manufacturers to produce better buses. Manufacturers had to constantly improve the quality and design of their products, which required a lot of capital.[72] Although the Patterson Company was a dominant force in the local bus industry during the 1920s, it failed in the 1930s in the face of competition and financial expense.

The Great Depression affected each individual bus manufacturer in a different way. It affected both large and small bus companies in one way or another. An example can be given of the two largest bus manufacturers in Ohio during the 1930s: Twin Coach and White. Between 1930 and 1938, Twin Coach continued to grow and built 5,384 buses. Business grew so rapidly that, in 1937, they expanded their factory size by 25% to meet the rising demand for their vehicles. Unlike Twin Coach, the White Motor Company had its share of difficulties during the Great Depression. In 1930, they suffered an $84,000 loss, and by the next year, the loss reached nearly $3 million. In 1932, they merged with Studebaker attempting to reorganize, but this was short-lived, and the merger broke apart in 1934. In 1937, White regained some ground when it switched to producing aluminum-bodied buses, which utilized die-stamped body panels, and eliminated over 1,500 individual body parts. They combined this with the heavy use of assembly line production in a newly constructed plant costing over $2 million.[73] This allowed the company to stay in competition and survive through the Great Depression era.

Of the small body manufacturers in Ohio during the era, perhaps the most successful was the Anderson Body Company. Anderson's success largely resulted from a 1931 merger with the Flxible Corporation, which combined the assets of the two large manufacturers and aided their survival. Anderson bodies were quite popular during the early 1930s and the growing demand for them required a 150,000 square feet expansion to the factory in 1932. Other body manufacturers were not so fortunate, and their business declined, just as the Patterson Company experienced. The Lang Body Company had been a leader in the local industry during the 1920s. Lang managed this through the adoption of assembly line production in the late 1920s, but despite its modern methods, Lang failed to survive the Depression. Another small company, the Fremont Metal Body Company, focused on manufacturing parlor style bodies for upscale city transit buses. As the Pattersons had done, the Fremont Company held on to using wood framed bodies in the late 1920s when the larger companies were making the transition to all-metal bodies even before it became a requirement. This, in combination with the low demand for expensive parlor style bodies during the Great Depression, led to their demise.[74] Many differing reasons led to the level of success each company achieved during the 1930s. While some companies were able to adapt to the changing industry and economic conditions, others, such as the Pattersons, could not, and failed.[75]

The 1930s were full of legal issues for the Pattersons as well. Many of these issues came because of the decline in business. During the Great Depression years, and the years leading up to it, many businesses did things that were a bit shady to survive; the Patterson Company was no different. The Pattersons began to double and triple up on mortgages in a desperate attempt to survive. The problem was that they could not pay them back. They mortgaged the same property to several different organizations and take the money without paying back the loan. Amazingly, likely due to their previous good credit standing, the lack of payments sometimes went on for years before any kind of court action was taken. During the 1930s, the Patterson properties began to disappear one at a time until there was eventually nothing left. The banks continued to work with the company and make deals with them to avoid foreclosure of the company. This happened in both 1935 and in 1937.[76]

One 1937 article discussed the foreclosure of the factory buildings. They were sold at Sheriff's sale to one of their mortgagers who made an agreement with the Pattersons for them to continue operations. Despite the potential negative attention caused by the article announcing the foreclosure, the article did make light of the situation by offering some praise for the Patterson Company stating that "Since founded many years ago by C.R. Patterson, the concern has conducted one of Greenfield's most prominent industries. With the passing of the horse-and-buggy era during which Patterson-built carriages were among the finest on the market, the firm changed to building of school bus and moving van bodies, and once even placed an automobile on the market. In addition to bus-body building, the company is now engaged in construction of trailers."[77]

Typically, the lower values were used until the second decade of the century, but after that, the amounts increased greatly. Between 1922 and 1926, the mortgage values jumped to the maximum values that the company could get, including the 1926 mortgage of both company buildings for $19,000. They had never taken this much money in one shot. Not only was this value extremely large, but these same properties were also mortgaged at the same time to two other institutions for $6,000 each. There was an earlier outstanding mortgage of $8,000 on these properties from another institution.[78] This placed the total mortgage value taken on these properties at $39,000 during a 5-year period. This money was never paid back, and the properties were eventually lost.

Since these mortgages were all taken out in the 1920s, the company had to find a new source of funding by the 1930s that was not local to the area where their credit reputation had been tarnished. They eventually found The Supreme Life Insurance Company out of Chicago to fill this void. This company was the largest Black-owned life insurance agency in the United States. This was a case of one Black business helping another, but by 1933 it had turned out to be a bad choice for the Chicago company and they had to take the Patterson Company to court. This, in addition to the company mortgages discussed above, indicates that the company's decline had already started by the mid-1920s, and they were making poor business decisions out of desperation.

Relocation to Gallipolis, Ohio – The Gallia Body Company

The Chamber of Commerce of the city of Gallipolis coaxed the Pattersons into relocating the bus building business there. They tried to bring new businesses to their city that would help them to grow economically by providing jobs in the community and attract outside commerce. According to Kathleen Patterson's autobiography, the members of the Chamber of Commerce had invited the family to tour the city and to look at the building offered to them to use for the business. The problem was that no one in the family, or any of the workers for that matter, wanted to make an uncertain move. As 1938 came into full swing, the family knew that all hope of the business surviving in Greenfield had been lost, so they reluctantly made the decision to move the company to Gallipolis. To make matters worse, the matriarch of the family, Josephine, died in this year.[79]

The announcement came in the *Gallia Times* that a "New Industry Comes to City at Once." The article gave a brief background of the company and stated that it would be operating in the city within 30 days. The Pattersons had apparently not made an official announcement yet in Greenfield, but a wire to the *Greenfield Republican* from the *Gallia Times* had confirmed the deal and the story ran in that week's Greenfield newspaper.[80] The article mentioned that no member of the Patterson family could be reached for comment.

The Pattersons moved into Gallipolis somewhat quietly with no fanfare or much attention to the new industry in town – just the simple brief article mentioned above. In contrast, the Hoy Company moved to Gallipolis just a few months after the Patterson Company and had large articles written about the company and their leadership. Additionally, in the same newspaper, 23 different local businesses posted congratulatory and welcome messages to the new company. It seems that besides the Chamber of Commerce of Gallipolis, there wasn't much welcome or support for the Patterson Company.

Photograph of Josephine Patterson in her later years. (R.P.)

When the time came to prepare the new building for installment of the machinery, the Pattersons hired two local men to aid in this task. To defray the costs of converting the building into a bus factory, the businesspeople of Gallipolis extended the Pattersons a loan. In exchange for this loan, the Pattersons renamed the business "The Gallia Body Company," but they still used the Patterson name in advertising, such as "Patterson Built."[81] The machinery, delivered by truck from Greenfield, was installed in the new building to prepare for business.

This move caused more legal problems for the Patterson Company. They had taken out a mortgage on some equipment at the shop in Greenfield from a farmer in Leesburg, Ohio. They had planned this one out wisely and had only mortgaged old equipment to the farmer, just in case they decided to make the move to Gallipolis. This protected the newer equipment that they needed to build the newer all-steel bodies for buses. When they made the move to Gallipolis, the farmer filed a claim against the Patterson Company that they had transported mortgaged equipment across county lines, which was not a misdemeanor, but rather a felony. This could have been big trouble for the Pattersons, but since they had planned for this possibility, they had left the mortgaged equipment behind in Greenfield. There were arrests made in the case, but when the list of mortgaged equipment was checked against the equipment left behind in Greenfield, the Pattersons were found not guilty.[82] This was a lucky break for the Patterson Company and the family.

The move to Gallipolis was not a successful venture. Many of the same problems followed them from Greenfield and several new ones arose. By moving to Gallipolis, the company couldn't leave the Great Depression behind or suddenly become competitive with the large bus companies who were better equipped to construct the all-steel bus bodies. Their supply routes were also compromised, as there were no major transportation networks set up between Gallipolis and their supplier's

Advertisement from the Patterson Company after relocating to their new location in Gallipolis and changing the company name. This is a good example of the later style all-metal construction buses. (Kathleen Patterson, 2003)

cities like they had enjoyed in Greenfield. New employees had to be trained, which proved to be a problem for the Pattersons. Some of the men learned slowly and held up operations. Postell began to travel once again to secure orders for bus bodies, but business came slowly. Although the company received a handful of orders, it did not amount to enough to feel that the move had given a fresh start to the business. As things got worse financially, some of the family wanted to quit, so Fred P., Bernice (his wife), and Estelline all left Gallipolis and moved to Bernice's mother's house in Lyndon, a small town just outside of Greenfield.[83] This left Postell and his wife Kathleen to run the business on their own.

Postell and Kathleen did what they could to keep the business alive. As things got worse, the two remaining Pattersons knew what they had to do; they had to give up. Kathleen stated in her autobiography "in a year of struggling, we found we just couldn't make it." They decided to close the business and move away. Kathleen talked about this sad event in her autobiography and stated, “It was very hard to give up the dream.”[84] After 74 years, the Patterson family’s business closed its doors for the final time in 1939.

CHAPTER 4 NOTES

[1] Craig Robert Semsel, "Built to Move Millions: Transit Vehicle Manufacture in Ohio, 1880-1938" (PhD Diss., Case Western Reserve University, 2001), 312-314.
[2] Ibid., 314.
[3] Ibid., 650.
[4] Ibid., 304, 567.
[5] *GR,* "Make Specialty of School Bus Bodies," June 23, 1921.
[6] Albert McGee Callon, "A Score Card and Typical Standards for School Bus Bodies" (MA Thesis, University of Cincinnati, 1930), 1; Dale J. Baughman, "A Survey of Pupil Transportation in the Consolidated School Districts of Hardin County" (MA Thesis, Ohio State University, 1933), 1; J. L. Clifton, *Handbook for Rural and Village Boards of Education and Clerks in Ohio* (Columbus, Ohio: F. J. Heer Printing Company, 1928), 59.
[7] Baughman, "Pupil Transportation," 1; Callon, "Scorecard," 1.
[8] Joseph Calvin Copeland, "Costs and the Measurements Applied in School Transportation" (MA Thesis, Ohio State University, 1931), 4.
[9] Clayton M. Barden, "School Bus Standards" (MA Thesis, Ohio State University, 1932), 4; Semsel, "Built to Move," 304.
[10] William T. Ashby et al., *The Motor Truck Red Book* (New York: Traffic Publishing Company, 1942), 11.
[11] Ibid., 11-12.
[12] James A. Wren and Genevieve J. Wren, *Motor Trucks of America: Milestones, Pioneers, RollCall, Highlights* (Ann Arbor, Michigan: The University of Michigan Press, 1979), 370.
[13] S. V. Norton, *The Motor Truck as an Aid to Business Profits* (Chicago: A. W. Shaw Company, 1918), 14.
[14] Norton, *Truck as an Aid,* 4; Ashby et al., *Red Book,* 5-6.
[15] *GR,* "Local Company Bought South Side Building: Greenfield Bus Body Company is Planning Expansion in 1928," January 12, 1928; Kathleen L. Patterson, *It's Been a Wonderful Life: One Day at a Time* (Dayton, Ohio: Self-published, 2001), 49.

[16] Letter from Frederick D. Patterson to W.E.B. DuBois, on C.R. Patterson and Sons Company letterhead, dated November 12, 1920; Harvey C. Smith, *Annual Report of the Secretary of State to the Governor and General Assembly of the State of Ohio for the Year Ending June 30, 1920* (Springfield, Ohio: Kelly-Springfield Printing Company, 1920), 22; Nemar Publishing Company, *Department Reports of the State of Ohio: Embracing Twenty-Six Issues Dating from October 16, 1919 to April 8, 1920* (Columbus, Ohio: Nemar Publishing Company, 1920), 88.
[17] *GR,* "Make Specialty of School Bus Bodies," June 23, 1921.
[18] "Big Demand for Bus Bodies," *Commercial Car Journal,* Vol. 18 (1922): 32.
[19] Ibid; "Do You Remember? A Nostalgic Look at School Buses of the Past," *School Bus Fleet,* October/November (1979): 9.
[20] Bill Naughton, "Former Slave, Descendants Were Early Vehicle Pioneers in Ohio," *Tri-State Trader,* July 5, 1980: 12; "Do You Remember? A Nostalgic Look at School Buses of the Past," *School Bus Fleet,* October/November (1979): 9.
[21] Callon, "Scorecard," 17-47, 90-91.
[22] Semsel, "Built to Move," 374.
[23] Baughman, "Pupil Transportation," 31.
[24] *GR* June 28, 1921; *GR* April 21, 1922; *GR* October 19, 1922; *GR* January 6, 1927; *GR* July 28, 1927; *GR* July 18, 1929; *GR* February 26, 1931; *Greenfield Times* January 22, 1982. According to many later articles about the Pattersons, it seems that Ford chassis were a favorite for use in the Patterson buses. This could help to explain a contract that the Patterson Company held with the Ford Motor Company. Several sources mention that the Patterson Company held a contract of some kind with Ford, but does not specify exactly what the contract consisted of. In an article by Harmon, the writer states that the company "has the patronage of the International Truck Company for building some of their special bodies and does some of the business of the Ford Manufacturing Company." According to an article posted on Ford's website, it becomes clear that this contract did exist. The

article was covering an awards ceremony in 2003 where "Ray Jensen, Director of Supplier Diversity Development of Ford Motor Company, will be accepting the Posthumous Enshrinement Award on behalf of the Patterson family, Ford Motor Company manufacturers from 1919 to 1938." The article does not give any other details about the contract. Other sources refer only to the relationship between Patterson and Ford as a contract. Considering the vast network of resources and major suppliers that Ford had at that time, it seems more likely that the contract was not a manufacturing contract, but rather a contract for chassis or some other necessary piece of equipment that the Pattersons needed for their business. What would Ford have needed from a low production, custom order type company? The time period of the contract, 1919 to 1938, coincided with the dates when the Pattersons built buses, and they had already given up on the automobile manufacturing venture. An attempt was made by the Ford archivists to locate the details of the contract, but it could not be located in the millions of documents that are housed at the Benson Ford Research Center. Naughton, "Former Slave," 12; Greenfield Historical Society, *Greenfield, Ohio: 1799-1999* (Paducah, Kentucky: Turner Publishing Company, 2000),36; Roderick Jenifer, "The Patterson Wheels," *The King Maker* Issue 9 (2008): 31; J. H. Harmon, "The Negro as a Local Business Man," *The Journal of Negro History* Vol. 14, No. 2 (1929): 136; Juliet E. K. Walker, *Encyclopedia of African American Business History* (Westport, Connecticut: Greenwood Press, 1999), 50; "Ford Motor Company and George Fraser Team Up Again to Bring Powernetworking 2nd Annual Conference," Ford Motor Company's media website, http://media.ford.com (2003).

[25] "Knock-Down Construction," *Motor Coach Transportation*, Vol. 1 (1923): 18.

[26] Charlotte Pack, *Time Travels: 200 Years of Highland County History Told Through Diaries, Letters, Stories, and Photos* (Fayetteville, Ohio: Chatfield Publishing Company, 2007), 232. A 1918 article in the *Greenfield Republican* described how the

Greenfield Board of Education had purchased a Ford truck from the local dealer, C. K. Middleton, and had it converted "into a carry-all for the purpose of bringing in the school children from the country districts, and the same is being tried at the present time." The article mentioned that if this test worked out, several more of these conveyances would be ordered. Although the article does not explicitly state that the Patterson Company performed the conversion, it is very likely that Patterson had done it, given the history of building the horse-drawn school wagons for the district and its close ties with the school.

[27] *GR,* "New School Trucks," July 28, 1921; *GR,* "New School Garage," January 5, 1922.

[28] *GR,* "Attracting Attention: Product of the Greenfield Bus Body Company is Center of Attraction at Chicago Meeting," February 8, 1923.

[29] Ibid; *GR,* "More Men Are Wanted," June 7, 1923.

[30] *Automotive Industries,* "Truck Plans of Car Makers Halted," Vol. 39., No. 13 (1918): 596.

[31] Wren, *Motor Trucks of America,* 74; *GR,* Patterson Ad, April 15 and April 22, 1915; *HD,* Patterson Ad, September 12, 1916.

[32] Harmon, "Negro Business," 136; Naughton, "Former Slave," 11; Patterson, *Wonderful Life,* 50-63.

[33] Harmon, "Negro Business," 136; *GR,* "Patterson Company Making Plans to 'Carry On,'" October 17, 1935.

[34] M. D. Hartsook, "A Survey of Bus Transportation to and from Ohio Schools, 1929-1930" (MA Thesis, Ohio State University, 1930), 8.

[35] CoachBuilt.com, "Wayne Works, Divco-Wayne Corporation, Wayne Corporation, Wayne Wheeled Vehicles," http://www.coachbuilt.com/bui/w/wayne/wayne.htm. (2004)

[36] Semsel, *Built to Move Millions,* 658, 552-571.

[37] Roe Lyell Johns, *State and Local Administration of School Transportation* (New York: Bureau of Publications, Columbia University, 1928), 85; Barden, "School Bus Standards," 8; Earl B. Kellmer, "The Transportation Factor in the Proposed Central-

---

ization of the High Schools of Lorain County" (MA Thesis, Ohio State University, 1929), 48.
[38] *GR,* "A Busy Place: C.R. Patterson and Sons Finishing Large Order of Top Bodies for School Trucks," April 21, 1921.
[39] Naughton, "Former Slave," 12; Leland J. Pennington, "The Patterson-Greenfield Automobile," *The Highland County Magazine,* Vol. 1, No. 3 (1984): 33; Walker, *African American Business,* 50; Pack, *Time Travels,* 232.
[40] Steve Konicki, "150 Built: Better than Model T," *Dayton Daily News,* March 21, 1976. Another distant family member, Henry May, went even further to say that the Pattersons "sold the first buses on the streets of Detroit." This would be quite ironic considering that the power of the Detroit automobile industry had crushed the dreams of many small automakers just a few years before including Frederick D. Patterson and his Patterson-Greenfield automobile. Ron Rollins, "Four-Wheel First," *Ohio Magazine,* Vol. 31, No. 5 (2008): 127.
[41] *GR,* "Deliver Buses," December 22, 1921; *GR,* "New Bus Completed," October 19, 1922.
[42] *Sabina News Record,* "New Bus Running Between Wilmington, Sabina, and Washington C. H.," December 29, 1921; Annotated Photo Scrapbook of Sabina located in the Sabina Public Library Genealogy room.
[43] *Greenfield Independent,* "Greenfield Products," June 30, 1921.
[44] "New Item," *The Automotive Manufacturer,* Vol. 68 (1926): 8.
[45] *GR,* "Turning Out Some Extra Fine Work," January 6, 1927.
[46] *GR,* "Furniture Company Bought New Truck," July 28, 1927; *GR,* "A New Truck," May 14, 1931; *GR,* "Furniture Truck is Mighty Pretty," January 11, 1934.
[47] Highland County Common Pleas Court, Case No. 11543, dated October 17, 1929.
[48] *GR,* "Bus Body Plant Damaged By Fire," February 25, 1926.
[49] *GR,* "Local Company Bought South Side Building: Greenfield Bus Body Company is Planning Expansion in 1928," January 12, 1928.

---

[50] Ibid.
[51] *GR,* "Bus Body Company Asset to Village," September 11, 1930.
[52] *GR*, "Highland County Bank Closed its Doors Wednesday Morning; Reserve Depleted, Assets Frozen," July 3, 1930; *GR*, "Depositors of H.C. Bank Hold Meeting," February 5, 1931; *GR*, "Bank Depositors to Hold Conference Friday Nite," February 12, 1931; *GR*, "Bank Will Pay Another 10 Percent Dividend," April 2, 1931; *GR*, "Stockholders of Highland Co. Bank Face Suits," April 30, 1931; *GR*, "Highland County Bank Will Pay Dividend December 22," December 10, 1931; *GR*, "Highland County Bank Will Pay 5 Percent Dividend Oct. 6," September 29, 1932; *GR*, "To Cut Melon at Highland County Bank," April 26, 1934; *Greenfield Daily Times*, "Final Bank Dividend Due Late in May," May 6, 1937.
[53] *GR,* "Fred D. Patterson, Prominent Colored Citizen, Died Monday," January 21, 1932.
[54] *GR,* "Get Five Orders for Truck Bodies," July 20, 1933. To make some inferences into how much a Patterson body may have cost during the early portion of the 1930s, some known figures can be used to approximate the price. Baughman gives the average price of a bus as $1,408 around the 1933 time frame. This was for a bus with a Ford or Chevrolet chassis. According to a report by the National Safety Council an average sized bus designed to carry between 25 and 35 passengers should utilize a 1 ½-ton chassis. The local newspaper, the *Greenfield Republican*, ran advertisements for the local Chevrolet dealer, the Motor Inn Garage in Greenfield. An advertisement in September of 1931 advertised a 1 ½-ton bare chassis for $520. Subtracting $520 for the chassis from the $1,408 average price for a total bus would leave approximately $888 for the cost of the body. Assuming that the Patterson buses were around average price, it can be inferred that their bodies brought in between $800 and $1,000 each. Of course, this price was for the older style bodies.

It is quite likely that the cost rose significantly with the advent of all-steel bodies.

[55] Patterson, *Wonderful Life,* 56, 65, 67.

[56] *GR,* "Pattersons Get Job," May 24, 1934; Patterson, *Wonderful Life,* 61.

[57] *GR,* "Local Buses May be Used in Port-Au-Prince: Greenfield Bus Body May Bood Order From Haiti," January 17, 1935.

[58] *GR,* "Greenfield Company Gets Order for Haiti Buses: Greenfield Bus Body Company Given Nice Contract," April 25, 1935. It is possible that Haiti ordered 3 more buses in 1937. In Kathleen Patterson's autobiography, she described how she helped to build and deliver 3 buses to New York after the company received an order from the Haitian government. Kathleen didn't return to Greenfield and marry Postell until 1935 and didn't begin working in the shop until 1936, so these dates suggest that it is possible that 3 more buses were built a couple of years later for the Haitian government.

[59] Thayer Watkins, "Political and Economic History of Haiti," www.sjsu.edu/faculty/watkins/haiti.htm.

[60] The Pattersons mortgaged most, if not all, of the properties that they owned at one point or another. There were certain properties that the Pattersons tended to mortgage more often than the others; typically these were the company properties. The west building was mortgaged 23 times between August 1895 and April 1926 with values ranging from $200 to $8,000. The east building was mortgaged 10 times between December 1900 and April 1926 and the values ranged from $1,000 to $6,000, but when both company buildings were mortgaged together, the values ranged from $6,000 to $19,000. They also had a storefront on East Washington Street downtown that they frequently mortgaged. This lot was mortgaged 10 times between July 1876 and June 1920 and had values ranging from $369.77 to $1,239.51.

[61] *GR,* "Patterson Company Has Fine Year," August 29, 1935; *GR,* "Patterson Company Making Plans to 'Carry On,'" October 17, 1935.

---

[62] *GR,* "New Manager at Bus Body Company," December 5, 1935.
[63] Patterson, *Wonderful Life,* 57.
[64] Ibid., 57.
[65] Kathleen Patterson, *It's Been a Wonderful Life: One Day at a Time* (Dayton, Ohio: Self-published, 2001), 61-62.
[66] *GR*, "Car Runs Into Church," April 18, 1935.
[67] Patterson, *Wonderful Life,* 58; Naughton, "Former Slave," 11; *Greenfield Times,* "Patterson Family Earned Niche in History of Automaking," July 14, 1980.
[68] Patterson, *Wonderful Life,* 58; Naughton, "Former Slave," 12.
[69] Clifton, *Handbook Boards of Education,* 61; Callon, "Score-card," 1; Walton B. Bliss, "The School Bus Situation in Ohio," *Ohio Schools,* Vol. 8, No. 3 (1930): 100.
[70] Patterson, *Wonderful Life,* 56; *GR,* "New Manager at Bus Body Company," December 5, 1935.
[71] Fred L. Prentiss, "Making Motor Bus Bodies of Steel," *The Iron Age*, December 26, 1929: 1732.
[72] Semsel, *Built to Move Millions,* 570, 595.
[73] Ibid., 552, 556, 557-563.
[74] Ibid., 569-571.
[75] In 1937, major financial trouble hit the Patterson Company once again. Despite whatever deal the Pattersons had made with the Fidelity Building and Loan Company in late 1935, a suit was brought and a sheriff's sale of the company was ordered once again in June of 1937. The article also stated that a new deal had already been agreed upon and the Patterson Company would continue operations of the bus building business. Even with this new arrangement with the bank, financial conditions continued to get worse. The Pattersons could not get ahead, so in March of 1938, they reluctantly accepted an offer from the Chamber of Commerce of the town of Gallipolis, Ohio, to relocate the business there.

---

[76] *GR,* "Patterson Company Making Plans to 'carry on'," October 17, 1935; *Greenfield Daily Times,* "Patterson Company to Continue Operations," June 4, 1937.
[77] *Greenfield Daily Times*, "Patterson Co. to Continue Operations," June 4, 1937.
[78] Highland County Court Case No. 12253. August 15, 1932.
[79] Patterson, *Wonderful Life,* 63. A March 3, 1938, *Gallia Times* article mentioned that the Pattersons had purchased the building to be used for the company, but according to Kathleen Patterson, the Chamber of Commerce had provided the building rent free to the company in exchange for them relocating there. The Second Avenue building, known at that time as the Dale Building, had been built for manufacturing purposes but had never been occupied. Patterson mentioned that the building contained a lot of rusted junk that had been stored there and had to be cleaned out to prepare it for the machinery to be installed.
[80] *The Gallia Times,* "New Industry Comes to City at Once," March 3, 1938; *GR,* "Pattersons to Move Body Plant to Gallipolis," March 7, 1938.
[81] Ibid., 67. Only one worker, Walter Byrd, came with the Pattersons from Greenfield. Byrd was a young crippled man, with one leg shorter than the other, and had worked for the company and shown his loyalty for several years. The company had to start over again with all new employees. John Simms helped dismantle the old shop in Greenfield and remove the equipment, but remained behind in Greenfield. He had been with the company for several years and had at times held major stock in the corporation.
[82] Highland County Court Case No. 3271. April 26, 1938.
[83] Ibid., 70-71. Business got so bad that all the employees quit except one local man named Sam and, of course, Walter Byrd, but eventually even Walter's family came from Greenfield to get him.
[84] Ibid., 63, 70-71.

5

SEVEN DECADES OF CHANGE, SUCCESS, AND FAILURE

The C.R. Patterson and Sons Company passed through three generations of the Patterson family over a period of 74 years from 1865 to 1939. Any business that survived this long should be considered a success. Although the Pattersons were in business for this length of time, they are historically remembered for becoming the first and only Blacks to manufacture automobiles; they did much more than that. Building the Patterson-Greenfield automobile occupied only three of the company's 74 years. This small portion was surrounded by many years of supplying other products and services. Today, in Greenfield, nothing remains of the structures that housed this company, as they have given way to recent construction. Some vehicles remain, but these consist of only a few carriages, since none of the motorized vehicles have survived. Little remains of this Black-owned company that survived in the vehicle industry by adapting to the changing demand for vehicles through time, making this a worthy subject for historical study.

The C.R. Patterson and Sons Company did meet success, but at times, there was also failure. The focus of the company changed several times to meet the demands of the public. Some of these changes were profitable and well timed, but others were ill timed and were not so successful. From the data available, it is easy to see that the carriage industry was a profitable venture for the Patterson Company. The company did not fail in this field; the carriage industry as a whole declined and gave way to the age of the automobile. With this new age, Frederick Patterson decided it would be wise to adapt to the times and he started producing automobiles. The Patterson-Greenfield came onto the scene too late. It was not an inferior car, but a newcomer just couldn't compete in the field as late as 1915 when the large corporations were already dominating the industry. Few Patterson-Greenfield automobiles were produced, and the company relied more on its automobile repair services to keep the company going financially. While it seems that the automobile phase of the Patterson Company should be considered a failure, it was actually an important step in the right direction for the next phase of the company, the manufacture of buses and trucks.

The manufacture of buses and trucks, although one long phase, has been evaluated as two separate phases. These two phases were separated into decades: the 1920s and the 1930s. Although late in the 1920s, the company had already been showing some signs of struggling financially, the decade should be considered a success. The Patterson Company acted as pioneers in this industry at the beginning of the decade and its products became quite popular throughout the midwestern States. Fortunately, figures exist from a 1930 inventory of school buses in Ohio that tell us just how popular the Patterson built buses were in the state. The Patterson Company had built an impressive 35.8% of school buses in Ohio. Considering the lack of standards at the time, which meant that buses could be used for a period of 10 years or more, many of this number had likely been built during the first half of the 1920s and were still in use. By the last half of the decade, several larger manufacturers were on the rise, but the Patterson Company had produced so many buses in the years before this rise that their buses had already saturated the schools of Ohio. The 1920s was undoubtedly a success for the Patterson Company.

The 1930s provided a different story. Sales slowed and they struggled financially as most small businesses of any kind did during the Great Depression. In the late 1920s and early 1930s, the Pattersons had misused their credit to obtain massive amounts of mortgages that they could not pay back, thus effectively ruining their credit worthiness. This affected them in a major way in 1935 when the new school bus standards were set that required buses to use an all-steel body. The Patterson Company did not have the money or access to money that would allow it to retool the shop properly to adapt to this change. The Pattersons produced a minimal number of buses since they could only purchase those new tools that they could afford. The Pattersons could not compete with the large manufacturers that were utilizing mass-production techniques at this time. Although not all the problems during the 1930s were due to financial conditions, the company could just no longer compete. The company had been a low production, yet high quality, manufacturer for its entire existence. Staying this way was not an option by the 1930s, so the failure to adapt to the current conditions in the industry was a major cause of the failure of the company to remain competitive.

Leadership had played an important role in the success of the company. C.R. was the founder of this company and led it to become a successful carriage building shop. His ability to build quality carriages attracted customers to purchase one from him and aided in the growth of the business and it became a popular builder by the turn of the century. The business grew to proportions that required additional properties for the company to use. When his son, Frederick, took over the leadership in 1910, the company was well established and sold vehicles across several states. By this time, the automobile had come onto the scene, and Frederick saw opportunity. He wisely added automobile repair to the company's services offered. Frederick's willingness to adapt to the times and begin a shift from horse-drawn vehicles to automobiles allowed the company to survive the decline in the carriage industry during the second decade of the century. Frederick's leadership took the company through several transitions from carriages to autos, then on to buses and trucks. He brought the company business in other ways, as well, by adding services such as the auto repair shop and building furniture. His leadership lasted until his death in 1932 when his sons, Frederick Postell and Postell, took charge of the company. Under their leadership, plagued by a host of complications, the company failed.

C.R. Patterson overcame adversity to start this company as the Civil War ended. He succeeded in business as a Black man much earlier than most Blacks were able to do. A 1910 book, *An Era of Progress and Promise,* written about the rise of Blacks from 1863 to 1910 talked about Abraham Lincoln for the first five pages. Lincoln's importance to the freedom and success of Blacks earned him a place in this book. In it was included the words of the Emancipation Proclamation. During this speech Lincoln stated, "I recommend to them that in all cases when allowed they labor faithfully for reasonable wages." A few pages later, a 1910 speech was quoted that had been given by Booker T. Washington about Lincoln and his role in the freedom of the Blacks. In this speech he stated that Lincoln "lives in the steady and unalterable determination of 10,000,000 Black citizens who continue to climb, year by year, the ladder of the highest usefulness and to perfect themselves in strong, robust character. For making all this possible, Lincoln lives." The very last entry in the book talks about C.R. and Frederick Patterson as if they were the best examples of Black success during this period. The entry about C.R. almost shadowed Washington's words and stated that he "was a man of usefulness and influence in the community, and, by reason of his mechanical skill, enjoyed opportunities not usually afforded thirty years or more ago to one of his race."[1] Through his hard work, skill, and faithful labor, C.R. was able to work side by side and in competition with White men.

C.R. Patterson started a business at a time when the Black man faced many obstacles. The Black businessman had to deal with economic isolation. That is, "while society is composed of Blacks and Whites…the financial and political worlds have been almost entirely White." This affected the Black businessman's ability to gain proper credit and other economic considerations. "Faced with the same obstacles as the Black man, many other businessmen would have closed their doors. But the Black man, fighting for the essential means of survival, learned techniques in his daily life that could be carried over into his business activities." When looking at the problems of going into business, "if the Black man had been entirely realistic, he would have been pessimistic about his chances of success in a White society; for he certainly encountered barriers of such resistance that much of the hope, and virtually all of the enthusiasm, understandably subsided."[2] C.R. Patterson pushed through these obstacles and took the chance to compete in the White man's world.

In starting any business, there were risks to be considered. For the Black businessman, these risks were inherently greater, because there were more obstacles for him to deal with. For this reason, the typical Black businessman was more cognizant of issues that did not seem important to the White businessman, such as credit, capital, and risks, because these many times impacted the Black businessman more strongly just because of his color. This caused the Black businessman to minimize his risks and, therefore, kept him out of major categories of the economic field, such as manufacturing. According to Carter G. Woodson, the "father of Black history," in 1890 only .02% of the total Black population aged 10 and over was actively engaged in manufacturing.[3] This amounts to 1,077 of the 5,328,972 total Black population at that time. Although 208,374 (3.9%) of this total population worked as laborers in manufacturing, those that owned manufacturing enterprises were few. A later historian, Vishnu V. Oak, evaluated the history of Black business and pointed out that "a prime category of business activity is missing from the account – manufacturing." He went on to say, "There were indeed some Black manufacturers, and the number would increase and decrease over the years, depending upon the state of the national economy. But it is a principal truth of the Black business situation that few Blacks have engaged successfully in manufacturing operations."[4] At the turn of the century, Black manufacturing began to slowly increase, but with the higher use of mass-production, the Black manufacturer was left behind by the wealthier White competition. "The competition was so great that hardly any Black manufacturing survived. So long as the old methods of production and distribution endured, the Black manufacturers had a chance of success. When nearly all business was small business, the Black manufacturer could keep pace…As the business world came to be dominated by fewer but larger manufacturers, the Black man, with his lack of capital, equipment, and skills, fell farther and farther behind."[5] According to Robert W. Fairlie and Alicia M. Robb, as it stood historically, the same trend still stands today in which "Black firms are underrepresented in construction , manufacturing, wholesale trade…and real estate relative to White firms."[6] Black and White firms tend to concentrate in different industries.

While some of these tendencies rang true for the Patterson Company, they defied others. It is true that some of the downfalls of the company were largely due to credit limitations and being pushed out by the larger factories capable of mass-production, but these had nothing to do with the Pattersons being Black. They had defied the norm for the Black businessman and gained credit worthiness early on, which lasted several decades into the business, but that was something that they lost on their own accord in the late 1920s. They also defied the idea that the Black businessman did not take risks. The Pattersons were not afraid to take chances and developed new products and attempted to be leaders in their field. A big risk was taken when Frederick attempted to produce automobiles. While this did not pay off directly, it did lead him to take a chance once again when he shifted to building buses and trucks, and for a while, this risk paid off.

All of this occurred during a time that Blacks faced many forms of opposition. Discrimination existed through several means. Segregation was common in most areas, and although Greenfield had been an active abolitionist town and Underground Railroad stop during slavery, the town was not fully free from later segregation. According to information collected from the 1900 Greenfield City Directory, Black residences tended to cluster along the eastern ends of Lafayette, North, and McClain Streets and the local A.M.E. church was built in that area as well. While this doesn't suggest segregation mandated housing in the town, it does suggest that Blacks did tend to cluster together in the community. On a larger scale, the Ku Klux Klan initially rose in the south just after the Civil War. Although their influence was mainly restricted to the south during that time, there was a larger rise of the KKK during the 1920s. This rise was on a national level and was the most powerful of the three periods of major activity (the third was during the Civil Rights Movement of the 1960s). While there was no apparent Klan activity in Greenfield, in the nearby city of Columbus, the Klan became active in the 1920s because of the massive migrations of Blacks to the northern cities during World War I. The first two episodes occurred during the Patterson Company's existence. While the KKK did not appear to have a direct effect on the Pattersons, the Klan's influence on many Whites may have caused potential customers or investors to discriminate against them as Black business owners. Many Whites that did not support the Klan were still prejudiced toward the Black community, and discrimination was common during the period.[7] This created more obstacles for the Pattersons and other Black business owners, but despite the discriminating social and business environment, the Patterson Company managed to accomplish numerous things during their 74 years in business.

The business passed through three generations of the Patterson family. There were many transitions during this company's history with some being more successful than others. In a 1965 interview, C.R.'s youngest daughter, Kate, spoke about her brother Frederick and the Patterson-Greenfield car. She stated "the Patterson-Greenfield automobile never reached the mass-production level due to credit limitations, but we are proud indeed. It was a credit not only to our race, but [the C.R. Patterson and Sons Company] was considered one of the prominent businesses of southern Ohio, meeting the needs of many people."[8] Frederick's son, Postell, spoke of his father in a 1976 interview and stated "there was little that could get in the man's way."[9] They should indeed be proud of their achievements, not only for Blacks, but also for small businesses in general.

Today, there are automobile industry awards that are named for the Patterson family. Two are given by Detroit-based *On Wheels* magazine. One is the "Patterson-Greenfield Company of the Year" award that is given to recognize an automotive company's commitment to diversifying its workforce, and relationships with African American and Latino supplies and dealers. The second is the "C.R. Patterson Golden Wheel Award" that is given yearly to an African American automobile executive for management leadership. The Patterson Company and family have also received some awards themselves. One was in 2003 when they received the "Posthumous Enshrinement Award" from the Ford Motor Company for their contract with Ford from 1919 to 1938. Another award was the "Distinguished Honoree Award" that was given to C.R. Patterson posthumously at a 1985 banquet of the Dayton, Ohio, chapter of the NAACP. The theme of the banquet was "A Salute to Transportation." Receiving awards and even having some named after you is a sign of recognized success and achievement.

Within the town of Greenfield, the C.R. Patterson and Sons Company is still a part of the community. The story of this company and family is told over and over among the residents of the town and to out-of-towners that come in for a visit. The annual Wheels of Progress Festival, now known as the Greene Countrie Towne Festival, in Greenfield gives many visitors an opportunity to learn about the Patterson Company and their achievements as they browse through displays and collections at the Greenfield Historical Society. It quickly becomes very apparent that this family and company meant a lot to the community, and it does as much now as it did back then. In 1910, when C.R. Patterson died, the town of Greenfield was still segregated, but when visiting the cemetery in Greenfield and seeing where C.R., Frederick, and many others of their family are buried, one realizes that this family was not seen for their color, but rather for their reputation in the community. C.R. and his family are not buried down in the "colored" cemetery, but rather in a place of honor on top of the hill. As one enters through the old wrought iron gates of the cemetery, and travels a short distance, there is a giant oval right in the middle of the cemetery. Almost in the exact center of this oval is the Patterson main monument. Aligned parallel to each side of this monument are laid the graves of C.R., Frederick, and eight others of the family. This is not the burial place of a "colored" family of the early 1900s, but rather the resting place of a well-respected family of the community.

In 2014, an Ohio Historical Marker was placed at the former location of the Patterson main factory. The marker was erected by The Historical Society of Greenfield, Ohio Southern Hills Community Bank, Shorter Chapel A. M. E. Church, and The Ohio Historical Society.

On July 22, 2021, C.R. Patterson and Frederick Douglass Patterson, the first two generations of family to run the Patterson Company, were inducted into the Automotive Hall of Fame in Detroit. This honor was bestowed on them for their contributions to the transportation industry and for Frederick being the first Black manufacturer of automobiles. This demonstrated that the Patterson contribution to history mattered and, while generally fallen by the wayside for 80 years after they closed the factory, a long-overdue recognition of their efforts has finally come to light and their story will live on.

C.R. and Frederick Patterson Class of 2021 induction into the Automotive Hall of Fame in Detroit.

Richard Patterson (center) accepting the induction of his great-grandfather C.R. and grandfather Frederick into the Class of 2021 of the Automotive Hall of Fame.

Overall, the C.R. Patterson and Sons Company must be considered to have been a successful business. The company existed for over seven decades and only truly failed to thrive during the 1930s amid the Great Depression when so many other small businesses failed along with many large ones, as well. They had one small setback during the second decade of the century for a few years when they attempted to enter the automobile-manufacturing field with the Patterson-Greenfield car. This should not be seen as a total failure since it provided valuable experience for them to be successful in their next phase of operations. They were always willing to provide services where there was a need. A 1914 article described the company as "an example of making the best of an opportunity. The chances are plentiful and only need hard, consistent, continuous labor and application to bring them to successful termination."[10] The Pattersons worked hard at everything they did, and it made them better businessmen as a result. A letter from the Patterson Company to their patrons perhaps sums up the method they used to build a successful business the best. The letter appeared as a 1912 advertisement but could easily be applied to their business belief system at any point during their 74-year history. Just as the *Greenfield Republican* newspaper did in 1915 when Frederick was allowed to introduce the Patterson-Greenfield automobile in his own words, it only seems fitting to conclude with the Patterson voice once again. A portion of the letter stated:

> Dear Reader, We are – all of us – right now at the beginning of a New Year, 1912. Every man owes it to himself – to his family – to his community, to do his level best.
>
> Every man worthy of name, husbands now his strength – maps out his plans and with faith and hope is eager to plunge into the very thick of doing things. The wise man banks on straight methods, honest hard labor and grit to

hold out to the end. He is a fool who depends on tricks and sharp practices, for at the best his success will be short-lived and his self-contempt infinite.

We belong to that class Who Do Things. We want to serve you, and by serving you thus serve ourselves.

Whatever in our line you need, if you bring it to us, You Can Depend Upon It – it will be done right to our utmost ability. No mistake but will be corrected. No defect but will be made good. Not a single hair's breadth misrepresentation…

Respectfully yours, C.R. Patterson & Sons, Greenfield, Ohio.[11]

An Open Letter

You Don't Have to Look After Things Personally---Send Them and Your Interests Will Be More Than Safeguarded.

C. R. Patterson & Sons

CARRIAGE BUILDERS

Greenfield, Ohio,
February 1912.

Dear Reader:

We are - all of us - right now at the beginning of a New Year, 1912. Every man owes it to himself - to his family - to his community, to do his level best.

Every Man worthy of name, husbands now his strength - maps out his plans and with faith and hope is eager to plunge into the very thick of Doing things. The wise man banks on straight methods, honest hard labor and grit to hold out to the end. He is a fool who depends on tricks and sharp practices, for at the best his success will be short-lived and his self-contempt infinite.

We belong to that class Who Do Things. We want to serve you, and by serving you thus serve ourselves.

Whatever in our line you need, if you bring it to us, You Can Depend Upon It - it will be done right to our utmost ability. No mistake but will be corrected. No defect but will be made good. Not a single hair's breadth misrepresentation.

Every service will be as cheap as our splendid equipment and full complete stock (by far the largest of its kind in the section of the State) can make it, and yet be honest with ourselves.

Remember Us - plow work - wagon work - buggy repair - forgings - buggy painting - rubber tire, all kinds. Bring them NOW and it shall be our most earnest struggle day by day to serve you faithfully and well.

We are very grateful for a very large and still growing patronage, and we shall continue to merit it by deserving to merit it.

Respectfully yours,

C. R. Patterson & Sons,

Greenfield, Ohio.

CHAPTER 5 NOTES

[1] William Newton Hartshorn, *An Era of Progress and Promise, 1863-1910: The Religious, Moral, and Educational Development of the American Negro Since His Emancipation* (Boston: Priscilla Publishing Company, 1910), 5, 9, 448.
[2] Edward H. Jones, *Blacks in Business* (New York: Grosset and Dunlap, 1971), 4,6.
[3] Lorenzo J. Greene, and Carter G. Woodson, *The Negro Wage Earner* (Berkeley Heights, New Jersey: Wildside Press, 2008), 37,42.
[4] Vishnu V. Oak, *The Negro's Adventure in General Business* (Westport, Connecticut: Negro Universities Press, 1970), 45-46.
[5] Edward H. Jones, *Blacks in Business,* 13,50, 52.
[6] Robert W. Fairlie and Alicia M. Robb, *Race and Entrepreneurial Success: Black-, Asian-, and White-Owned Businesses in the United States* (Cambridge: MIT Press, 2008), 120.
[7] J. S. Himes, Jr., "Forty Years of Negro Life in Columbus, Ohio," *The Journal of Negro History,* Vol. 27, No. 2 (1942): 150; James W. Loewen, *Lies Across America: What Our Historic Sites Get Wrong* (New York: Simon and Schuster, 2007), 147.
[8] *Courier (Cleveland),* "Negro Family Made 'First' Cars," December 10, 1965.
[9] Steve Konicki, "Better Than a Model T," *Dayton Daily News,* March 21, 1976.
[10] *JNMA,* "Books, Lay Press, etc.," Vol. 6, No. 4 (1914): 269.
[11] *HD,* Patterson Ad, February 13, 1912.

BIBLIOGRAPHY

Archival/Public Records

Antique Automobile Club of America, Hershey, Pennsylvania
Benson Ford Research Center, Dearborn, Michigan
Carriage Museum of America, Lexington, Kentucky
Central Michigan University Archives, Mount Pleasant, Michigan
Columbus Metropolitan Library, Columbus, Ohio
Greenfield Historical Society, Greenfield, Ohio
Greenfield Public Library, Greenfield, Ohio
Highland County Courthouse, Hillsboro, Ohio
Icabod Flewellen Collection at the East Cleveland Public Library in Cleveland, Ohio
National Afro-American Museum and Cultural Center, Wilberforce, Ohio
National Automobile History Collection, Detroit, Michigan
National Automobile Museum/The Harrah Collection, Reno, Nevada
Ohio Historical Society/State Archives of Ohio, Columbus, Ohio
Ohio State University Archives, Columbus, Ohio
Ohio University Archives, Athens, Ohio
State Library of Ohio, Columbus, Ohio
Western Reserve Historical Society, Cleveland, Ohio

Newspapers/Periodicals

Annals of the American Academy of Political and Social Science
Automotive Industries
Business and Economic History
Carriage Monthly
Cleveland Advocate
Columbus Dispatch
Commercial Car Journal
Courier (Cleveland)
Dayton Daily News
Gallia Times
Greenfield Independent
Greenfield Journal
Greenfield Republican (GR)
Greenfield Times
Highland County Magazine
Hillsboro Dispatch (HD)
Horseless Age
Iron Age
Journal of Negro History
Journal of the National Medical Association (JNMA)
Leesburg Citizen
Motor Age
Motor Coach Transportation
Motor Record
New Pittsburgh Courier
New York Times
Ohio Farmer
Ohio Magazine
Ohio Schools
Ohio State University Monthly
Old Cars Weekly
Sabina News Record
The Automobile
The Automotive Manufacturer

The Crisis (NAACP)
Tri-State Trader
Ward's Auto World

Articles, Books, Government Documents, and Unpublished Works

American Technical Society. *Automobile Engineering.* Chicago: American Technical Society, 1917.

Ashby, William T., Walter E. Aebischer, John J. Powelson, Walter W. Weller, and Charles J. Fagg. *The Motor Truck Red Book.* New York: Traffic Publishing Company, 1942.

Babson, Steve. *Working Detroit: The Making of a Union Town.* Detroit: Wayne State University Press, 1986.

Banham, Russ. *The Ford Century: Ford Motor Company and the Innovations that Shaped the World.* New York: Artisan Books, 2002.

Barden, Clayton M. "School Bus Standards." MA Thesis, Ohio State University, 1932.

Baughman, Dale J. "A Survey of Pupil Transportation in the Consolidated School Districts of Hardin County." MA Thesis, Ohio State University, 1933.

Beach, Frederick Converse. *The Encyclopedia Americana.* New York: The Scientific American, 1903.

Berkebile, Don H. *American Carriages, Sleighs, Sulkies, and Carts.* New York: Dover Publications, 1977.

Bliss, Walton B. "The School Bus Situation in Ohio," *Ohio Schools,* Vol. 8, No. 3 (1930).

Borg, Kevin L. *Auto Mechanics: Technology and Expertise in Twentieth-Century America.* Baltimore: The Johns Hopkins University Press, 2007.

Brooke, Lindsay. *Ford Model T: The Car that Put the World on Wheels.* Minneapolis: Motorbooks, 2008.

Burrows, John Howard. "The Necessity of Myth: A History of the National Negro Business League, 1900-1945." PhD diss., Auburn University, 1977.

Burton, Thomas William. *What Experience Has Taught Me.* Cincinnati, Ohio: Jennings and Graham, 1910.

Callon, Albert McGee. "A Score Card and Typical Standards for School Bus Bodies." MA Thesis, University of Cincinnati, 1930.

Casey, Robert. *The Model T: A Centennial History.* Baltimore: The Johns Hopkins University Press, 2008.

Clifton, J. L. *Handbook for Rural and Village Boards of Education and Clerks in Ohio.* Columbus, Ohio: F. J. Heer Printing Company, 1928.

Columbus Carriage and Harness Company. *Thirty-third Annual Catalogue, The Columbus Carriage and Harness Company.* Cincinnati, Ohio: C. J. Krehbiel, 1910.

Copeland, Joseph Calvin. "Costs and the Measurements Applied in School Transportation." MA Thesis, Ohio State University, 1931.

Courier, July 31, 1996.

Cox, Joseph Mason Andrew. *Great Black Men of Masonry, 1723-1982.* Bronx, New York: Blue Diamond Press, 1982.

Downes, Randolph C. *The Rise of Warren Gamaliel Harding, 1865-1920.* Columbus, Ohio: Ohio State University Press, 1970.

Dyke, Andrew Lee. *Dyke's Automobile Encyclopedia.* St. Louis, Missouri: A. L. Dyke, 1911.

Fairlie, Robert W., and Alicia M. Robb. *Race and Entrepreneurial Success: Black-, Asian-, and White-Owned Businesses in the United States.* Cambridge, Massachusetts, The MIT Press, 2008.

Flink, James J. *America Adopts the Automobile, 1895-1910.* Cambridge, Massachusetts: The MIT Press, 1970.

Flink, James J. *The Car Culture.* Cambridge, Massachusetts: The MIT Press, 1975.

Freeman, Larry. *The Merry Old Mobiles.* Watkins Glen, New York: Century House, 1949.

Galbreath, Charles B. *History of Ohio.* Chicago: The American Historical Society, 1925.

Georgano, G. N. *The Beaulieu Encyclopedia of the Automobile,* Volume 2. Chicago: Fitzroy Dearborn Publishers, 2000.

Georgano, G. N. *The New Encyclopedia of Motor Cars, 1885 to the Present.* New York: E. P. Dutton, 1982.

Gibbs, Mifflin Wistar. *Shadow of Light: An Autobiography with Reminiscences of the Last and Present Century.* Washington D. C., 1902.

Greeley, Horace. *The Great Industries of the United States.* Hartford, Connecticut: J. B. Burr and Hyde, 1873.

Greene, Lorenzo J., and Carter G. Woodson. *The Negro Wage Earner.* Wildside Press, Berkeley Heights, New Jersey, 2008.

Greenfield Historical Society. *Greenfield, Ohio: 1799-1999.* Paducah, Kentucky: Turner Publishing Company, 2000.

Harmon, J. H. "The Negro as a Local Business Man," *The Journal of Negro History* Vol. 14, No. 2 (1929).

Hartshorn, William Newton. *An Era of Progress and Promise, 1863-1910: The Religious, Moral, and Educational Development of the American Negro Since His Emancipation.* Boston: Priscilla Publishing Company, 1910.

Hartsook, M. D. "A Survey of Bus Transportation to and from Ohio Schools, 1929-1930." MA Thesis, Ohio State University, 1930.

Hebb, David. *Wheels on the Road: A History of the Automobile from the Steam Engine to the Car of Tomorrow.* New York: Crowell-Collier Press, 1966.

Heitmann, John Alfred. *The Automobile and American Life.* McFarland, Jefferson, North Carolina, 2009.

Himes, J. S., "Forty Years of Negro Life in Columbus, Ohio," *The Journal of Negro History,* Vol. 27, No. 2 (1942).

Hochfelder, David, and Susan Helper. "Suppliers and Product Development in the Early American Automobile Industry," *Business and Economic History,* Vol. 25, No. 2 (1996).

Homans, James E. *Self-Propelled Vehicles: A Practical Treatise on the Theory, Construction, Operation, Care, and Management of all Forms of Automobiles.* New York: Theo. Audel and Company, 1903.

Hounshell, David. *From the American System to Mass Production, 1800-1932: The Development of Manufacturing Technology in the United States.* Baltimore: The Johns Hopkins University Press, 1984.

Howe, Henry. *Historical Collections of Ohio,* Cincinnati, Ohio: C. J. Krehbiel & Company, 1904.

Hunter, Bob. "Athletics Not Always an Open Field," *Columbus Dispatch* January 29, 1992.

Hunter, H. W. *New Atlas of Highland County, Ohio*. Hillsboro, Ohio: H. W. Hunter, 1916.

Hutchins, C. D. "Pupil Transportation Equipment," *Review of Educational Research,* Vol. 8, No. 4 (1938).

Hyde, Charles K. *The Dodge Brothers: The Men, the Motor Cars, and the Legacy.* Detroit: Wayne State University Press, 2005.

Jaynes, Gerald D. *Encyclopedia of African American Society.* Thousand Oaks, California: Sage Publications, 2005.

Jenifer, Roderick. "The Patterson Wheels," *The King Maker* Issue 9 (2008).

Johns, Roe Lyell. *State and Local Administration of School Transportation*. New York: Bureau of Publications, Columbia University, 1928.

Joiner, W. A. *A Half Century of Freedom of the Negro in Ohio.* Xenia, Ohio: Smith Advertising Company, 1915.

Jones, Edward H. *Blacks in Business.* New York: Grosset and Dunlap, 1971.

Kellmer, Earl B. "The Transportation Factor in the Proposed Centralization of the High Schools of Lorain County." MA Thesis, Ohio State University, 1929.

Kelsey, Carl. "The Evolution of Negro Labor," *Annals of the American Academy of Political and Social Science,* Vol. 21 (1903).

Kelsey, Carl. "The Evolution of Negro Labor," *Annals of the American Academy of Political and Social Science,* Vol. 21 (1903).

Kimes, Beverly Rae. *Standard Catalog of American Cars, 1805-1942*. Iola, Wisconsin: Krause Publications, 1989.

Kinney, Thomas. *The Carriage Trade: Making Horse-Drawn Vehicles in America.* Baltimore: The Johns Hopkins University Press, 2004.

Klise, J. W. *The County of Highland: A History of Highland County, Ohio, From the Earliest Days, with Special Chapters on the Bench and Bar, Medical Profession, Educational Development, Industry and Agriculture, and Biographical Sketches.* Madison, Wisconsin: Northwestern Historical Association, 1902.

Konicki, Steve. "150 Built: Better than Model T," *Dayton Daily News,* March 21, 1976.

Lake, D. J. *Atlas of Highland County, Ohio.* Philadelphia: C. O. Titus, 1871.

Lapchick, Richard. *100 Pioneers: African-Americans Who Broke Color Barriers in Sport*. Morgantown, West Virginia: West Virginia University, 2008.

Larrie, Reginald. "From Slave to Auto Manufacturer: The Black Family Who Built Cars," *Ward's Auto World* Vol. 17, No. 4 (1981): 60-61.

Loewen, James W. *Lies Across America: What Our Historic Sites Get Wrong.* Simon and Schuster, New York, 2007.

Mather, Frank Lincoln. *Who's Who of the Colored Race: A General Biographical Dictionary of Men and Women of African Descent.* Chicago, 1915.

May, Henry A. *First Black Autos: The Charles Richard "C.R." Patterson & Sons Company: African-American Automobile Manufacturer of Patterson-Greenfield Motorcars, Buses, and Trucks.* Mount Vernon, New York: Stalwart Publications, 2006.

McBride, David. *Common Pleas Court Records of Highland County, Ohio, 1805-1860.* Hillsboro, Ohio: Southern Ohio Genealogical Society, 1959.

McBride, David. *Personal Property Taxpayers of Highland County, Ohio.* Hillsboro, Ohio: D. N. McBride, 1980.

Meyer, Henry W. *Memories of the Buggy Days.* Cincinnati, Ohio: Brinker, Inc., 1965.

Myers, William E. *Book of the Republican National Convention, Cleveland, Ohio, June 10th, 1924.* Cleveland, Ohio: Allied Printing, 1924.

Napper, Charles W. "Concretionary Forms in the Greenfield Limestone," (paper presented for the Geological Section of the Ohio Academy of Science Meeting at Columbus, Ohio, on April 7, 1917).

Napper, Charles W. "Occurrence of Carbonaceous Material in the Greenfield Member of the Monroe Formation," *The Knowledge Bank at O.S.U.* Vol. 16, No. 4 (1916).

National Safety Council. *School Buses: Their Safe Design and Operation.* Chicago: National Safety Council, 1938.

Naughton, Bill. "Former Slave, Descendants Were Early Vehicle Pioneers in Ohio," *Tri-State Trader*, July 5, 1980.

Naughton, Bill. "Patterson-Greenfield Cars," *Old Cars Weekly*, June 19, 1980.

Nemar Publishing Company. *Department Reports of the State of Ohio: Embracing Twenty-Six Issues Dating from October 16, 1919 to April 8, 1920.* Columbus, Ohio: Nemar Publishing Company, 1920.

Norton, S. V. *The Motor Truck as an Aid to Business Profits.* Chicago: A. W. Shaw Company, 1918.

Oak, Vishnu V. *The Negro's Adventure in General Business.* Westport, Connecticut: Negro Universities Press, 1970.

Ohio Carriage Manufacturing Company. *Split Hickory Vehicles & Ohio Oak Tanned Harness, Sixteenth Annual Catalogue.* Columbus, Ohio: Ohio Carriage Manufacturing Company, 1916.

Ohio State University. *Register of Graduates and Members of the Ohio State University Association, 1878-1917.* Columbus, Ohio: Ohio State University, 1917.

Pack, Charlotte. *Time Travels: 200 Years of Highland County History Told Through Diaries, Letters, Stories, and Photos.* Fayetteville, Ohio: Chatfield Publishing Company, 2007.

Parham, William Hartwell. *An Official History of the Most Worshipful Grand Lodge Free and Accepted Masons for the State of Ohio.* Columbus, Ohio: Grand Lodge for the State of Ohio, 1906.

Patterson, Kathleen L. *It's Been a Wonderful Life: One Day at a Time.* Dayton, Ohio: Self-published, 2001.

Pennington, Leland J. "The Patterson-Greenfield Automobile," *The Highland County Magazine,* Vol. 1, No. 3 (1984).

Perris, George Herbert. *The Industrial History of Modern England.* New York: Henry Holt and Company, 1914.

Perry, Brandon A. "Black History Profile: Early Automaker Frederick D. Patterson Built Cars to Last," *Indianapolis Recorder* February 22, 2002.

Pfau, Hugo. *The Custom Body Era.* New York: A. S. Barnes and Company, 1970.

Prentiss, Fred L. "Making Motor Bus Bodies of Steel," *The Iron Age,* December 26, 1929.

Rae, John B. *The American Automobile Industry.* Boston: Twayne Publishers, 1984.

Reed, William. "100 Years of Car Making: Blacks Then and Now," *New Pittsburgh Courier*, July 31, 1996.

Rollins, Ron. "Four-Wheel First," *Ohio Magazine,* Vol. 13, No. 5 (2008).

Salzman, Jack, David Lionel Smith, and Cornel West. *Encyclopedia of African-American Culture and History.* New York: Macmillan Library Reference USA, 1996.

Schmidt, Harold. "State of Ohio on Relation of C.R. Patterson vs The Board of Education of the Incorporated Village of Greenfield, Ohio, and W. G. Moler as Superintendent," Greenfield Historical Society, http://www.greenfieldhistoricalsociety.org/PattersonvsBd.pdf (Accessed December 3, 2008)

Sears, Stephen W. *The American Heritage History of the Automobile in America.* New York: American Heritage Publishing Company, 1977.

Semsel, Craig Robert. "Built to Move Millions: Transit Vehicle Manufacture in Ohio, 1880-1938." PhD Diss., Case Western Reserve University, 2001.

Sinclair, Bruce. *Technology and the African-American Experience: Needs and Opportunities for Study.* MIT Press, Cambridge, 2004.

Smith, Harvey C. *Annual Report of the Secretary of State to the Governor and General Assembly of the State of Ohio for the Year Ending June 30, 1920.* Springfield, Ohio: Kelly-Springfield Printing Company, 1920.

Smith, Jessie Carney. *Black Firsts: 4,000 Ground-Breaking and Pioneering Historical Events.* Detroit: Visible Ink Press, 2003.

Snider, Wayne L. *All in the Same Spaceship: Portions of American Negro History Illustrated in Highland County, Ohio, U. S. A.* New York: Vantage Press, 1974.

The Industrial Commission of Ohio. *Directory of Ohio Manufacturers.* Columbus: F. J. Heer Printing Company, 1918.

The Museums at Stony Brook. *19th Century American Carriages: Their Manufacture, Decoration, and Use.* Stony Brook, New York: The Museums at Stony Brook, 1987.

U. S. Committee on Industrial Arts and Expositions. *Celebration of the Semicentennial Anniversary of the Act of Emancipation.* Washington D. C.: Government Printing Office, 1914. (Bill HR15733 presented to the 63rd United States Congress on May 27, 1914)

Varassi, John. "Wheeling and Dealing: The Inventiveness of Engineers Made the Auto Trade Big Business in the Early 20th Century," *Mechanical Engineering* April (2005).

Walker, Juliet E. K. *Encyclopedia of African American Business History.* Westport, Connecticut: Greenwood Press, 1999.

Washington, Booker T. *Booker T. Washington's Own Story of His Life and Work.* Naperville, Illinois: J. L. Nichols & Company, 1901.

Williams, W. W. *History of Ross and Highland Counties, Ohio, With Illustrations and Biographical Sketches.* Cleveland, Ohio: Williams Brothers, 1880.

Work, Monroe N. "The Negro in Business and the Professions," *Annals of the American Academy of Political and Social Science,* Vol. 140 (1928).

Wren, James A., and Genevieve J. Wren. *Motor Trucks of America: Milestones, Pioneers, RollCall, Highlights.* Ann Arbor, Michigan: The University of Michigan Press, 1979.

Wynkoop Hallenbeck Crawford Company. *Second Annual Report of the Department of Labor of the State of Michigan.* Lansing, Michigan: Wynkoop Hallenbeck Crawford Company, 1911.

Appendix A:

Patterson Company Advertisements Selected for this Study

Greenfield Republican and *Hillsboro Dispatch* ran the same advertisements simultaneously. To prevent duplication, the ads from only one or the other is documented here.

Month	Year	Vehicle Type	Specs	Prices	Sale	Other Services	Slogan Only	Promotions	TESTIMONIAL	Negro ID	Audience	Ad Source
1	1902	NO. 60 CUT-UNDER STANHOPE									ALL	GREENFIELD REPUBLICAN
1	1902	NO. 1 END SPRING PIANO BUGGY									ALL	GREENFIELD REPUBLICAN
1	1902	NO. 2 END SPRING PIANO BUGGY									ALL	GREENFIELD REPUBLICAN
1	1902	NO. 1 END SPRING PIANO BUGGY									ALL	GREENFIELD REPUBLICAN
2	1902	NO. 1 END SPRING PIANO BUGGY									ALL	GREENFIELD REPUBLICAN
2	1902	DOCTOR'S BUGGY	Y								PHYSICIANS	GREENFIELD REPUBLICAN
2	1902	NO. 1 END SPRING PIANO BUGGY		Y							ALL	GREENFIELD REPUBLICAN
2	1902	NO. 3 END SPRING PIANO BUGGY		Y							ALL	GREENFIELD REPUBLICAN
3	1902	NO. 3 END SPRING PIANO BUGGY		Y							ALL	GREENFIELD REPUBLICAN
3	1902	NO. 1 END SPRING PIANO BUGGY									ALL	GREENFIELD REPUBLICAN
3	1902	NO. 20 BOULEVARD									ALL	GREENFIELD REPUBLICAN
3	1902	NO. 1 END SPRING PIANO BUGGY									ALL	GREENFIELD REPUBLICAN
4	1902	NO. 3 END SPRING PIANO BUGGY		Y							ALL	GREENFIELD REPUBLICAN
4	1902	NO. 119 ELLIPTIC SPRING BIKE WAGON									ALL	GREENFIELD REPUBLICAN
5	1902	NO. 1 END SPRING PIANO BUGGY									ALL	GREENFIELD REPUBLICAN
5	1902	NO. 60 CUT-UNDER STANHOPE									ALL	GREENFIELD REPUBLICAN
5	1902	NO. 61 CUT-UNDER STANHOPE									ALL	GREENFIELD REPUBLICAN
5	1902				Y						ALL	GREENFIELD REPUBLICAN
5	1902	NO. 1 END SPRING PIANO BUGGY									ALL	GREENFIELD REPUBLICAN
6	1902	NO. 1 END SPRING PIANO BUGGY									ALL	GREENFIELD REPUBLICAN
6	1902	NO. 119 ELLIPTIC SPRING BIKE WAGON	Y								ALL	GREENFIELD REPUBLICAN
6	1902	NO. 1 END SPRING PIANO BUGGY									ALL	GREENFIELD REPUBLICAN
7	1902	NO. 1 END SPRING PIANO BUGGY									ALL	GREENFIELD REPUBLICAN
9	1902			Y	Y						ALL	GREENFIELD REPUBLICAN
3	1903	NO. 1 END SPRING PIANO BUGGY									ALL	GREENFIELD REPUBLICAN
3	1903	NO. 1 END SPRING PIANO BUGGY									ALL	GREENFIELD REPUBLICAN

Month	Year	Vehicle Type	Specs	Prices	Sale	Other Services	Slogan Only	Promotions	TESTIMONIAL	Negro ID	Audience	Ad Source
4	1903	NO. 1 END SPRING PIANO BUGGY									ALL	GREENFIELD REPUBLICAN
5	1903	NO. 1 END SPRING PIANO BUGGY									ALL	GREENFIELD REPUBLICAN
5	1903	NO. 1 END SPRING PIANO BUGGY									ALL	GREENFIELD REPUBLICAN
3	1904	NO. 1 PHAETON SEATED BUGGY									ALL	GREENFIELD REPUBLICAN
3	1904	NO. 1 PHAETON SEATED BUGGY									ALL	GREENFIELD REPUBLICAN
3	1904	NO. 1 PHAETON SEATED BUGGY									ALL	GREENFIELD REPUBLICAN
3	1904	NO. 1 PHAETON SEATED BUGGY									ALL	GREENFIELD REPUBLICAN
3	1904	NO. 1 PHAETON SEATED BUGGY									ALL	GREENFIELD REPUBLICAN
4	1904	NO. 1 PHAETON SEATED BUGGY									ALL	GREENFIELD REPUBLICAN
4	1904	NO. 1 PHAETON SEATED BUGGY									ALL	GREENFIELD REPUBLICAN
4	1904	NO. 1 PHAETON SEATED BUGGY									ALL	GREENFIELD REPUBLICAN
4	1904	NO. 1 PHAETON SEATED BUGGY									ALL	GREENFIELD REPUBLICAN
5	1904	NO. 1 PHAETON SEATED BUGGY									ALL	GREENFIELD REPUBLICAN
5	1904	NO. 1 PHAETON SEATED BUGGY									ALL	GREENFIELD REPUBLICAN
5	1904	NO. 1 PHAETON SEATED BUGGY									ALL	GREENFIELD REPUBLICAN
5	1904	NO. 1 PHAETON SEATED BUGGY									ALL	GREENFIELD REPUBLICAN
6	1904	NO. 1 PHAETON SEATED BUGGY									ALL	GREENFIELD REPUBLICAN
6	1904	NO. 1 PHAETON SEATED BUGGY									ALL	GREENFIELD REPUBLICAN
6	1904	NO. 1 PHAETON SEATED BUGGY									ALL	GREENFIELD REPUBLICAN
6	1904	NO. 1 PHAETON SEATED BUGGY									ALL	GREENFIELD REPUBLICAN
8	1904			Y	Y						ALL	GREENFIELD REPUBLICAN
11	1904										ALL	GREENFIELD REPUBLICAN
12	1904										ALL	GREENFIELD REPUBLICAN
12	1904										ALL	GREENFIELD REPUBLICAN
12	1904										ALL	GREENFIELD REPUBLICAN

Month	Year	Vehicle Type	Specs	Prices	Sale	Other Services	Slogan Only	Promotions	TESTIMONIAL	Negro ID	Audience	Ad Source
12	1904	PORTLAND SLEIGHS AND CUTTER SLEIGHS		Y	Y						ALL	GREENFIELD REPUBLICAN
12	1904										ALL	GREENFIELD REPUBLICAN
1	1905										ALL	GREENFIELD REPUBLICAN
1	1905										ALL	GREENFIELD REPUBLICAN
1	1905					Y					FARMERS	GREENFIELD REPUBLICAN
1	1905										ALL	GREENFIELD REPUBLICAN
2	1905					Y					FARMERS	GREENFIELD REPUBLICAN
2	1905										ALL	GREENFIELD REPUBLICAN
2	1905										ALL	GREENFIELD REPUBLICAN
2	1905		Y			Y					ALL	GREENFIELD REPUBLICAN
3	1905					Y					ALL	GREENFIELD REPUBLICAN
3	1905						Y				ALL	GREENFIELD REPUBLICAN
4	1905										ALL	GREENFIELD REPUBLICAN
4	1905			Y	Y						ALL	GREENFIELD REPUBLICAN
4	1905						Y				ALL	GREENFIELD REPUBLICAN
5	1905					Y					ALL	GREENFIELD REPUBLICAN
5	1905						Y				ALL	GREENFIELD REPUBLICAN
6	1905						Y				ALL	GREENFIELD REPUBLICAN
6	1905						Y				ALL	GREENFIELD REPUBLICAN
6	1905										ALL	GREENFIELD REPUBLICAN
7	1905										ALL	GREENFIELD REPUBLICAN
8	1905										ALL	GREENFIELD REPUBLICAN
9	1905				Y						ALL	GREENFIELD REPUBLICAN
10	1905						Y				ALL	GREENFIELD REPUBLICAN
10	1905						Y				ALL	GREENFIELD REPUBLICAN

Month	Year	Vehicle Type	Specs	Prices	Sale	Other Services	Slogan Only	Promotions	TESTIMONIAL	Negro ID	Audience	Ad Source
11	1905										ALL	GREENFIELD REPUBLICAN
12	1905						Y				ALL	GREENFIELD REPUBLICAN
12	1905			Y	Y						ALL	GREENFIELD REPUBLICAN
12	1905			Y	Y						ALL	GREENFIELD REPUBLICAN
12	1905	NO. 1 PIANO BUGGY	Y								ALL	GREENFIELD REPUBLICAN
1	1906	NO. 1 PIANO BUGGY	Y	Y							ALL	GREENFIELD REPUBLICAN
1	1906	NO. 1 PIANO BUGGY		Y							ALL	GREENFIELD REPUBLICAN
1	1906	NO. 1 PIANO BUGGY									ALL	GREENFIELD REPUBLICAN
1	1906	NO. 1 PIANO BUGGY									ALL	GREENFIELD REPUBLICAN
1	1906	NO. 10									ALL	GREENFIELD REPUBLICAN
2	1906	NO. 1 PIANO BUGGY									ALL	GREENFIELD REPUBLICAN
2	1906	NO. 1 PIANO BUGGY									ALL	GREENFIELD REPUBLICAN
2	1906	NO. 1 PIANO BUGGY				Y					ALL	GREENFIELD REPUBLICAN
3	1906	NO. 1 PIANO BUGGY				Y					ALL	GREENFIELD REPUBLICAN
3	1906	NO. 1 PIANO BUGGY				Y					ALL	GREENFIELD REPUBLICAN
3	1906	SURREY				Y					ALL	GREENFIELD REPUBLICAN
3	1906	NO. 1 PIANO BUGGY				Y					ALL	GREENFIELD REPUBLICAN
3	1906	NO. 1 PIANO BUGGY				Y					ALL	GREENFIELD REPUBLICAN
4	1906	NO. 1 PIANO BUGGY				Y					ALL	GREENFIELD REPUBLICAN
4	1906	NO. 1 PIANO BUGGY				Y					ALL	GREENFIELD REPUBLICAN
4	1906	NO. 1 PIANO BUGGY				Y					ALL	GREENFIELD REPUBLICAN
5	1906	NO. 1 PIANO BUGGY				Y					ALL	GREENFIELD REPUBLICAN
5	1906	STANHOPE				Y					ALL	GREENFIELD REPUBLICAN
5	1906	NO. 1 PIANO BUGGY				Y					ALL	GREENFIELD REPUBLICAN
5	1906	NO. 1 FINE STATION WAGON				Y					ALL	GREENFIELD REPUBLICAN

Month	Year	Vehicle Type	Specs	Prices	Sale	Other Services	Slogan Only	Promotions	TESTIMONIAL	Negro ID	Audience	Ad Source
5	1906	NO. 41 BIKE RUNABOUT				Y					ALL	GREENFIELD REPUBLICAN
5	1906	STANHOPE				Y					ALL	GREENFIELD REPUBLICAN
6	1906	NO. 1 PIANO BUGGY				Y					ALL	GREENFIELD REPUBLICAN
6	1906	NO. 1 PIANO BUGGY				Y					ALL	GREENFIELD REPUBLICAN
6	1906	NO. 1 PIANO BUGGY				Y					ALL	GREENFIELD REPUBLICAN
7	1906	FINE PHAETON				Y					ALL	GREENFIELD REPUBLICAN
7	1906	FINE STRAIGHT SILL SURREY				Y					ALL	GREENFIELD REPUBLICAN
7	1906	NO. 1 PIANO BUGGY				Y					ALL	GREENFIELD REPUBLICAN
7	1906	FINE PHAETON				Y					ALL	GREENFIELD REPUBLICAN
8	1906	NO. 1 PIANO BUGGY				Y					ALL	GREENFIELD REPUBLICAN
8	1906	NO. 1 PIANO BUGGY				Y					ALL	GREENFIELD REPUBLICAN
8	1906	FINE LIGHT THREE-QUARTER SURREY				Y					ALL	GREENFIELD REPUBLICAN
8	1906	NO. 1 PIANO BUGGY				Y					ALL	GREENFIELD REPUBLICAN
8	1906	NO. 1 PIANO BUGGY				Y					ALL	GREENFIELD REPUBLICAN
9	1906	NO. 1 PIANO BUGGY				Y					ALL	GREENFIELD REPUBLICAN
9	1906				Y						ALL	GREENFIELD REPUBLICAN
9	1906				Y						ALL	GREENFIELD REPUBLICAN
9	1906				Y						ALL	GREENFIELD REPUBLICAN
10	1906				Y						ALL	GREENFIELD REPUBLICAN
10	1906				Y						ALL	GREENFIELD REPUBLICAN
10	1906	VERY FINE THREE QUARTER SURREY									ALL	GREENFIELD REPUBLICAN
11	1906				Y						ALL	GREENFIELD REPUBLICAN
11	1906				Y						ALL	GREENFIELD REPUBLICAN
11	1906				Y						ALL	GREENFIELD REPUBLICAN
11	1906				Y						ALL	GREENFIELD REPUBLICAN

Month	Year	Vehicle Type	Specs	Prices	Sale	Other Services	Slogan Only	Promotions	TESTIMONIAL	Negro ID	Audience	Ad Source
11	1906				Y						ALL	GREENFIELD REPUBLICAN
12	1906			Y	Y						ALL	GREENFIELD REPUBLICAN
12	1906			Y	Y						ALL	GREENFIELD REPUBLICAN
12	1906										ALL	GREENFIELD REPUBLICAN
1	1907										ALL	GREENFIELD REPUBLICAN
1	1907										ALL	GREENFIELD REPUBLICAN
2	1907					Y					ALL	GREENFIELD REPUBLICAN
2	1907	NO. 1 PIANO BUGGY									ALL	GREENFIELD REPUBLICAN
3	1907	NO. 1 PIANO BUGGY									ALL	GREENFIELD REPUBLICAN
4	1907	NO. 1 PIANO BUGGY									ALL	GREENFIELD REPUBLICAN
5	1907	NO. 1 PIANO BUGGY	Y								ALL	GREENFIELD REPUBLICAN
5	1907	NO. 1 PIANO BUGGY	Y								ALL	GREENFIELD REPUBLICAN
5	1907	CURTAIN STATION WAGON									ALL	GREENFIELD REPUBLICAN
5	1907	NO. 40 STRAIGHT SILL SURREY									ALL	GREENFIELD REPUBLICAN
6	1907	NO. 1 PIANO BUGGY									ALL	GREENFIELD REPUBLICAN
6	1907	NO. 12 FINE AUTO SEAT RUNABOUT									ALL	GREENFIELD REPUBLICAN
7	1907	NO. 12 FINE AUTO SEAT RUNABOUT									ALL	GREENFIELD REPUBLICAN
7	1907	NO. 1 PIANO BUGGY									ALL	GREENFIELD REPUBLICAN
7	1907	NO. 100 DOCTOR'S PIANO BUGGY									ALL	GREENFIELD REPUBLICAN
7	1907	NO. 1 PIANO BUGGY									ALL	GREENFIELD REPUBLICAN
7	1907	NO. 1 PIANO BUGGY			Y						ALL	GREENFIELD REPUBLICAN
8	1907				Y						ALL	GREENFIELD REPUBLICAN
9	1907	FARM WAGON	Y		Y						FARMERS	GREENFIELD REPUBLICAN
10	1907				Y						ALL	GREENFIELD REPUBLICAN
10	1907							Y			ALL	GREENFIELD REPUBLICAN

Month	Year	Vehicle Type	Specs	Prices	Sale	Other Services	Slogan Only	Promotions	TESTIMONIAL	Negro ID	Audience	Ad Source
10	1907							Y			ALL	GREENFIELD REPUBLICAN
10	1907							Y			ALL	GREENFIELD REPUBLICAN
11	1907							Y			ALL	GREENFIELD REPUBLICAN
11	1907							Y			ALL	GREENFIELD REPUBLICAN
11	1907							Y			ALL	GREENFIELD REPUBLICAN
11	1907							Y			ALL	GREENFIELD REPUBLICAN
12	1907							Y			ALL	GREENFIELD REPUBLICAN
12	1907				Y						ALL	GREENFIELD REPUBLICAN
12	1907							Y			ALL	GREENFIELD REPUBLICAN
12	1907							Y			ALL	GREENFIELD REPUBLICAN
12	1907							Y			ALL	GREENFIELD REPUBLICAN
1	1908							Y			ALL	GREENFIELD REPUBLICAN
1	1908										ALL	GREENFIELD REPUBLICAN
1	1908										ALL	GREENFIELD REPUBLICAN
1	1908										ALL	GREENFIELD REPUBLICAN
1	1908										ALL	GREENFIELD REPUBLICAN
2	1908										ALL	GREENFIELD REPUBLICAN
2	1908										ALL	GREENFIELD REPUBLICAN
2	1908										ALL	GREENFIELD REPUBLICAN
2	1908					Y					ALL	GREENFIELD REPUBLICAN
3	1908					Y					ALL	GREENFIELD REPUBLICAN
3	1908				Y						ALL	GREENFIELD REPUBLICAN
3	1908				Y						ALL	GREENFIELD REPUBLICAN
3	1908				Y						ALL	GREENFIELD REPUBLICAN
4	1908				Y						ALL	GREENFIELD REPUBLICAN

Month	Year	Vehicle Type	Specs	Prices	Sale	Other Services	Slogan Only	Promotions	TESTIMONIAL	Negro ID	Audience	Ad Source
4	1908				Y						ALL	GREENFIELD REPUBLICAN
4	1908				Y						ALL	GREENFIELD REPUBLICAN
4	1908				Y						ALL	GREENFIELD REPUBLICAN
4	1908				Y						ALL	GREENFIELD REPUBLICAN
5	1908				Y						ALL	GREENFIELD REPUBLICAN
5	1908				Y						ALL	GREENFIELD REPUBLICAN
5	1908				Y						ALL	GREENFIELD REPUBLICAN
6	1908				Y						ALL	GREENFIELD REPUBLICAN
6	1908				Y						ALL	GREENFIELD REPUBLICAN
6	1908				Y						ALL	GREENFIELD REPUBLICAN
6	1908				Y						ALL	GREENFIELD REPUBLICAN
7	1908	LIGHT SURREY									ALL	GREENFIELD REPUBLICAN
7	1908	LIGHT SURREY									ALL	GREENFIELD REPUBLICAN
7	1908				Y						ALL	GREENFIELD REPUBLICAN
7	1908	LIGHT SURREY		Y							ALL	GREENFIELD REPUBLICAN
7	1908	LIGHT SURREY		Y							ALL	GREENFIELD REPUBLICAN
8	1908	LIGHT SURREY		Y							ALL	GREENFIELD REPUBLICAN
8	1908	LIGHT SURREY		Y							ALL	GREENFIELD REPUBLICAN
8	1908	LIGHT SURREY		Y							ALL	GREENFIELD REPUBLICAN
8	1908										ALL	GREENFIELD REPUBLICAN
9	1908										ALL	GREENFIELD REPUBLICAN
9	1908										ALL	GREENFIELD REPUBLICAN
9	1908				Y						ALL	GREENFIELD REPUBLICAN
9	1908										ALL	GREENFIELD REPUBLICAN
10	1908						Y				ALL	GREENFIELD REPUBLICAN

Month	Year	Vehicle Type	Specs	Prices	Sale	Other Services	Slogan Only	Promotions	TESTIMONIAL	Negro ID	Audience	Ad Source
10	1908				Y						ALL	GREENFIELD REPUBLICAN
10	1908					Y					FARMERS	GREENFIELD REPUBLICAN
11	1908										ALL	GREENFIELD REPUBLICAN
11	1908										ALL	GREENFIELD REPUBLICAN
12	1908	STORM BUGGY	Y								ALL	GREENFIELD REPUBLICAN
12	1908										ALL	GREENFIELD REPUBLICAN
12	1908										ALL	GREENFIELD REPUBLICAN
12	1908				Y						ALL	GREENFIELD REPUBLICAN
12	1908								Y		ALL	GREENFIELD REPUBLICAN
1	1912										FARMERS	THE OHIO FARMER
1	1912										FARMERS	THE OHIO FARMER
2	1912										FARMERS	THE OHIO FARMER
2	1912										FARMERS	THE OHIO FARMER
3	1912										FARMERS	THE OHIO FARMER
3	1912										FARMERS	THE OHIO FARMER
4	1912										FARMERS	THE OHIO FARMER
4	1912										FARMERS	THE OHIO FARMER
1	1909	STORM BUGGY			Y						ALL	GREENFIELD REPUBLICAN
`	1909	STORM BUGGY									ALL	GREENFIELD REPUBLICAN
1	1909					Y					ALL	GREENFIELD REPUBLICAN
2	1909				Y						ALL	GREENFIELD REPUBLICAN
2	1909				Y				Y		ALL	GREENFIELD REPUBLICAN
2	1909								Y		ALL	GREENFIELD REPUBLICAN
2	1909	RUNABOUT							Y		ALL	GREENFIELD REPUBLICAN
3	1909				Y						ALL	GREENFIELD REPUBLICAN

Month	Year	Vehicle Type	Specs	Prices	Sale	Other Services	Slogan Only	Promotions	TESTIMONIAL	Negro ID	Audience	Ad Source
3	1909								Y		ALL	GREENFIELD REPUBLICAN
3	1909				Y						ALL	GREENFIELD REPUBLICAN
3	1909								Y		ALL	GREENFIELD REPUBLICAN
4	1909						Y				ALL	GREENFIELD REPUBLICAN
4	1909						Y				ALL	GREENFIELD REPUBLICAN
4	1909										ALL	GREENFIELD REPUBLICAN
5	1914					Y					ALL	GREENFIELD REPUBLICAN
5	1914	NO. 4 GREENFIELD BUGGY	Y								ALL	GREENFIELD REPUBLICAN
5	1914	NO. 2 END SPRING PIANO BUGGY		Y							ALL	GREENFIELD REPUBLICAN
5	1914	PHAETON		Y							ALL	GREENFIELD REPUBLICAN
6	1914	AUTO SEAT BUGGY									ALL	GREENFIELD REPUBLICAN
6	1914	SURREY		Y							ALL	GREENFIELD REPUBLICAN
6	1914	SPRING WAGON		Y							ALL	GREENFIELD REPUBLICAN
6	1914										ALL	GREENFIELD REPUBLICAN
7	1914										ALL	GREENFIELD REPUBLICAN
7	1914						Y				ALL	GREENFIELD REPUBLICAN
7	1914			Y							FARMERS	GREENFIELD REPUBLICAN
7	1914	RUNABOUT							Y		ALL	GREENFIELD REPUBLICAN
7	1914				Y						ALL	GREENFIELD REPUBLICAN
7	1914			Y		Y					FARMERS	GREENFIELD REPUBLICAN
8	1914					Y					FARMERS	GREENFIELD REPUBLICAN
8	1914										FARMERS	GREENFIELD REPUBLICAN
8	1914										FARMERS	GREENFIELD REPUBLICAN
8	1914										FARMERS	GREENFIELD REPUBLICAN
9	1914	STORM BUGGY									ALL	GREENFIELD REPUBLICAN

Month	Year	Vehicle Type	Specs	Prices	Sale	Other Services	Slogan Only	Promotions	TESTIMONIAL	Negro ID	Audience	Ad Source
9	1914	STORM BUGGY									ALL	GREENFIELD REPUBLICAN
10	1914	STORM BUGGY		Y							ALL	GREENFIELD REPUBLICAN
10	1914	STORM BUGGY		Y							ALL	GREENFIELD REPUBLICAN
10	1914	STORM BUGGY		Y							ALL	GREENFIELD REPUBLICAN
10	1914	STORM BUGGY		Y							ALL	GREENFIELD REPUBLICAN
10	1914	STORM BUGGY		Y							ALL	GREENFIELD REPUBLICAN
11	1914	STORM BUGGY		Y							ALL	GREENFIELD REPUBLICAN
11	1914	STORM BUGGY		Y							ALL	GREENFIELD REPUBLICAN
11	1914	STORM BUGGY		Y							ALL	GREENFIELD REPUBLICAN
11	1914	STORM BUGGY		Y							ALL	GREENFIELD REPUBLICAN
12	1914	STORM BUGGY		Y							ALL	GREENFIELD REPUBLICAN
12	1914	STORM BUGGY		Y							ALL	GREENFIELD REPUBLICAN
12	1914	STORM BUGGY		Y							ALL	GREENFIELD REPUBLICAN
12	1914	STORM BUGGY		Y							ALL	GREENFIELD REPUBLICAN
12	1914	STORM BUGGY		Y							ALL	GREENFIELD REPUBLICAN
1	1915	STORM BUGGY		Y							ALL	GREENFIELD REPUBLICAN
1	1915	STORM BUGGY		Y							ALL	GREENFIELD REPUBLICAN
1	1915	STORM BUGGY		Y							ALL	GREENFIELD REPUBLICAN
1	1915	STORM BUGGY		Y							ALL	GREENFIELD REPUBLICAN
2	1915	STORM BUGGY		Y							ALL	GREENFIELD REPUBLICAN
2	1915	STORM BUGGY		Y							ALL	GREENFIELD REPUBLICAN
2	1915					Y					ALL	GREENFIELD REPUBLICAN
2	1915			Y		Y					ALL	GREENFIELD REPUBLICAN
3	1915	NO. 4 GREENFIELD BUGGY		Y							ALL	GREENFIELD REPUBLICAN
3	1915	NO. 4 GREENFIELD BUGGY		Y							ALL	GREENFIELD REPUBLICAN

Month	Year	Vehicle Type	Specs	Prices	Sale	Other Services	Slogan Only	Promotions	TESTIMONIAL	Negro ID	Audience	Ad Source
3	1915			Y							ALL	GREENFIELD REPUBLICAN
3	1915	NO. 4 GREENFIELD BUGGY		Y							ALL	GREENFIELD REPUBLICAN
4	1915			Y		Y					ALL	GREENFIELD REPUBLICAN
4	1915			Y		Y					ALL	GREENFIELD REPUBLICAN
4	1915	AUTO REPAIR				Y					ALL	GREENFIELD REPUBLICAN
4	1915	AUTO REPAIR				Y					ALL	GREENFIELD REPUBLICAN
4	1915	NO. 4 GREENFIELD BUGGY		Y		Y					ALL	GREENFIELD REPUBLICAN
4	1915	NO. 1 END SPRING PIANO BUGGY		Y		Y					ALL	GREENFIELD REPUBLICAN
4	1915	NO. 37 PHAETON		Y		Y					ALL	GREENFIELD REPUBLICAN
4	1915	NO. 33 SURREY		Y		Y					ALL	GREENFIELD REPUBLICAN
5	1915	NO. 4 GREENFIELD BUGGY		Y		Y					ALL	GREENFIELD REPUBLICAN
5	1915	PHAETON		Y		Y					ALL	GREENFIELD REPUBLICAN
5	1915	SURREY		Y		Y					ALL	GREENFIELD REPUBLICAN
5	1915	SPRING WAGON		Y		Y					ALL	GREENFIELD REPUBLICAN
6	1915	AUTO / BUGGY REPAIR				Y					ALL	GREENFIELD REPUBLICAN
6	1915	PONY VEHICLE		Y		Y					ALL	GREENFIELD REPUBLICAN
6	1915	AUTO / BUGGY REPAIR				Y					ALL	GREENFIELD REPUBLICAN
6	1915	SPRING WAGON		Y		Y					ALL	GREENFIELD REPUBLICAN
7	1915	PHAETON		Y		Y					ALL	GREENFIELD REPUBLICAN
7	1915	AUTO / BUGGY REPAIR		Y		Y					ALL	GREENFIELD REPUBLICAN
7	1915	SPRING WAGON		Y		Y					ALL	GREENFIELD REPUBLICAN
7	1915	AUTO REPAIR									ALL	GREENFIELD REPUBLICAN
7	1915	DOCTOR'S BUGGY		Y					Y		PHYSICIANS	GREENFIELD REPUBLICAN
8	1915								Y		FARMERS	GREENFIELD REPUBLICAN
8	1915					Y					ALL	GREENFIELD REPUBLICAN

Month	Year	Vehicle Type	Specs	Prices	Sale	Other Services	Slogan Only	Promotions	TESTIMONIAL	Negro ID	Audience	Ad Source
8	1915						Y				ALL	GREENFIELD REPUBLICAN
8	1915						Y				ALL	GREENFIELD REPUBLICAN
9	1915						Y				ALL	GREENFIELD REPUBLICAN
9	1915						Y				ALL	GREENFIELD REPUBLICAN
9	1915						Y				ALL	GREENFIELD REPUBLICAN
9	1915						Y				ALL	GREENFIELD REPUBLICAN
9	1915	PATTERSON GREENFIELD / BUGGY				Y					ALL	GREENFIELD REPUBLICAN
10	1915	STORM BUGGY		Y		Y					ALL	GREENFIELD REPUBLICAN
10	1915	PATTERSON GREENFIELD									ALL	GREENFIELD REPUBLICAN
10	1915	STORM BUGGY		Y		Y					ALL	GREENFIELD REPUBLICAN
10	1915	PATTERSON GREENFIELD									ALL	GREENFIELD REPUBLICAN
10	1915	STORM BUGGY		Y		Y					ALL	GREENFIELD REPUBLICAN
10	1915	PATTERSON GREENFIELD									ALL	GREENFIELD REPUBLICAN
10	1915	STORM BUGGY		Y		Y					ALL	GREENFIELD REPUBLICAN
10	1915	PATTERSON GREENFIELD									ALL	GREENFIELD REPUBLICAN
11	1915	STORM BUGGY		Y		Y					ALL	GREENFIELD REPUBLICAN
11	1915	PATTERSON GREENFIELD									ALL	GREENFIELD REPUBLICAN
11	1915	STORM BUGGY		Y		Y					ALL	GREENFIELD REPUBLICAN
11	1915	PATTERSON GREENFIELD									ALL	GREENFIELD REPUBLICAN
11	1915	STORM BUGGY		Y		Y					ALL	GREENFIELD REPUBLICAN
11	1915	PATTERSON GREENFIELD									ALL	GREENFIELD REPUBLICAN
11	1915	STORM BUGGY		Y		Y					ALL	GREENFIELD REPUBLICAN
11	1915	PATTERSON GREENFIELD									ALL	GREENFIELD REPUBLICAN
12	1915	STORM BUGGY		Y		Y					ALL	GREENFIELD REPUBLICAN
12	1915	PATTERSON GREENFIELD									ALL	GREENFIELD REPUBLICAN

Month	Year	Vehicle Type	Specs	Prices	Sale	Other Services	Slogan Only	Promotions	TESTIMONIAL	Negro ID	Audience	Ad Source
12	1915	STORM BUGGY		Y		Y					ALL	GREENFIELD REPUBLICAN
12	1915	PATTERSON GREENFIELD									ALL	GREENFIELD REPUBLICAN
12	1915	STORM BUGGY		Y		Y					ALL	GREENFIELD REPUBLICAN
12	1915	PATTERSON GREENFIELD									ALL	GREENFIELD REPUBLICAN
12	1915	STORM BUGGY		Y		Y					ALL	GREENFIELD REPUBLICAN
12	1915	PATTERSON GREENFIELD									ALL	GREENFIELD REPUBLICAN
1	1917	AUTO / BUGGY REPAIR				Y					ALL	GREENFIELD REPUBLICAN
1	1917	AUTO / BUGGY REPAIR				Y					ALL	GREENFIELD REPUBLICAN
2	1917	AUTO / BUGGY REPAIR				Y					ALL	GREENFIELD REPUBLICAN
2	1917	AUTO / BUGGY REPAIR				Y					ALL	GREENFIELD REPUBLICAN
2	1917	AUTO / BUGGY REPAIR				Y					ALL	GREENFIELD REPUBLICAN
2	1917	AUTO / BUGGY REPAIR				Y					ALL	GREENFIELD REPUBLICAN
3	1917	AUTO / BUGGY REPAIR				Y					ALL	GREENFIELD REPUBLICAN
3	1917	PATTERSON GREENFIELD				Y					ALL	GREENFIELD REPUBLICAN
3	1917	AUTO / BUGGY REPAIR									ALL	GREENFIELD REPUBLICAN
3	1917	AUTO / BUGGY REPAIR									ALL	GREENFIELD REPUBLICAN
3	1917	PATTERSON GREENFIELD		Y		Y					ALL	GREENFIELD REPUBLICAN
4	1917	PATTERSON GREENFIELD		Y		Y					ALL	GREENFIELD REPUBLICAN
4	1917	PATTERSON GREENFIELD		Y		Y					ALL	GREENFIELD REPUBLICAN
4	1917	PATTERSON GREENFIELD		Y		Y					ALL	GREENFIELD REPUBLICAN
4	1917	PATTERSON GREENFIELD		Y		Y					ALL	GREENFIELD REPUBLICAN
5	1917	PATTERSON GREENFIELD		Y		Y					ALL	GREENFIELD REPUBLICAN
5	1917	PATTERSON GREENFIELD		Y		Y					ALL	GREENFIELD REPUBLICAN
5	1917	PATTERSON GREENFIELD		Y		Y					ALL	GREENFIELD REPUBLICAN
5	1917	PATTERSON GREENFIELD		Y		Y					ALL	GREENFIELD REPUBLICAN

Month	Year	Vehicle Type	Specs	Prices	Sale	Other Services	Slogan Only	Promotions	TESTIMONIAL	Negro ID	Audience	Ad Source
5	1917	PATTERSON GREENFIELD		Y		Y					ALL	GREENFIELD REPUBLICAN
6	1917	PATTERSON GREENFIELD		Y		Y					ALL	GREENFIELD REPUBLICAN
6	1917	PATTERSON GREENFIELD		Y		Y					ALL	GREENFIELD REPUBLICAN
6	1917	PATTERSON GREENFIELD		Y		Y					ALL	GREENFIELD REPUBLICAN
6	1917	PATTERSON GREENFIELD		Y		Y					ALL	GREENFIELD REPUBLICAN
7	1917	PATTERSON GREENFIELD		Y		Y					ALL	GREENFIELD REPUBLICAN
7	1917	PATTERSON GREENFIELD		Y		Y					ALL	GREENFIELD REPUBLICAN
7	1917	AUTO REPAIR									ALL	GREENFIELD REPUBLICAN
7	1917	AUTO / BUGGY REPAIR									ALL	GREENFIELD REPUBLICAN
8	1917	AUTO / BUGGY REPAIR									ALL	GREENFIELD REPUBLICAN
8	1917	AUTO / BUGGY REPAIR									ALL	GREENFIELD REPUBLICAN
8	1917	AUTO / BUGGY REPAIR									ALL	GREENFIELD REPUBLICAN
8	1917	AUTO / BUGGY REPAIR									ALL	GREENFIELD REPUBLICAN
9	1917	AUTO / BUGGY REPAIR									ALL	GREENFIELD REPUBLICAN
9	1917	AUTO / BUGGY REPAIR									ALL	GREENFIELD REPUBLICAN
9	1917	AUTO / BUGGY REPAIR									ALL	GREENFIELD REPUBLICAN
9	1917	AUTO / BUGGY REPAIR									ALL	GREENFIELD REPUBLICAN
10	1917	AUTO / BUGGY REPAIR									ALL	GREENFIELD REPUBLICAN
10	1917	AUTO / BUGGY REPAIR									ALL	GREENFIELD REPUBLICAN
10	1917	AUTO / BUGGY REPAIR									ALL	GREENFIELD REPUBLICAN
10	1917	AUTO / BUGGY REPAIR									ALL	GREENFIELD REPUBLICAN
11	1917	AUTO / BUGGY REPAIR									ALL	GREENFIELD REPUBLICAN
11	1917	AUTO / BUGGY REPAIR									ALL	GREENFIELD REPUBLICAN
11	1917	AUTO / BUGGY REPAIR									ALL	GREENFIELD REPUBLICAN
11	1917	AUTO / BUGGY REPAIR									ALL	GREENFIELD REPUBLICAN

Month	Year	Vehicle Type	Specs	Prices	Sale	Other Services	Slogan Only	Promotions	TESTIMONIAL	Negro ID	Audience	Ad Source
12	1917	AUTO / BUGGY REPAIR									ALL	GREENFIELD REPUBLICAN
12	1917	AUTO / BUGGY REPAIR									ALL	GREENFIELD REPUBLICAN
12	1917	AUTO / BUGGY REPAIR									ALL	GREENFIELD REPUBLICAN
12	1917	AUTO / BUGGY REPAIR									ALL	GREENFIELD REPUBLICAN
12	1917	AUTO / BUGGY REPAIR									ALL	GREENFIELD REPUBLICAN
1	1918	AUTO / BUGGY REPAIR									ALL	GREENFIELD REPUBLICAN
1	1918	AUTO / BUGGY REPAIR									ALL	GREENFIELD REPUBLICAN
1	1918	AUTO / BUGGY REPAIR									ALL	GREENFIELD REPUBLICAN
1	1918	AUTO / BUGGY REPAIR									ALL	GREENFIELD REPUBLICAN
1	1918	AUTO / BUGGY REPAIR									ALL	GREENFIELD REPUBLICAN
2	1918	AUTO / BUGGY REPAIR									ALL	GREENFIELD REPUBLICAN
2	1918	AUTO / BUGGY REPAIR									ALL	GREENFIELD REPUBLICAN
2	1918	AUTO / BUGGY REPAIR									ALL	GREENFIELD REPUBLICAN
2	1918	AUTO / BUGGY REPAIR									ALL	GREENFIELD REPUBLICAN
3	1918	AUTO / BUGGY REPAIR									ALL	GREENFIELD REPUBLICAN
3	1918	AUTO / BUGGY REPAIR									ALL	GREENFIELD REPUBLICAN
3	1918	AUTO / BUGGY REPAIR									ALL	GREENFIELD REPUBLICAN
3	1918	AUTO / BUGGY REPAIR									ALL	GREENFIELD REPUBLICAN
3	1918	AUTO / BUGGY REPAIR									ALL	GREENFIELD REPUBLICAN
3	1918	AUTO / BUGGY REPAIR									ALL	GREENFIELD REPUBLICAN
4	1918	AUTO / BUGGY REPAIR									ALL	GREENFIELD REPUBLICAN
4	1918	AUTO / BUGGY REPAIR									ALL	GREENFIELD REPUBLICAN
4	1918	AUTO / BUGGY REPAIR									ALL	GREENFIELD REPUBLICAN
4	1918	AUTO / BUGGY REPAIR									ALL	GREENFIELD REPUBLICAN
5	1918	AUTO / BUGGY REPAIR									ALL	GREENFIELD REPUBLICAN

Month	Year	Vehicle Type	Specs	Prices	Sale	Other Services	Slogan Only	Promotions	TESTIMONIAL	Negro ID	Audience	Ad Source
5	1918	AUTO / BUGGY REPAIR									ALL	GREENFIELD REPUBLICAN
5	1918	AUTO / BUGGY REPAIR									ALL	GREENFIELD REPUBLICAN
5	1918	AUTO / BUGGY REPAIR									ALL	GREENFIELD REPUBLICAN
5	1918	AUTO / BUGGY REPAIR									ALL	GREENFIELD REPUBLICAN
6	1918	AUTO / BUGGY REPAIR									ALL	GREENFIELD REPUBLICAN
6	1918	AUTO / BUGGY REPAIR									ALL	GREENFIELD REPUBLICAN
6	1918	AUTO / BUGGY REPAIR									ALL	GREENFIELD REPUBLICAN
6	1918	AUTO / BUGGY REPAIR									ALL	GREENFIELD REPUBLICAN
6	1918	AUTO / BUGGY REPAIR									ALL	GREENFIELD REPUBLICAN
7	1918	AUTO / BUGGY REPAIR									ALL	GREENFIELD REPUBLICAN
7	1918	AUTO / BUGGY REPAIR									ALL	GREENFIELD REPUBLICAN
7	1918	AUTO / BUGGY REPAIR									ALL	GREENFIELD REPUBLICAN
7	1918	AUTO / BUGGY REPAIR									ALL	GREENFIELD REPUBLICAN
8	1918	AUTO / BUGGY REPAIR									ALL	GREENFIELD REPUBLICAN
8	1918	AUTO / BUGGY REPAIR									ALL	GREENFIELD REPUBLICAN
8	1918	AUTO / BUGGY REPAIR									ALL	GREENFIELD REPUBLICAN
8	1918	AUTO REPAIR									ALL	GREENFIELD REPUBLICAN
8	1918	AUTO REPAIR									ALL	GREENFIELD REPUBLICAN
9	1918	AUTO / BUGGY REPAIR									ALL	GREENFIELD REPUBLICAN
9	1918	AUTO / BUGGY REPAIR									ALL	GREENFIELD REPUBLICAN
9	1918	AUTO / BUGGY REPAIR									ALL	GREENFIELD REPUBLICAN
9	1918	AUTO / BUGGY REPAIR									ALL	GREENFIELD REPUBLICAN
10	1918	AUTO / BUGGY REPAIR									ALL	GREENFIELD REPUBLICAN
10	1918	AUTO / BUGGY REPAIR									ALL	GREENFIELD REPUBLICAN
10	1918	AUTO / BUGGY REPAIR									ALL	GREENFIELD REPUBLICAN

Month	Year	Vehicle Type	Specs	Prices	Sale	Other Services	Slogan Only	Promotions	TESTIMONIAL	Negro ID	Audience	Ad Source
10	1918	AUTO / BUGGY REPAIR									ALL	GREENFIELD REPUBLICAN
11	1918	AUTO / BUGGY REPAIR									ALL	GREENFIELD REPUBLICAN
11	1918	AUTO / BUGGY REPAIR									ALL	GREENFIELD REPUBLICAN
11	1918	AUTO / BUGGY REPAIR									ALL	GREENFIELD REPUBLICAN
11	1918	AUTO / BUGGY REPAIR									ALL	GREENFIELD REPUBLICAN
12	1918	AUTO / BUGGY REPAIR									ALL	GREENFIELD REPUBLICAN
12	1918	AUTO / BUGGY REPAIR									ALL	GREENFIELD REPUBLICAN
12	1918	AUTO / BUGGY REPAIR									ALL	GREENFIELD REPUBLICAN
12	1918	AUTO / BUGGY REPAIR									ALL	GREENFIELD REPUBLICAN
1	1919	AUTO / BUGGY REPAIR									ALL	GREENFIELD REPUBLICAN
1	1919	AUTO / BUGGY REPAIR									ALL	GREENFIELD REPUBLICAN
1	1919	AUTO / BUGGY REPAIR									ALL	GREENFIELD REPUBLICAN
1	1919	AUTO / BUGGY REPAIR									ALL	GREENFIELD REPUBLICAN
2	1919	AUTO / BUGGY REPAIR									ALL	GREENFIELD REPUBLICAN
2	1919	AUTO / BUGGY REPAIR									ALL	GREENFIELD REPUBLICAN
2	1919	AUTO / BUGGY REPAIR									ALL	GREENFIELD REPUBLICAN
2	1919	AUTO / BUGGY REPAIR									ALL	GREENFIELD REPUBLICAN
3	1919	AUTO / BUGGY REPAIR									ALL	GREENFIELD REPUBLICAN
3	1919	AUTO / BUGGY REPAIR									ALL	GREENFIELD REPUBLICAN
3	1919	AUTO / BUGGY REPAIR									ALL	GREENFIELD REPUBLICAN
3	1919	AUTO / BUGGY REPAIR									ALL	GREENFIELD REPUBLICAN
4	1919	AUTO / BUGGY REPAIR									ALL	GREENFIELD REPUBLICAN
4	1919	BRISCOE AUTOMOBILE		Y		Y					ALL	GREENFIELD REPUBLICAN
4	1919	BRISCOE AUTOMOBILE		Y		Y					ALL	GREENFIELD REPUBLICAN
4	1919	BRISCOE AUTOMOBILE		Y		Y					ALL	GREENFIELD REPUBLICAN

Month	Year	Vehicle Type	Specs	Prices	Sale	Other Services	Slogan Only	Promotions	TESTIMONIAL	Negro ID	Audience	Ad Source
5	1919	AUTO / BUGGY REPAIR									ALL	GREENFIELD REPUBLICAN
5	1919	AUTO / BUGGY REPAIR									ALL	GREENFIELD REPUBLICAN
5	1919	AUTO / BUGGY REPAIR									ALL	GREENFIELD REPUBLICAN
5	1919	AUTO / BUGGY REPAIR									ALL	GREENFIELD REPUBLICAN
6	1919					Y					FARMERS	GREENFIELD REPUBLICAN
6	1919					Y					FARMERS	GREENFIELD REPUBLICAN
7	1919					Y					FARMERS	GREENFIELD REPUBLICAN
7	1919	AUTO / BUGGY REPAIR									ALL	GREENFIELD REPUBLICAN
7	1919	AUTO / BUGGY REPAIR									ALL	GREENFIELD REPUBLICAN
7	1919	AUTO / BUGGY REPAIR									ALL	GREENFIELD REPUBLICAN
8	1919	AUTO / BUGGY REPAIR									ALL	GREENFIELD REPUBLICAN
8	1919	AUTO / BUGGY REPAIR									ALL	GREENFIELD REPUBLICAN
8	1919	AUTO / BUGGY REPAIR									ALL	GREENFIELD REPUBLICAN
8	1919	AUTO / BUGGY REPAIR									ALL	GREENFIELD REPUBLICAN
8	1919	AUTO / BUGGY REPAIR									ALL	GREENFIELD REPUBLICAN
9	1919	AUTO / BUGGY REPAIR									ALL	GREENFIELD REPUBLICAN
9	1919	AUTO / BUGGY REPAIR									ALL	GREENFIELD REPUBLICAN
9	1919	AUTO / BUGGY REPAIR									ALL	GREENFIELD REPUBLICAN
9	1919	AUTO / BUGGY REPAIR									ALL	GREENFIELD REPUBLICAN
10	1919	AUTO / BUGGY REPAIR									ALL	GREENFIELD REPUBLICAN
10	1919	AUTO / BUGGY REPAIR									ALL	GREENFIELD REPUBLICAN
10	1919	AUTO / BUGGY REPAIR									ALL	GREENFIELD REPUBLICAN
10	1919	AUTO / BUGGY REPAIR									ALL	GREENFIELD REPUBLICAN
10	1919	AUTO / BUGGY REPAIR									ALL	GREENFIELD REPUBLICAN
11	1919	AUTO / BUGGY REPAIR									ALL	GREENFIELD REPUBLICAN

Month	Year	Vehicle Type	Specs	Prices	Sale	Other Services	Slogan Only	Promotions	TESTIMONIAL	Negro ID	Audience	Ad Source
11	1919	AUTO / BUGGY REPAIR									ALL	GREENFIELD REPUBLICAN
11	1919	AUTO / BUGGY REPAIR									ALL	GREENFIELD REPUBLICAN
11	1919	AUTO / BUGGY REPAIR									ALL	GREENFIELD REPUBLICAN
12	1919	AUTO / BUGGY REPAIR									ALL	GREENFIELD REPUBLICAN
12	1919	AUTO / BUGGY REPAIR									ALL	GREENFIELD REPUBLICAN
12	1919	AUTO / BUGGY REPAIR									ALL	GREENFIELD REPUBLICAN
12	1919	AUTO / BUGGY REPAIR									ALL	GREENFIELD REPUBLICAN
12	1919	AUTO / BUGGY REPAIR									ALL	GREENFIELD REPUBLICAN
1	1920	AUTO / BUGGY REPAIR									ALL	GREENFIELD REPUBLICAN
1	1920										ALL	GREENFIELD REPUBLICAN
1	1920										ALL	GREENFIELD REPUBLICAN
2	1920										ALL	GREENFIELD REPUBLICAN
2	1920										ALL	GREENFIELD REPUBLICAN
2	1920										ALL	GREENFIELD REPUBLICAN
2	1920										ALL	GREENFIELD REPUBLICAN
3	1920	BATTERY SHOP				Y					ALL	GREENFIELD REPUBLICAN
3	1920										ALL	GREENFIELD REPUBLICAN
3	1920										ALL	GREENFIELD REPUBLICAN
3	1920										ALL	GREENFIELD REPUBLICAN
3	1920										ALL	GREENFIELD REPUBLICAN
4	1920	BATTERY SHOP				Y					ALL	GREENFIELD REPUBLICAN
4	1920	BATTERY SHOP				Y					ALL	GREENFIELD REPUBLICAN
4	1920	BATTERY SHOP				Y					ALL	GREENFIELD REPUBLICAN
4	1920	BATTERY SHOP				Y					ALL	GREENFIELD REPUBLICAN
4	1920	BATTERY SHOP				Y					ALL	GREENFIELD REPUBLICAN

Month	Year	Vehicle Type	Specs	Prices	Sale	Other Services	Slogan Only	Promotions	TESTIMONIAL	Negro ID	Audience	Ad Source
4	1920	BATTERY SHOP				Y					ALL	GREENFIELD REPUBLICAN
5	1920	AUTO REPAIR				Y					ALL	GREENFIELD REPUBLICAN
5	1920	AUTO REPAIR				Y					ALL	GREENFIELD REPUBLICAN
5	1920	AUTO REPAIR				Y					ALL	GREENFIELD REPUBLICAN
5	1920	CAR WASH									ALL	GREENFIELD REPUBLICAN
5	1920	AUTO REPAIR				Y					ALL	GREENFIELD REPUBLICAN
6	1920	AUTO REPAIR				Y					ALL	GREENFIELD REPUBLICAN
6	1920	AUTO REPAIR				Y					ALL	GREENFIELD REPUBLICAN
6	1920	AUTO REPAIR				Y					ALL	GREENFIELD REPUBLICAN
6	1920	AUTO REPAIR				Y					ALL	GREENFIELD REPUBLICAN
7	1920	AUTO REPAIR				Y					ALL	GREENFIELD REPUBLICAN
7	1920	AUTO REPAIR				Y					ALL	GREENFIELD REPUBLICAN
7	1920	AUTO REPAIR				Y					ALL	GREENFIELD REPUBLICAN
7	1920	BATTERY SHOP									ALL	GREENFIELD REPUBLICAN
7	1920	AUTO REPAIR				Y					ALL	GREENFIELD REPUBLICAN
8	1920	AUTO REPAIR				Y					ALL	GREENFIELD REPUBLICAN
8	1920	AUTO REPAIR				Y					ALL	GREENFIELD REPUBLICAN
8	1920	BATTERY SHOP									ALL	GREENFIELD REPUBLICAN
8	1920	AUTO REPAIR				Y					ALL	GREENFIELD REPUBLICAN
8	1920	BATTERY SHOP									ALL	GREENFIELD REPUBLICAN
8	1920	AUTO REPAIR				Y					ALL	GREENFIELD REPUBLICAN
8	1920	BATTERY SHOP									ALL	GREENFIELD REPUBLICAN
8	1920	AUTO REPAIR				Y					ALL	GREENFIELD REPUBLICAN
8	1920	BATTERY SHOP									ALL	GREENFIELD REPUBLICAN
9	1920	AUTO REPAIR				Y					ALL	GREENFIELD REPUBLICAN

Month	Year	Vehicle Type	Specs	Prices	Sale	Other Services	Slogan Only	Promotions	TESTIMONIAL	Negro ID	Audience	Ad Source
9	1920	BATTERY SHOP									ALL	GREENFIELD REPUBLICAN
9	1920	AUTO REPAIR				Y					ALL	GREENFIELD REPUBLICAN
9	1920	BATTERY SHOP									ALL	GREENFIELD REPUBLICAN
9	1920	AUTO REPAIR				Y					ALL	GREENFIELD REPUBLICAN
9	1920	BATTERY SHOP									ALL	GREENFIELD REPUBLICAN
9	1920	AUTO REPAIR				Y					ALL	GREENFIELD REPUBLICAN
10	1920	BATTERY SHOP									ALL	GREENFIELD REPUBLICAN
10	1920	AUTO REPAIR				Y					ALL	GREENFIELD REPUBLICAN
10	1920	USED CARS				Y					ALL	GREENFIELD REPUBLICAN
10	1920	BATTERY SHOP				Y					ALL	GREENFIELD REPUBLICAN
10	1920	BATTERY SHOP				Y					ALL	GREENFIELD REPUBLICAN
10	1920	BATTERY SHOP				Y					ALL	GREENFIELD REPUBLICAN
11	1920	BATTERY SHOP				Y					ALL	GREENFIELD REPUBLICAN
11	1920	BATTERY SHOP				Y					ALL	GREENFIELD REPUBLICAN
11	1920	BATTERY SHOP				Y					ALL	GREENFIELD REPUBLICAN
11	1920	BATTERY SHOP				Y					ALL	GREENFIELD REPUBLICAN
12	1920	BATTERY SHOP				Y					ALL	GREENFIELD REPUBLICAN
12	1920	BATTERY SHOP				Y					ALL	GREENFIELD REPUBLICAN
12	1920	BATTERY SHOP				Y					ALL	GREENFIELD REPUBLICAN
12	1920	BATTERY SHOP				Y					ALL	GREENFIELD REPUBLICAN
12	1920	BATTERY SHOP				Y					ALL	GREENFIELD REPUBLICAN
1	1921	BATTERY SHOP				Y					ALL	GREENFIELD REPUBLICAN
1	1921	BATTERY SHOP				Y					ALL	GREENFIELD REPUBLICAN
1	1921	BATTERY SHOP				Y					ALL	GREENFIELD REPUBLICAN
1	1921	BATTERY SHOP				Y					ALL	GREENFIELD REPUBLICAN

Month	Year	Vehicle Type	Specs	Prices	Sale	Other Services	Slogan Only	Promotions	TESTIMONIAL	Negro ID	Audience	Ad Source
1	1921	BATTERY SHOP				Y					ALL	GREENFIELD REPUBLICAN
2	1921	BATTERY SHOP				Y					ALL	GREENFIELD REPUBLICAN
2	1921	BATTERY SHOP				Y					ALL	GREENFIELD REPUBLICAN
2	1921	BATTERY SHOP				Y					ALL	GREENFIELD REPUBLICAN
2	1921	BATTERY SHOP				Y					ALL	GREENFIELD REPUBLICAN
3	1921	BATTERY SHOP				Y					ALL	GREENFIELD REPUBLICAN
3	1921	BATTERY SHOP				Y					ALL	GREENFIELD REPUBLICAN
3	1921	BATTERY SHOP				Y					ALL	GREENFIELD REPUBLICAN
3	1921	BATTERY SHOP				Y					ALL	GREENFIELD REPUBLICAN
3	1921	BATTERY SHOP				Y					ALL	GREENFIELD REPUBLICAN
4	1921	BATTERY SHOP				Y					ALL	GREENFIELD REPUBLICAN
4	1921	BATTERY SHOP				Y					ALL	GREENFIELD REPUBLICAN
4	1921	BATTERY SHOP				Y					ALL	GREENFIELD REPUBLICAN
4	1921	BATTERY SHOP				Y					ALL	GREENFIELD REPUBLICAN
5	1921	BATTERY SHOP				Y					ALL	GREENFIELD REPUBLICAN
5	1921	BATTERY SHOP				Y					ALL	GREENFIELD REPUBLICAN
5	1921	BATTERY SHOP				Y					ALL	GREENFIELD REPUBLICAN
5	1921	BATTERY SHOP				Y					ALL	GREENFIELD REPUBLICAN
6	1921	BATTERY SHOP				Y					ALL	GREENFIELD REPUBLICAN
6	1921	BATTERY SHOP				Y					ALL	GREENFIELD REPUBLICAN
6	1921	AUTO / BUGGY REPAIR									ALL	GREENFIELD REPUBLICAN
6	1921	AUTO / BUGGY REPAIR									ALL	GREENFIELD REPUBLICAN
7	1921	AUTO / BUGGY REPAIR									ALL	GREENFIELD REPUBLICAN
7	1921	AUTO / BUGGY REPAIR									ALL	GREENFIELD REPUBLICAN
7	1921	AUTO / BUGGY REPAIR									ALL	GREENFIELD REPUBLICAN

Month	Year	Vehicle Type	Specs	Prices	Sale	Other Services	Slogan Only	Promotions	TESTIMONIAL	Negro ID	Audience	Ad Source
7	1921	AUTO / BUGGY REPAIR									ALL	GREENFIELD REPUBLICAN
8	1921	AUTO / BUGGY REPAIR									ALL	GREENFIELD REPUBLICAN
8	1921	AUTO / BUGGY REPAIR									ALL	GREENFIELD REPUBLICAN
8	1921	AUTO / BUGGY REPAIR									ALL	GREENFIELD REPUBLICAN
8	1921	AUTO / BUGGY REPAIR									ALL	GREENFIELD REPUBLICAN
9	1921	AUTO / BUGGY REPAIR									ALL	GREENFIELD REPUBLICAN
9	1921	AUTO / BUGGY REPAIR									ALL	GREENFIELD REPUBLICAN
9	1921	AUTO / BUGGY REPAIR									ALL	GREENFIELD REPUBLICAN
9	1921	AUTO / BUGGY REPAIR									ALL	GREENFIELD REPUBLICAN
9	1921	AUTO / BUGGY REPAIR									ALL	GREENFIELD REPUBLICAN
10	1921	AUTO / BUGGY REPAIR									ALL	GREENFIELD REPUBLICAN
10	1921	AUTO / BUGGY REPAIR									ALL	GREENFIELD REPUBLICAN
10	1921	AUTO / BUGGY REPAIR									ALL	GREENFIELD REPUBLICAN
10	1921	AUTO / BUGGY REPAIR									ALL	GREENFIELD REPUBLICAN
11	1921	AUTO / BUGGY REPAIR									ALL	GREENFIELD REPUBLICAN
11	1921	AUTO / BUGGY REPAIR									ALL	GREENFIELD REPUBLICAN
11	1921	AUTO / BUGGY REPAIR									ALL	GREENFIELD REPUBLICAN
11	1921	AUTO / BUGGY REPAIR									ALL	GREENFIELD REPUBLICAN
12	1921	AUTO / BUGGY REPAIR									ALL	GREENFIELD REPUBLICAN
12	1921	AUTO / BUGGY REPAIR									ALL	GREENFIELD REPUBLICAN
12	1921	AUTO / BUGGY REPAIR									ALL	GREENFIELD REPUBLICAN
12	1921	AUTO / BUGGY REPAIR									ALL	GREENFIELD REPUBLICAN
12	1921	AUTO / BUGGY REPAIR									ALL	GREENFIELD REPUBLICAN
1	1922	AUTO / BUGGY REPAIR									ALL	GREENFIELD REPUBLICAN
1	1922	AUTO / BUGGY REPAIR									ALL	GREENFIELD REPUBLICAN

Month	Year	Vehicle Type	Specs	Prices	Sale	Other Services	Slogan Only	Promotions	TESTIMONIAL	Negro ID	Audience	Ad Source
1	1922	AUTO / BUGGY REPAIR									ALL	GREENFIELD REPUBLICAN
1	1922	AUTO / BUGGY REPAIR									ALL	GREENFIELD REPUBLICAN
2	1922	AUTO / BUGGY REPAIR									ALL	GREENFIELD REPUBLICAN
2	1922	AUTO / BUGGY REPAIR									ALL	GREENFIELD REPUBLICAN
2	1922	AUTO / BUGGY REPAIR									ALL	GREENFIELD REPUBLICAN
10	1929	AUTO REPAIR				Y					ALL	GREENFIELD REPUBLICAN
10	1929	AUTO REPAIR				Y					ALL	GREENFIELD REPUBLICAN
10	1929	AUTO REPAIR				Y					ALL	GREENFIELD REPUBLICAN
11	1929	AUTO REPAIR				Y					ALL	GREENFIELD REPUBLICAN
11	1929	AUTO REPAIR				Y					ALL	GREENFIELD REPUBLICAN
11	1929	AUTO REPAIR				Y					ALL	GREENFIELD REPUBLICAN
11	1929	AUTO REPAIR				Y					ALL	GREENFIELD REPUBLICAN
12	1929	AUTO REPAIR				Y					ALL	GREENFIELD REPUBLICAN
12	1929	AUTO REPAIR				Y					ALL	GREENFIELD REPUBLICAN
12	1929	AUTO REPAIR				Y					ALL	GREENFIELD REPUBLICAN
12	1929	AUTO REPAIR				Y					ALL	GREENFIELD REPUBLICAN
12	1929	AUTO REPAIR				Y					ALL	GREENFIELD REPUBLICAN
1	1930	AUTO REPAIR				Y					ALL	GREENFIELD REPUBLICAN
1	1930	AUTO REPAIR				Y					ALL	GREENFIELD REPUBLICAN
1	1930	AUTO REPAIR				Y					ALL	GREENFIELD REPUBLICAN
1	1930	AUTO REPAIR				Y					ALL	GREENFIELD REPUBLICAN
1	1930	AUTO REPAIR				Y					ALL	GREENFIELD REPUBLICAN
2	1930	AUTO REPAIR				Y					ALL	GREENFIELD REPUBLICAN
2	1930	AUTO REPAIR				Y					ALL	GREENFIELD REPUBLICAN
2	1930	AUTO REPAIR				Y					ALL	GREENFIELD REPUBLICAN

Month	Year	Vehicle Type	Specs	Prices	Sale	Other Services	Slogan Only	Promotions	TESTIMONIAL	Negro ID	Audience	Ad Source
2	1930	AUTO REPAIR				Y					ALL	GREENFIELD REPUBLICAN
3	1930	AUTO REPAIR				Y					ALL	GREENFIELD REPUBLICAN
3	1930	AUTO REPAIR				Y					ALL	GREENFIELD REPUBLICAN
3	1930	AUTO REPAIR				Y					ALL	GREENFIELD REPUBLICAN
3	1930	AUTO REPAIR				Y					ALL	GREENFIELD REPUBLICAN
4	1930	AUTO REPAIR				Y					ALL	GREENFIELD REPUBLICAN
4	1930	AUTO REPAIR				Y					ALL	GREENFIELD REPUBLICAN
4	1930	AUTO REPAIR				Y					ALL	GREENFIELD REPUBLICAN
4	1930	AUTO REPAIR				Y					ALL	GREENFIELD REPUBLICAN
4	1930	AUTO REPAIR				Y					ALL	GREENFIELD REPUBLICAN
5	1930	AUTO REPAIR				Y					ALL	GREENFIELD REPUBLICAN
5	1930	AUTO REPAIR				Y					ALL	GREENFIELD REPUBLICAN
5	1930	AUTO REPAIR				Y					ALL	GREENFIELD REPUBLICAN
5	1930	AUTO REPAIR				Y					ALL	GREENFIELD REPUBLICAN
6	1930	AUTO REPAIR				Y					ALL	GREENFIELD REPUBLICAN
6	1930	AUTO REPAIR				Y					ALL	GREENFIELD REPUBLICAN
6	1930	AUTO REPAIR				Y					ALL	GREENFIELD REPUBLICAN
6	1930	AUTO REPAIR				Y					ALL	GREENFIELD REPUBLICAN
7	1930	AUTO REPAIR				Y					ALL	GREENFIELD REPUBLICAN
7	1930	AUTO REPAIR				Y					ALL	GREENFIELD REPUBLICAN
7	1930	AUTO REPAIR				Y					ALL	GREENFIELD REPUBLICAN
7	1930	AUTO REPAIR				Y					ALL	GREENFIELD REPUBLICAN
7	1930	AUTO REPAIR				Y					ALL	GREENFIELD REPUBLICAN
8	1930	AUTO REPAIR				Y					ALL	GREENFIELD REPUBLICAN
8	1930	AUTO REPAIR				Y					ALL	GREENFIELD REPUBLICAN

Month	Year	Vehicle Type	Specs	Prices	Sale	Other Services	Slogan Only	Promotions	TESTIMONIAL	Negro ID	Audience	Ad Source
8	1930	AUTO REPAIR				Y					ALL	GREENFIELD REPUBLICAN
8	1930	AUTO REPAIR				Y					ALL	GREENFIELD REPUBLICAN
9	1930	AUTO REPAIR				Y					ALL	GREENFIELD REPUBLICAN
9	1930	AUTO REPAIR				Y					ALL	GREENFIELD REPUBLICAN
9	1930	AUTO REPAIR				Y					ALL	GREENFIELD REPUBLICAN
9	1930	AUTO REPAIR				Y					ALL	GREENFIELD REPUBLICAN
10	1930	AUTO REPAIR				Y					ALL	GREENFIELD REPUBLICAN
10	1930	AUTO REPAIR				Y					ALL	GREENFIELD REPUBLICAN
10	1930	AUTO REPAIR				Y					ALL	GREENFIELD REPUBLICAN
10	1930	AUTO REPAIR				Y					ALL	GREENFIELD REPUBLICAN
10	1930	AUTO REPAIR				Y					ALL	GREENFIELD REPUBLICAN
11	1930	AUTO REPAIR				Y					ALL	GREENFIELD REPUBLICAN
11	1930	AUTO REPAIR				Y					ALL	GREENFIELD REPUBLICAN
11	1930	AUTO REPAIR				Y					ALL	GREENFIELD REPUBLICAN
11	1930	AUTO REPAIR				Y					ALL	GREENFIELD REPUBLICAN
12	1930	AUTO REPAIR				Y					ALL	GREENFIELD REPUBLICAN
12	1930	AUTO REPAIR				Y					ALL	GREENFIELD REPUBLICAN
12	1930	AUTO REPAIR				Y					ALL	GREENFIELD REPUBLICAN
12	1930	AUTO REPAIR				Y					ALL	GREENFIELD REPUBLICAN
12	1930	AUTO REPAIR				Y					ALL	GREENFIELD REPUBLICAN
1	1931	AUTO REPAIR				Y					ALL	GREENFIELD REPUBLICAN
1	1931	AUTO REPAIR				Y					ALL	GREENFIELD REPUBLICAN
1	1931	AUTO REPAIR				Y					ALL	GREENFIELD REPUBLICAN
1	1931	AUTO REPAIR				Y					ALL	GREENFIELD REPUBLICAN
1	1931	AUTO REPAIR				Y					ALL	GREENFIELD REPUBLICAN

Month	Year	Vehicle Type	Specs	Prices	Sale	Other Services	Slogan Only	Promotions	TESTIMONIAL	Negro ID	Audience	Ad Source
2	1931	AUTO REPAIR				Y					ALL	GREENFIELD REPUBLICAN
2	1931	AUTO REPAIR				Y					ALL	GREENFIELD REPUBLICAN
2	1931	AUTO REPAIR				Y					ALL	GREENFIELD REPUBLICAN
2	1931	AUTO REPAIR				Y					ALL	GREENFIELD REPUBLICAN
3	1931	AUTO REPAIR				Y					ALL	GREENFIELD REPUBLICAN
3	1931	AUTO REPAIR				Y					ALL	GREENFIELD REPUBLICAN
3	1931	AUTO REPAIR				Y					ALL	GREENFIELD REPUBLICAN
3	1931	AUTO REPAIR				Y					ALL	GREENFIELD REPUBLICAN
4	1931	AUTO REPAIR				Y					ALL	GREENFIELD REPUBLICAN
4	1931	AUTO REPAIR				Y					ALL	GREENFIELD REPUBLICAN
4	1931	AUTO REPAIR				Y					ALL	GREENFIELD REPUBLICAN
4	1931	AUTO REPAIR				Y					ALL	GREENFIELD REPUBLICAN
4	1931	AUTO REPAIR				Y					ALL	GREENFIELD REPUBLICAN
5	1931	AUTO REPAIR				Y					ALL	GREENFIELD REPUBLICAN
5	1931	AUTO REPAIR				Y					ALL	GREENFIELD REPUBLICAN
5	1931	AUTO REPAIR				Y					ALL	GREENFIELD REPUBLICAN
5	1931	AUTO REPAIR				Y					ALL	GREENFIELD REPUBLICAN
6	1931	AUTO REPAIR				Y					ALL	GREENFIELD REPUBLICAN
6	1931	AUTO REPAIR				Y					ALL	GREENFIELD REPUBLICAN
6	1931	AUTO REPAIR				Y					ALL	GREENFIELD REPUBLICAN
6	1931	AUTO REPAIR				Y					ALL	GREENFIELD REPUBLICAN
7	1931	AUTO REPAIR				Y					ALL	GREENFIELD REPUBLICAN
7	1931	AUTO REPAIR				Y					ALL	GREENFIELD REPUBLICAN
7	1931	AUTO REPAIR				Y					ALL	GREENFIELD REPUBLICAN
7	1931	AUTO REPAIR				Y					ALL	GREENFIELD REPUBLICAN

Month	Year	Vehicle Type	Specs	Prices	Sale	Other Services	Slogan Only	Promotions	TESTIMONIAL	Negro ID	Audience	Ad Source
7	1931	AUTO REPAIR				Y					ALL	GREENFIELD REPUBLICAN
8	1931	AUTO REPAIR				Y					ALL	GREENFIELD REPUBLICAN
8	1931	AUTO REPAIR				Y					ALL	GREENFIELD REPUBLICAN
8	1931	AUTO REPAIR				Y					ALL	GREENFIELD REPUBLICAN
8	1931	AUTO REPAIR				Y					ALL	GREENFIELD REPUBLICAN
9	1931	AUTO REPAIR				Y					ALL	GREENFIELD REPUBLICAN
9	1931	AUTO REPAIR				Y					ALL	GREENFIELD REPUBLICAN
9	1931	AUTO REPAIR				Y					ALL	GREENFIELD REPUBLICAN
9	1931	AUTO REPAIR				Y					ALL	GREENFIELD REPUBLICAN
10	1931	AUTO REPAIR				Y					ALL	GREENFIELD REPUBLICAN
10	1931	AUTO REPAIR				Y					ALL	GREENFIELD REPUBLICAN
10	1931	AUTO REPAIR				Y					ALL	GREENFIELD REPUBLICAN
10	1931	AUTO REPAIR				Y					ALL	GREENFIELD REPUBLICAN
10	1931	AUTO REPAIR				Y					ALL	GREENFIELD REPUBLICAN
11	1931	AUTO REPAIR				Y					ALL	GREENFIELD REPUBLICAN
11	1931	AUTO REPAIR				Y					ALL	GREENFIELD REPUBLICAN
11	1931	AUTO REPAIR				Y					ALL	GREENFIELD REPUBLICAN
11	1931	AUTO REPAIR				Y					ALL	GREENFIELD REPUBLICAN
12	1931	AUTO REPAIR				Y					ALL	GREENFIELD REPUBLICAN
12	1931	AUTO REPAIR				Y					ALL	GREENFIELD REPUBLICAN
12	1931	AUTO REPAIR				Y					ALL	GREENFIELD REPUBLICAN
12	1931	AUTO REPAIR				Y					ALL	GREENFIELD REPUBLICAN
1	1932	AUTO REPAIR				Y					ALL	GREENFIELD REPUBLICAN
1	1932	AUTO REPAIR				Y					ALL	GREENFIELD REPUBLICAN
1	1932	AUTO REPAIR				Y					ALL	GREENFIELD REPUBLICAN

Month	Year	Vehicle Type	Specs	Prices	Sale	Other Services	Slogan Only	Promotions	TESTIMONIAL	Negro ID	Audience	Ad Source
1	1932	AUTO REPAIR				Y					ALL	GREENFIELD REPUBLICAN
1	1932	AUTO REPAIR				Y					ALL	GREENFIELD REPUBLICAN
2	1932	AUTO REPAIR				Y					ALL	GREENFIELD REPUBLICAN
2	1932	AUTO REPAIR				Y					ALL	GREENFIELD REPUBLICAN
2	1932	AUTO REPAIR				Y					ALL	GREENFIELD REPUBLICAN
2	1932	AUTO REPAIR									ALL	GREENFIELD REPUBLICAN
3	1932	AUTO REPAIR									ALL	GREENFIELD REPUBLICAN
3	1932	AUTO REPAIR									ALL	GREENFIELD REPUBLICAN
3	1932	AUTO REPAIR									ALL	GREENFIELD REPUBLICAN
3	1932	AUTO REPAIR									ALL	GREENFIELD REPUBLICAN
4	1932	AUTO REPAIR									ALL	GREENFIELD REPUBLICAN
4	1932	AUTO REPAIR									ALL	GREENFIELD REPUBLICAN
4	1932	AUTO REPAIR									ALL	GREENFIELD REPUBLICAN
4	1932	AUTO REPAIR									ALL	GREENFIELD REPUBLICAN
4	1932	AUTO REPAIR									ALL	GREENFIELD REPUBLICAN
5	1932	AUTO REPAIR									ALL	GREENFIELD REPUBLICAN
5	1932	AUTO REPAIR									ALL	GREENFIELD REPUBLICAN
5	1932	AUTO REPAIR									ALL	GREENFIELD REPUBLICAN
5	1932	AUTO REPAIR									ALL	GREENFIELD REPUBLICAN
6	1932	AUTO REPAIR									ALL	GREENFIELD REPUBLICAN
6	1932	AUTO REPAIR									ALL	GREENFIELD REPUBLICAN
6	1932	AUTO REPAIR									ALL	GREENFIELD REPUBLICAN
6	1932	AUTO REPAIR									ALL	GREENFIELD REPUBLICAN
7	1932	AUTO REPAIR									ALL	GREENFIELD REPUBLICAN
7	1932	AUTO REPAIR									ALL	GREENFIELD REPUBLICAN

Month	Year	Vehicle Type	Specs	Prices	Sale	Other Services	Slogan Only	Promotions	TESTIMONIAL	Negro ID	Audience	Ad Source
7	1932	AUTO REPAIR									ALL	GREENFIELD REPUBLICAN
7	1932	AUTO REPAIR									ALL	GREENFIELD REPUBLICAN
7	1932	AUTO REPAIR									ALL	GREENFIELD REPUBLICAN
8	1932	AUTO REPAIR									ALL	GREENFIELD REPUBLICAN
8	1932	AUTO REPAIR									ALL	GREENFIELD REPUBLICAN
8	1932	AUTO REPAIR									ALL	GREENFIELD REPUBLICAN
8	1932	AUTO REPAIR									ALL	GREENFIELD REPUBLICAN
9	1932	AUTO REPAIR									ALL	GREENFIELD REPUBLICAN
9	1932	AUTO REPAIR									ALL	GREENFIELD REPUBLICAN
9	1932	AUTO REPAIR									ALL	GREENFIELD REPUBLICAN
9	1932	AUTO REPAIR									ALL	GREENFIELD REPUBLICAN
10	1932	AUTO REPAIR									ALL	GREENFIELD REPUBLICAN
10	1932	AUTO REPAIR									ALL	GREENFIELD REPUBLICAN
10	1932	AUTO REPAIR									ALL	GREENFIELD REPUBLICAN
10	1932	AUTO REPAIR									ALL	GREENFIELD REPUBLICAN
10	1932	AUTO REPAIR									ALL	GREENFIELD REPUBLICAN
11	1932	AUTO REPAIR									ALL	GREENFIELD REPUBLICAN
11	1932	AUTO REPAIR									ALL	GREENFIELD REPUBLICAN
11	1932	AUTO REPAIR									ALL	GREENFIELD REPUBLICAN
12	1909	WINTER BUGGY		Y							ALL	HILLSBORO DISPATCH
12	1909	WINTER BUGGY		Y					Y		ALL	HILLSBORO DISPATCH
1	1910	WINTER BUGGY		Y					Y		ALL	HILLSBORO DISPATCH
1	1910	WINTER BUGGY		Y					Y		ALL	HILLSBORO DISPATCH
1	1910	WINTER BUGGY		Y					Y		ALL	HILLSBORO DISPATCH
2	1910	WINTER BUGGY		Y							ALL	HILLSBORO DISPATCH

Month	Year	Vehicle Type	Specs	Prices	Sale	Other Services	Slogan Only	Promotions	TESTIMONIAL	Negro ID	Audience	Ad Source
2	1910	BUGGY TIRES	Y	Y							ALL	HILLSBORO DISPATCH
2	1910	BUGGY TIRES	Y	Y							ALL	HILLSBORO DISPATCH
2	1910	BUGGY TIRES	Y	Y							ALL	HILLSBORO DISPATCH
3	1910	BUGGY TIRES	Y	Y							ALL	HILLSBORO DISPATCH
3	1910				Y				Y		ALL	HILLSBORO DISPATCH
3	1910				Y				Y		ALL	HILLSBORO DISPATCH
3	1910				Y				Y		ALL	HILLSBORO DISPATCH
3	1910				Y				Y		ALL	HILLSBORO DISPATCH
4	1910				Y				Y		ALL	HILLSBORO DISPATCH
4	1910				Y				Y		ALL	HILLSBORO DISPATCH
4	1910										FARMERS	HILLSBORO DISPATCH
4	1910				Y				Y		ALL	HILLSBORO DISPATCH
5	1910				Y				Y		ALL	HILLSBORO DISPATCH
5	1910				Y				Y		ALL	HILLSBORO DISPATCH
5	1910				Y				Y		ALL	HILLSBORO DISPATCH
5	1910				Y				Y		ALL	HILLSBORO DISPATCH
5	1910				Y				Y		ALL	HILLSBORO DISPATCH
6	1910	BUGGY REPAIR				Y					ALL	HILLSBORO DISPATCH
6	1910	BUGGY REPAIR				Y					ALL	HILLSBORO DISPATCH
6	1910	BUGGY REPAIR				Y					ALL	HILLSBORO DISPATCH
6	1910			Y	Y						ALL	HILLSBORO DISPATCH
7	1910			Y	Y						ALL	HILLSBORO DISPATCH
7	1910				Y						ALL	HILLSBORO DISPATCH
7	1910				Y						ALL	HILLSBORO DISPATCH
8	1910				Y						ALL	HILLSBORO DISPATCH

Month	Year	Vehicle Type	Specs	Prices	Sale	Other Services	Slogan Only	Promotions	TESTIMONIAL	Negro ID	Audience	Ad Source
8	1910				Y						ALL	HILLSBORO DISPATCH
8	1910				Y						ALL	HILLSBORO DISPATCH
8	1910				Y						ALL	HILLSBORO DISPATCH
8	1910				Y						ALL	HILLSBORO DISPATCH
9	1910				Y						ALL	HILLSBORO DISPATCH
9	1910				Y						ALL	HILLSBORO DISPATCH
9	1910				Y						ALL	HILLSBORO DISPATCH
9	1910				Y						ALL	HILLSBORO DISPATCH
10	1910	SURREY		Y	Y						ALL	HILLSBORO DISPATCH
10	1910	SURREY		Y	Y						ALL	HILLSBORO DISPATCH
10	1910	SURREY		Y	Y						ALL	HILLSBORO DISPATCH
10	1910	SURREY		Y	Y						ALL	HILLSBORO DISPATCH
11	1910	WINTER BUGGY									ALL	HILLSBORO DISPATCH
11	1910	WINTER BUGGY									ALL	HILLSBORO DISPATCH
11	1910	WINTER BUGGY	Y	Y							ALL	HILLSBORO DISPATCH
11	1910	WINTER BUGGY	Y	Y							ALL	HILLSBORO DISPATCH
11	1910	WINTER BUGGY	Y	Y	Y						ALL	HILLSBORO DISPATCH
11	1910	WINTER BUGGY	Y	Y	Y						ALL	HILLSBORO DISPATCH
12	1910	WINTER BUGGY									ALL	HILLSBORO DISPATCH
12	1910	WINTER BUGGY	Y	Y							ALL	HILLSBORO DISPATCH
12	1910	WINTER BUGGY	Y	Y							ALL	HILLSBORO DISPATCH
12	1910	WINTER BUGGY	Y	Y							ALL	HILLSBORO DISPATCH
1	1911	WINTER BUGGY		Y	Y						ALL	HILLSBORO DISPATCH
1	1911	WINTER BUGGY		Y							ALL	HILLSBORO DISPATCH
1	1911	WINTER BUGGY		Y					Y		ALL	HILLSBORO DISPATCH

Month	Year	Vehicle Type	Specs	Prices	Sale	Other Services	Slogan Only	Promotions	TESTIMONIAL	Negro ID	Audience	Ad Source
1	1911	WINTER BUGGY		Y							ALL	HILLSBORO DISPATCH
1	1911	WINTER BUGGY									ALL	HILLSBORO DISPATCH
2	1911				Y						ALL	HILLSBORO DISPATCH
2	1911				Y						ALL	HILLSBORO DISPATCH
2	1911				Y						ALL	HILLSBORO DISPATCH
2	1911	BUGGY REPAIR		Y		Y					ALL	HILLSBORO DISPATCH
3	1911	SURREY									ALL	HILLSBORO DISPATCH
3	1911	SURREY									ALL	HILLSBORO DISPATCH
3	1911	SURREY									ALL	HILLSBORO DISPATCH
3	1911	BUGGY REPAIR		Y							ALL	HILLSBORO DISPATCH
4	1911	BUGGY REPAIR									ALL	HILLSBORO DISPATCH
4	1911	BUGGY REPAIR									ALL	HILLSBORO DISPATCH
4	1911	BUGGY REPAIR									ALL	HILLSBORO DISPATCH
4	1911	BUGGY REPAIR									ALL	HILLSBORO DISPATCH
5	1911	BUGGY REPAIR									ALL	HILLSBORO DISPATCH
5	1911	NO. 4 BUGGY		Y							ALL	HILLSBORO DISPATCH
5	1911				Y						ALL	HILLSBORO DISPATCH
5	1911				Y						ALL	HILLSBORO DISPATCH
5	1911				Y						FARMERS	HILLSBORO DISPATCH
6	1911	NO. 1 BUGGY	Y	Y							ALL	HILLSBORO DISPATCH
6	1911	NO. 2 BUGGY	Y	Y							ALL	HILLSBORO DISPATCH
6	1911	NO. 4 BUGGY		Y							ALL	HILLSBORO DISPATCH
6	1911										ALL	HILLSBORO DISPATCH
7	1911				Y						ALL	HILLSBORO DISPATCH
7	1911			Y	Y						ALL	HILLSBORO DISPATCH

Month	Year	Vehicle Type	Specs	Prices	Sale	Other Services	Slogan Only	Promotions	TESTIMONIAL	Negro ID	Audience	Ad Source
7	1911	NO. 1 BUGGY	Y	Y							ALL	HILLSBORO DISPATCH
7	1911	SURREY		Y							ALL	HILLSBORO DISPATCH
8	1911	SURREY		Y							ALL	HILLSBORO DISPATCH
8	1911	SURREY		Y							ALL	HILLSBORO DISPATCH
8	1911	SURREY		Y							ALL	HILLSBORO DISPATCH
8	1911	SURREY		Y							ALL	HILLSBORO DISPATCH
8	1911	SURREY		Y							ALL	HILLSBORO DISPATCH
9	1911			Y	Y						ALL	HILLSBORO DISPATCH
9	1911			Y	Y						ALL	HILLSBORO DISPATCH
9	1911			Y	Y						ALL	HILLSBORO DISPATCH
9	1911	WINTER BUGGY			Y						ALL	HILLSBORO DISPATCH
10	1911	WINTER BUGGY		Y							ALL	HILLSBORO DISPATCH
10	1911	WINTER BUGGY		Y							ALL	HILLSBORO DISPATCH
10	1911	WINTER BUGGY		Y							ALL	HILLSBORO DISPATCH
10	1911	WINTER BUGGY		Y							ALL	HILLSBORO DISPATCH
10	1911	WINTER BUGGY		Y							ALL	HILLSBORO DISPATCH
11	1911	WINTER BUGGY		Y							ALL	HILLSBORO DISPATCH
11	1911	WINTER BUGGY		Y							ALL	HILLSBORO DISPATCH
11	1911	WINTER BUGGY		Y							ALL	HILLSBORO DISPATCH
11	1911	WINTER BUGGY		Y							ALL	HILLSBORO DISPATCH
12	1911	WINTER BUGGY		Y							ALL	HILLSBORO DISPATCH
12	1911	WINTER BUGGY		Y							ALL	HILLSBORO DISPATCH
12	1911	WINTER BUGGY		Y					Y		ALL	HILLSBORO DISPATCH
12	1911	WINTER BUGGY		Y							ALL	HILLSBORO DISPATCH
1	1912	WINTER BUGGY		Y							ALL	HILLSBORO DISPATCH

Month	Year	Vehicle Type	Specs	Prices	Sale	Other Services	Slogan Only	Promotions	TESTIMONIAL	Negro ID	Audience	Ad Source
1	1912	WINTER BUGGY		Y							ALL	HILLSBORO DISPATCH
1	1912	WINTER BUGGY		Y							ALL	HILLSBORO DISPATCH
1	1912	WINTER BUGGY		Y							ALL	HILLSBORO DISPATCH
2	1912				Y						FARMERS	HILLSBORO DISPATCH
2	1912	BUGGY REPAIR				Y					ALL	HILLSBORO DISPATCH
2	1912				Y						ALL	HILLSBORO DISPATCH
2	1912	BUGGY REPAIR				Y					FARMERS	HILLSBORO DISPATCH
3	1912	NO. 33 LIGHT SURREY	Y	Y							ALL	HILLSBORO DISPATCH
3	1912	BUGGY REPAIR		Y							ALL	HILLSBORO DISPATCH
3	1912	BUGGY TIRES	Y	Y							ALL	HILLSBORO DISPATCH
3	1912	NO. 1 BUGGY	Y	Y							ALL	HILLSBORO DISPATCH
4	1912	SPRING WAGON	Y	Y							FARMERS	HILLSBORO DISPATCH
4	1912	SPRING WAGON	Y	Y							FARMERS	HILLSBORO DISPATCH
4	1912	RUNABOUT		Y							ALL	HILLSBORO DISPATCH
4	1912	CONCORD		Y							ALL	HILLSBORO DISPATCH
5	1912	BUGGY REPAIR		Y		Y					ALL	HILLSBORO DISPATCH
5	1912	BUGGY REPAIR		Y							ALL	HILLSBORO DISPATCH
5	1912			Y							ALL	HILLSBORO DISPATCH
5	1912			Y	Y						ALL	HILLSBORO DISPATCH
6	1912			Y	Y						ALL	HILLSBORO DISPATCH
6	1912			Y	Y						ALL	HILLSBORO DISPATCH
6	1912						Y				ALL	HILLSBORO DISPATCH
6	1912	NO. 1 BUGGY		Y							ALL	HILLSBORO DISPATCH
6	1912	NO. 2 BUGGY		Y							ALL	HILLSBORO DISPATCH
6	1912	NO. 3 BUGGY		Y							ALL	HILLSBORO DISPATCH

Month	Year	Vehicle Type	Specs	Prices	Sale	Other Services	Slogan Only	Promotions	TESTIMONIAL	Negro ID	Audience	Ad Source
7	1912	BUGGY REPAIR									FARMERS	HILLSBORO DISPATCH
7	1912	BUGGY REPAIR		Y							ALL	HILLSBORO DISPATCH
7	1912	NO. 4 BUGGY		Y							ALL	HILLSBORO DISPATCH
7	1912	NO. 1 BUGGY		Y							ALL	HILLSBORO DISPATCH
7	1912	NO. 1 BUGGY		Y							ALL	HILLSBORO DISPATCH
7	1912	NO. 1 BUGGY		Y							ALL	HILLSBORO DISPATCH
8	1912	NO. 1 BUGGY		Y							ALL	HILLSBORO DISPATCH
8	1912	NO. 1 BUGGY		Y							ALL	HILLSBORO DISPATCH
8	1912	NO. 1 BUGGY									ALL	HILLSBORO DISPATCH
8	1912	BUGGY TIRES									ALL	HILLSBORO DISPATCH
9	1912				Y						ALL	HILLSBORO DISPATCH
9	1912				Y						ALL	HILLSBORO DISPATCH
9	1912				Y						ALL	HILLSBORO DISPATCH
10	1912	NO. 176 WINTER BUGGY		Y							ALL	HILLSBORO DISPATCH
10	1912	NO. 176 WINTER BUGGY		Y							ALL	HILLSBORO DISPATCH
10	1912	NO. 176 WINTER BUGGY		Y							ALL	HILLSBORO DISPATCH
10	1912	NO. 176 WINTER BUGGY		Y							ALL	HILLSBORO DISPATCH
10	1912	NO. 176 WINTER BUGGY		Y							ALL	HILLSBORO DISPATCH
11	1912	NO. 176 WINTER BUGGY		Y							ALL	HILLSBORO DISPATCH
11	1912	NO. 176 WINTER BUGGY		Y							ALL	HILLSBORO DISPATCH
11	1912	NO. 176 WINTER BUGGY		Y							ALL	HILLSBORO DISPATCH
11	1912	NO. 176 WINTER BUGGY		Y							ALL	HILLSBORO DISPATCH
12	1912	NO. 176 WINTER BUGGY		Y							ALL	HILLSBORO DISPATCH
12	1912	NO. 176 WINTER BUGGY		Y							ALL	HILLSBORO DISPATCH
12	1912	NO. 176 WINTER BUGGY		Y							ALL	HILLSBORO DISPATCH

Month	Year	Vehicle Type	Specs	Prices	Sale	Other Services	Slogan Only	Promotions	TESTIMONIAL	Negro ID	Audience	Ad Source
12	1912	NO. 176 WINTER BUGGY		Y							ALL	HILLSBORO DISPATCH
12	1912	NO. 176 WINTER BUGGY		Y							ALL	HILLSBORO DISPATCH
1	1913	NO. 176 WINTER BUGGY		Y							ALL	HILLSBORO DISPATCH
1	1913	NO. 176 WINTER BUGGY		Y							ALL	HILLSBORO DISPATCH
1	1913	NO. 176 WINTER BUGGY		Y							ALL	HILLSBORO DISPATCH
1	1913	NO. 176 WINTER BUGGY		Y							ALL	HILLSBORO DISPATCH
2	1913	NO. 176 WINTER BUGGY		Y							ALL	HILLSBORO DISPATCH
2	1913	AUTO REPAIR									ALL	HILLSBORO DISPATCH
2	1913	AUTO REPAIR									ALL	HILLSBORO DISPATCH
2	1913	AUTO REPAIR		Y							ALL	HILLSBORO DISPATCH
3	1913	BUGGY TIRES	Y	Y							ALL	HILLSBORO DISPATCH
3	1913	BUGGY REPAIR		Y		Y					ALL	HILLSBORO DISPATCH
3	1913	BUGGY REPAIR									ALL	HILLSBORO DISPATCH
3	1913	BUGGY REPAIR		Y		Y					FARMERS	HILLSBORO DISPATCH
4	1913		Y								ALL	HILLSBORO DISPATCH
4	1913		Y								ALL	HILLSBORO DISPATCH
4	1913		Y								ALL	HILLSBORO DISPATCH
4	1913		Y								ALL	HILLSBORO DISPATCH
4	1913		Y						Y		ALL	HILLSBORO DISPATCH
5	1913		Y								ALL	HILLSBORO DISPATCH
5	1913		Y								ALL	HILLSBORO DISPATCH
5	1913		Y								ALL	HILLSBORO DISPATCH
5	1913		Y								ALL	HILLSBORO DISPATCH
6	1913		Y								ALL	HILLSBORO DISPATCH
6	1913	BUGGY REPAIR		Y							ALL	HILLSBORO DISPATCH

Month	Year	Vehicle Type	Specs	Prices	Sale	Other Services	Slogan Only	Promotions	TESTIMONIAL	Negro ID	Audience	Ad Source
12	1913	COMPETITOR WINTER BUGGY		Y	Y						ALL	HILLSBORO DISPATCH
12	1913	NO. 176 WINTER BUGGY		Y	Y						ALL	HILLSBORO DISPATCH
1	1914	NO. 176 WINTER BUGGY		Y	Y						ALL	HILLSBORO DISPATCH
1	1914	NO. 176 WINTER BUGGY		Y	Y						ALL	HILLSBORO DISPATCH
1	1914	NO. 176 WINTER BUGGY		Y	Y						ALL	HILLSBORO DISPATCH
1	1914	NO. 176 WINTER BUGGY		Y	Y						ALL	HILLSBORO DISPATCH
2	1914	NO. 176 WINTER BUGGY		Y	Y						ALL	HILLSBORO DISPATCH
2	1914	BUGGY REPAIR									ALL	HILLSBORO DISPATCH
2	1914	BUGGY REPAIR									ALL	HILLSBORO DISPATCH
2	1914	SPRING WAGON		Y							ALL	HILLSBORO DISPATCH
3	1914				Y						ALL	HILLSBORO DISPATCH
3	1914				Y						ALL	HILLSBORO DISPATCH
3	1914	BUGGY REPAIR									ALL	HILLSBORO DISPATCH
3	1914	BUGGY REPAIR									ALL	HILLSBORO DISPATCH
3	1914				Y				Y		ALL	HILLSBORO DISPATCH
4	1914	BUGGY REPAIR									ALL	HILLSBORO DISPATCH
4	1914	BUGGY REPAIR	Y								ALL	HILLSBORO DISPATCH
4	1914	BUGGY REPAIR		Y							ALL	HILLSBORO DISPATCH
4	1914	AUTO / BUGGY REPAIR									ALL	HILLSBORO DISPATCH
5	1914	AUTO / BUGGY REPAIR	Y								ALL	HILLSBORO DISPATCH
5	1914	NO. 1 BUGGY		Y							ALL	HILLSBORO DISPATCH
5	1914	NO. 2 BUGGY		Y							ALL	HILLSBORO DISPATCH
5	1914	PHAETON		Y							ALL	HILLSBORO DISPATCH
6	1914	AUTO SEAT BUGGY									ALL	HILLSBORO DISPATCH
1	1916	STORM BUGGY		Y		Y					ALL	HILLSBORO DISPATCH

Month	Year	Vehicle Type	Specs	Prices	Sale	Other Services	Slogan Only	Promotions	TESTIMONIAL	Negro ID	Audience	Ad Source
1	1916	PATTERSON GREENFIELD									ALL	HILLSBORO DISPATCH
1	1916	STORM BUGGY		Y		Y					ALL	HILLSBORO DISPATCH
1	1916	PATTERSON GREENFIELD									ALL	HILLSBORO DISPATCH
1	1916	STORM BUGGY		Y		Y					ALL	HILLSBORO DISPATCH
1	1916	PATTERSON GREENFIELD									ALL	HILLSBORO DISPATCH
1	1916	STORM BUGGY		Y		Y					ALL	HILLSBORO DISPATCH
1	1916	PATTERSON GREENFIELD									ALL	HILLSBORO DISPATCH
2	1916	AUTO / BUGGY REPAIR		Y		Y					ALL	HILLSBORO DISPATCH
2	1916	PATTERSON GREENFIELD									ALL	HILLSBORO DISPATCH
2	1916	AUTO / BUGGY REPAIR		Y		Y					ALL	HILLSBORO DISPATCH
2	1916	PATTERSON GREENFIELD									ALL	HILLSBORO DISPATCH
2	1916	AUTO / BUGGY REPAIR		Y		Y					ALL	HILLSBORO DISPATCH
2	1916	PATTERSON GREENFIELD									ALL	HILLSBORO DISPATCH
2	1916	AUTO / BUGGY REPAIR		Y		Y					ALL	HILLSBORO DISPATCH
2	1916	PATTERSON GREENFIELD									ALL	HILLSBORO DISPATCH
3	1916	AUTO / BUGGY REPAIR				Y					ALL	HILLSBORO DISPATCH
3	1916	PATTERSON GREENFIELD									ALL	HILLSBORO DISPATCH
3	1916	AUTO / BUGGY REPAIR				Y					ALL	HILLSBORO DISPATCH
3	1916	PATTERSON GREENFIELD									ALL	HILLSBORO DISPATCH
3	1916	AUTO / BUGGY REPAIR				Y					ALL	HILLSBORO DISPATCH
4	1916	AUTO / BUGGY REPAIR				Y					ALL	HILLSBORO DISPATCH
4	1916	PATTERSON GREENFIELD									ALL	HILLSBORO DISPATCH
4	1916	AUTO / BUGGY REPAIR				Y					ALL	HILLSBORO DISPATCH
4	1916	PATTERSON GREENFIELD									ALL	HILLSBORO DISPATCH
4	1916	AUTO / BUGGY REPAIR				Y					ALL	HILLSBORO DISPATCH

Month	Year	Vehicle Type	Specs	Prices	Sale	Other Services	Slogan Only	Promotions	TESTIMONIAL	Negro ID	Audience	Ad Source
5	1916	BUGGY TIRES		Y							ALL	HILLSBORO DISPATCH
5	1916	NO. 1 BUGGY									ALL	HILLSBORO DISPATCH
5	1916	NO. 1 BUGGY		Y							ALL	HILLSBORO DISPATCH
6	1916	PATTERSON GREENFIELD				Y					ALL	HILLSBORO DISPATCH
6	1916	NO. 1 BUGGY		Y							ALL	HILLSBORO DISPATCH
6	1916	PATTERSON GREENFIELD				Y					ALL	HILLSBORO DISPATCH
6	1916	NO. 1 BUGGY		Y							ALL	HILLSBORO DISPATCH
6	1916	PATTERSON GREENFIELD		Y		Y					ALL	HILLSBORO DISPATCH
7	1916	SURREY		Y							ALL	HILLSBORO DISPATCH
7	1916	PATTERSON GREENFIELD		Y		Y					ALL	HILLSBORO DISPATCH
7	1916	SURREY		Y							ALL	HILLSBORO DISPATCH
7	1916	PATTERSON GREENFIELD		Y		Y					ALL	HILLSBORO DISPATCH
8	1916	SURREY		Y							ALL	HILLSBORO DISPATCH
8	1916	PATTERSON GREENFIELD				Y					ALL	HILLSBORO DISPATCH
8	1916	SURREY		Y							ALL	HILLSBORO DISPATCH
8	1916	PATTERSON GREENFIELD				Y					ALL	HILLSBORO DISPATCH
8	1916	AUTO / BUGGY REPAIR									ALL	HILLSBORO DISPATCH
8	1916	SCHOOL BUGGIES		Y	Y						ALL	HILLSBORO DISPATCH
8	1916	AUTO REPAIR									ALL	HILLSBORO DISPATCH
8	1916	PATTERSON GREENFIELD									ALL	HILLSBORO DISPATCH
8	1916	AUTO REPAIR			Y	Y					ALL	HILLSBORO DISPATCH
8	1916	PATTERSON GREENFIELD									ALL	HILLSBORO DISPATCH
9	1916	AUTO REPAIR			Y	Y					ALL	HILLSBORO DISPATCH
9	1916	PATTERSON GREENFIELD									ALL	HILLSBORO DISPATCH
9	1916	AUTO REPAIR			Y	Y					ALL	HILLSBORO DISPATCH

Month	Year	Vehicle Type	Specs	Prices	Sale	Other Services	Slogan Only	Promotions	TESTIMONIAL	Negro ID	Audience	Ad Source
9	1916	PATTERSON GREENFIELD									ALL	HILLSBORO DISPATCH
9	1916	AUTO REPAIR				Y					ALL	HILLSBORO DISPATCH
9	1916	PATTERSON GREENFIELD									ALL	HILLSBORO DISPATCH
9	1916	AUTO / BUGGY REPAIR				Y					ALL	HILLSBORO DISPATCH
9	1916	PATTERSON GREENFIELD									ALL	HILLSBORO DISPATCH
10	1916	AUTO REPAIR							Y		ALL	HILLSBORO DISPATCH
10	1916	NO. 176 WINTER BUGGY		Y							ALL	HILLSBORO DISPATCH
10	1916	AUTO REPAIR				Y					ALL	HILLSBORO DISPATCH
11	1916	AUTO / BUGGY REPAIR		Y		Y					ALL	HILLSBORO DISPATCH
11	1916	PATTERSON GREENFIELD		Y							ALL	HILLSBORO DISPATCH
11	1916	AUTO REPAIR				Y			Y		ALL	HILLSBORO DISPATCH
12	1916	AUTO REPAIR		Y		Y					ALL	HILLSBORO DISPATCH
12	1916	AUTO REPAIR		Y		Y					ALL	HILLSBORO DISPATCH
3	1910				Y				Y		ALL	LEESBURG WEEKLY BUCKEYE
3	1910				Y				Y		ALL	LEESBURG WEEKLY BUCKEYE
3	1910				Y				Y		ALL	LEESBURG WEEKLY BUCKEYE
3	1910				Y				Y		ALL	LEESBURG WEEKLY BUCKEYE
4	1910				Y				Y		ALL	LEESBURG WEEKLY BUCKEYE
4	1910				Y				Y		ALL	LEESBURG WEEKLY BUCKEYE
4	1910										FARMERS	LEESBURG WEEKLY BUCKEYE
4	1910				Y				Y		ALL	LEESBURG WEEKLY BUCKEYE
5	1910				Y				Y		ALL	LEESBURG WEEKLY BUCKEYE
5	1910				Y				Y		ALL	LEESBURG WEEKLY BUCKEYE
5	1910				Y				Y		ALL	LEESBURG WEEKLY BUCKEYE
5	1910				Y				Y		ALL	LEESBURG WEEKLY BUCKEYE

Month	Year	Vehicle Type	Specs	Prices	Sale	Other Services	Slogan Only	Promotions	TESTIMONIAL	Negro ID	Audience	Ad Source
5	1910				Y				Y		ALL	LEESBURG WEEKLY BUCKEYE
6	1910	BUGGY REPAIR				Y					ALL	LEESBURG WEEKLY BUCKEYE
6	1910	BUGGY REPAIR				Y					ALL	LEESBURG WEEKLY BUCKEYE
6	1910	BUGGY REPAIR				Y					ALL	LEESBURG WEEKLY BUCKEYE
6	1910			Y	Y						ALL	LEESBURG WEEKLY BUCKEYE
7	1910			Y	Y						ALL	LEESBURG WEEKLY BUCKEYE
7	1910				Y						ALL	LEESBURG WEEKLY BUCKEYE
7	1910				Y						ALL	LEESBURG WEEKLY BUCKEYE
8	1910				Y						ALL	LEESBURG WEEKLY BUCKEYE
8	1910				Y						ALL	LEESBURG WEEKLY BUCKEYE
8	1910				Y						ALL	LEESBURG WEEKLY BUCKEYE
8	1910				Y						ALL	LEESBURG WEEKLY BUCKEYE
8	1910				Y						ALL	LEESBURG WEEKLY BUCKEYE
9	1910				Y						ALL	LEESBURG WEEKLY BUCKEYE
9	1910				Y						ALL	LEESBURG WEEKLY BUCKEYE
9	1910				Y						ALL	LEESBURG WEEKLY BUCKEYE
9	1910				Y						ALL	LEESBURG WEEKLY BUCKEYE
10	1910	SURREY		Y	Y						ALL	LEESBURG WEEKLY BUCKEYE
10	1910	SURREY		Y	Y						ALL	LEESBURG WEEKLY BUCKEYE
10	1910	SURREY		Y	Y						ALL	LEESBURG WEEKLY BUCKEYE
10	1910	SURREY		Y	Y						ALL	LEESBURG WEEKLY BUCKEYE
11	1910	WINTER BUGGY									ALL	LEESBURG WEEKLY BUCKEYE
11	1910	WINTER BUGGY									ALL	LEESBURG WEEKLY BUCKEYE
11	1910	WINTER BUGGY	Y	Y							ALL	LEESBURG WEEKLY BUCKEYE
11	1910	WINTER BUGGY	Y	Y							ALL	LEESBURG WEEKLY BUCKEYE

Month	Year	Vehicle Type	Specs	Prices	Sale	Other Services	Slogan Only	Promotions	TESTIMONIAL	Negro ID	Audience	Ad Source
11	1910	WINTER BUGGY	Y	Y	Y						ALL	LEESBURG WEEKLY BUCKEYE
11	1910	WINTER BUGGY	Y	Y	Y						ALL	LEESBURG WEEKLY BUCKEYE
12	1910	WINTER BUGGY									ALL	LEESBURG WEEKLY BUCKEYE
12	1910	WINTER BUGGY	Y	Y							ALL	LEESBURG WEEKLY BUCKEYE
12	1910	WINTER BUGGY	Y	Y							ALL	LEESBURG WEEKLY BUCKEYE
12	1910	WINTER BUGGY	Y	Y							ALL	LEESBURG WEEKLY BUCKEYE
1	1911	WINTER BUGGY		Y	Y						ALL	LEESBURG WEEKLY BUCKEYE
1	1911	WINTER BUGGY		Y							ALL	LEESBURG WEEKLY BUCKEYE
1	1911	WINTER BUGGY		Y					Y		ALL	LEESBURG WEEKLY BUCKEYE
1	1911	WINTER BUGGY		Y							ALL	LEESBURG WEEKLY BUCKEYE
1	1911	WINTER BUGGY									ALL	LEESBURG WEEKLY BUCKEYE
2	1911				Y				Y		ALL	LEESBURG WEEKLY BUCKEYE
2	1911				Y				Y		ALL	LEESBURG WEEKLY BUCKEYE
2	1911				Y				Y		ALL	LEESBURG WEEKLY BUCKEYE
2	1911	BUGGY REPAIR		Y		Y			Y		ALL	LEESBURG WEEKLY BUCKEYE
2	1913	AUTO REPAIR				Y					ALL	OHIO STATE UNIVERSITY MONTHLY
10	1913	AUTO REPAIR				Y					ALL	OHIO STATE UNIVERSITY MONTHLY
11	1913	AUTO REPAIR				Y					ALL	OHIO STATE UNIVERSITY MONTHLY
12	1910	NO. 24 SOUTHERN BUGGY		Y						Y	NEGROES	THE CRISIS (NAACP)
1	1911	NO. 4 BUGGY		Y						Y	NEGROES	THE CRISIS (NAACP)
2	1911	NO. 4 BUGGY		Y						Y	NEGROES	THE CRISIS (NAACP)
3	1911	NO. 4 BUGGY		Y						Y	NEGROES	THE CRISIS (NAACP)
4	1911	NO. 4 BUGGY		Y						Y	NEGROES	THE CRISIS (NAACP)
6	1911	NO. 4 BUGGY		Y						Y	NEGROES	THE CRISIS (NAACP)
7	1911	NO. 4 BUGGY		Y						Y	NEGROES	THE CRISIS (NAACP)

Month	Year	Vehicle Type	Specs	Prices	Sale	Other Services	Slogan Only	Promotions	TESTIMONIAL	Negro ID	Audience	Ad Source
7	1909	DOCTOR'S BUGGY							Y	Y	PHYSICIANS	JOURNAL NATIONAL MEDICAL ASSN.
7	1909	NO. 1 BUGGY							Y		PHYSICIANS	JOURNAL NATIONAL MEDICAL ASSN.
1	1910	GODDARD PHAETON							Y		PHYSICIANS	JOURNAL NATIONAL MEDICAL ASSN.
4	1910	DOCTOR'S BUGGY							Y		PHYSICIANS	JOURNAL NATIONAL MEDICAL ASSN.
7	1910	DOCTOR'S BUGGY							Y	Y	PHYSICIANS	JOURNAL NATIONAL MEDICAL ASSN.
10	1910	DOCTOR'S BUGGY							Y	Y	PHYSICIANS	JOURNAL NATIONAL MEDICAL ASSN.
1	1911	NO. 65 STANHOPE							Y	Y	PHYSICIANS	JOURNAL NATIONAL MEDICAL ASSN.
4	1911	NO. 70 DOCTOR'S WAGON							Y	Y	PHYSICIANS	JOURNAL NATIONAL MEDICAL ASSN.
7	1911	DOCTOR'S BUGGY							Y	Y	PHYSICIANS	JOURNAL NATIONAL MEDICAL ASSN.
10	1911	STORM BUGGY							Y	Y	PHYSICIANS	JOURNAL NATIONAL MEDICAL ASSN.
1	1912	NO. 65 STANHOPE							Y	Y	PHYSICIANS	JOURNAL NATIONAL MEDICAL ASSN.
4	1912	NO. 4 BUGGY							Y		PHYSICIANS	JOURNAL NATIONAL MEDICAL ASSN.
7	1912	NO. 4 BUGGY							Y		PHYSICIANS	JOURNAL NATIONAL MEDICAL ASSN.
10	1912	NO. 176 WINTER BUGGY	Y								PHYSICIANS	JOURNAL NATIONAL MEDICAL ASSN.
10	1912	NO. 176 WINTER BUGGY		Y					Y		PHYSICIANS	JOURNAL NATIONAL MEDICAL ASSN.
1	1913	NO. 4 BUGGY	Y	Y							PHYSICIANS	JOURNAL NATIONAL MEDICAL ASSN.
4	1913	NO. 4 BUGGY		Y						Y	PHYSICIANS	JOURNAL NATIONAL MEDICAL ASSN.
7	1913	NO. 176 WINTER BUGGY		Y						Y	PHYSICIANS	JOURNAL NATIONAL MEDICAL ASSN.
10	1913	NO. 176 WINTER BUGGY	Y	Y						Y	PHYSICIANS	JOURNAL NATIONAL MEDICAL ASSN.
10	1913	BUGGY ACCESSORIES	Y	Y					Y		PHYSICIANS	JOURNAL NATIONAL MEDICAL ASSN.
1	1914	DOCTOR'S BUGGY								Y	PHYSICIANS	JOURNAL NATIONAL MEDICAL ASSN.
4	1914										PHYSICIANS	JOURNAL NATIONAL MEDICAL ASSN.
7	1914	BUGGY ACCESSORIES		Y						Y	PHYSICIANS	JOURNAL NATIONAL MEDICAL ASSN.
10	1914	BUGGY ACCESSORIES		Y							PHYSICIANS	JOURNAL NATIONAL MEDICAL ASSN.
10	1914	NO. 176 WINTER BUGGY									PHYSICIANS	JOURNAL NATIONAL MEDICAL ASSN.

Month	Year	Vehicle Type	Specs	Prices	Sale	Other Services	Slogan Only	Promotions	TESTIMONIAL	Negro ID	Audience	Ad Source
8	1911	NO. 4 BUGGY		Y						Y	NEGROES	THE CRISIS (NAACP)
9	1911	NO. 4 BUGGY		Y						Y	NEGROES	THE CRISIS (NAACP)
10	1911	NO. 4 BUGGY		Y						Y	NEGROES	THE CRISIS (NAACP)
11	1911	NO. 4 BUGGY		Y						Y	NEGROES	THE CRISIS (NAACP)
12	1911	NO. 4 BUGGY		Y						Y	NEGROES	THE CRISIS (NAACP)
1	1912	NO. 4 BUGGY		Y						Y	NEGROES	THE CRISIS (NAACP)
2	1912	NO. 4 BUGGY		Y						Y	NEGROES	THE CRISIS (NAACP)
3	1912	NO. 4 BUGGY		Y						Y	NEGROES	THE CRISIS (NAACP)
4	1912	NO. 4 BUGGY		Y						Y	NEGROES	THE CRISIS (NAACP)
5	1912	NO. 4 BUGGY		Y						Y	NEGROES	THE CRISIS (NAACP)
6	1912	NO. 4 BUGGY		Y						Y	NEGROES	THE CRISIS (NAACP)
7	1912	NO. 4 BUGGY		Y						Y	NEGROES	THE CRISIS (NAACP)
8	1912	NO. 4 BUGGY		Y						Y	NEGROES	THE CRISIS (NAACP)
9	1912	NO. 4 BUGGY		Y						Y	NEGROES	THE CRISIS (NAACP)
10	1912	NO. 4 BUGGY		Y						Y	NEGROES	THE CRISIS (NAACP)
4	1913	NO. 4 BUGGY		Y						Y	NEGROES	THE CRISIS (NAACP)
5	1913	NO. 4 BUGGY		Y						Y	NEGROES	THE CRISIS (NAACP)
9	1913	NO. 24 SOUTHERN BUGGY		Y						Y	NEGROES	THE CRISIS (NAACP)
10	1913	NO. 24 SOUTHERN BUGGY		Y						Y	NEGROES	THE CRISIS (NAACP)
5	1915	NO. 24 SOUTHERN BUGGY		Y						Y	NEGROES	THE CRISIS (NAACP)
12	1916	AUTO AND BUGGY MAKERS								Y	NEGROES	THE CRISIS (NAACP)
1	1917	AUTO AND BUGGY MAKERS								Y	NEGROES	THE CRISIS (NAACP)
1	1909	DOCTOR'S BUGGY							Y	Y	PHYSICIANS	JOURNAL NATIONAL MEDICAL ASSN.
4	1909	DOCTOR'S BUGGY							Y	Y	PHYSICIANS	JOURNAL NATIONAL MEDICAL ASSN.
4	1909	NO. 1 BUGGY							Y		PHYSICIANS	JOURNAL NATIONAL MEDICAL ASSN.

Month	Year	Vehicle Type	Specs	Prices	Sale	Other Services	Slogan Only	Promotions	TESTIMONIAL	Negro ID	Audience	Ad Source
1	1915	NO. 4 BUGGY		Y						Y	PHYSICIANS	JOURNAL NATIONAL MEDICAL ASSN.
1	1915	NO. 2 BUGGY		Y						Y	PHYSICIANS	JOURNAL NATIONAL MEDICAL ASSN.
1	1915	NO. 44 SURREY		Y						Y	PHYSICIANS	JOURNAL NATIONAL MEDICAL ASSN.
4	1915	NO. 4 BUGGY	Y	Y						Y	PHYSICIANS	JOURNAL NATIONAL MEDICAL ASSN.
4	1915	AUTO REPAIR		Y							PHYSICIANS	JOURNAL NATIONAL MEDICAL ASSN.
7	1915	NO. 4 BUGGY	Y	Y						Y	PHYSICIANS	JOURNAL NATIONAL MEDICAL ASSN.
10	1915	PATTERSON GREENFIELD	Y	Y						Y	PHYSICIANS	JOURNAL NATIONAL MEDICAL ASSN.
10	1915	PATTERSON GREENFIELD		Y							PHYSICIANS	JOURNAL NATIONAL MEDICAL ASSN.
1	1916	PATTERSON GREENFIELD		Y					Y	Y	PHYSICIANS	JOURNAL NATIONAL MEDICAL ASSN.
4	1916	PATTERSON GREENFIELD	Y						Y	Y	PHYSICIANS	JOURNAL NATIONAL MEDICAL ASSN.
7	1916	PATTERSON GREENFIELD	Y	Y						Y	PHYSICIANS	JOURNAL NATIONAL MEDICAL ASSN.
10	1916	PATTERSON GREENFIELD	Y	Y							PHYSICIANS	JOURNAL NATIONAL MEDICAL ASSN.
1	1917	PATTERSON GREENFIELD	Y	Y						Y	PHYSICIANS	JOURNAL NATIONAL MEDICAL ASSN.
4	1917	PATTERSON GREENFIELD		Y							PHYSICIANS	JOURNAL NATIONAL MEDICAL ASSN.
4	1917	PATTERSON GREENFIELD	Y								PHYSICIANS	JOURNAL NATIONAL MEDICAL ASSN.
10	1917	AUTO ACCESSORIES		Y							PHYSICIANS	JOURNAL NATIONAL MEDICAL ASSN.
10	1917	AUTO ACCESSORIES									PHYSICIANS	JOURNAL NATIONAL MEDICAL ASSN.
1	1918	AUTO ACCESSORIES		Y							PHYSICIANS	JOURNAL NATIONAL MEDICAL ASSN.
1	1912										FARMERS	THE OHIO FARMER
1	1912										FARMERS	THE OHIO FARMER
2	1912										FARMERS	THE OHIO FARMER
2	1912										FARMERS	THE OHIO FARMER
3	1912										FARMERS	THE OHIO FARMER
3	1912										FARMERS	THE OHIO FARMER
4	1912										FARMERS	THE OHIO FARMER

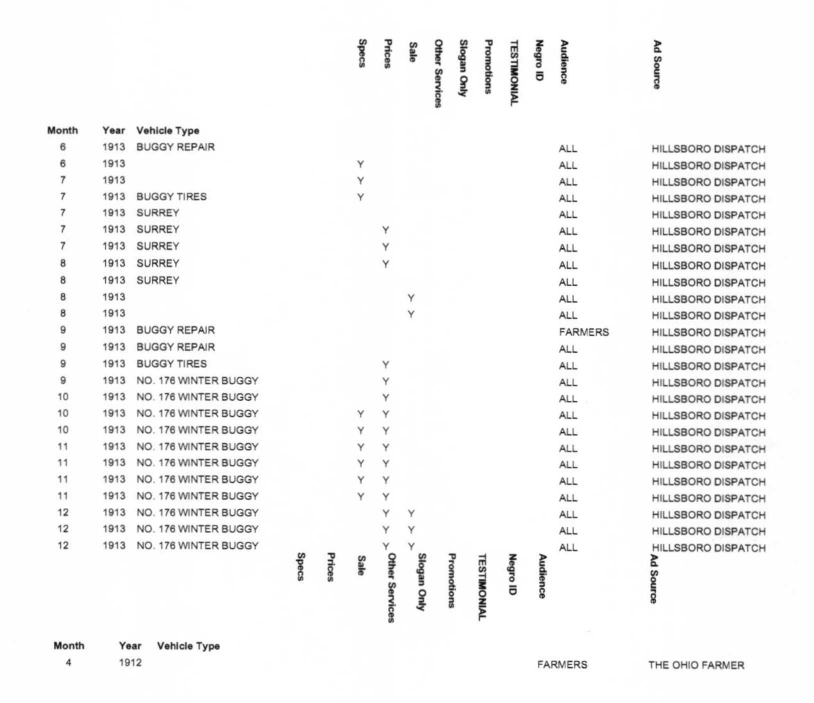

Month	Year	Vehicle Type	Specs	Prices	Sale	Other Services	Slogan Only	Promotions	TESTIMONIAL	Negro ID	Audience	Ad Source
6	1913	BUGGY REPAIR									ALL	HILLSBORO DISPATCH
6	1913		Y								ALL	HILLSBORO DISPATCH
7	1913		Y								ALL	HILLSBORO DISPATCH
7	1913	BUGGY TIRES	Y								ALL	HILLSBORO DISPATCH
7	1913	SURREY									ALL	HILLSBORO DISPATCH
7	1913	SURREY		Y							ALL	HILLSBORO DISPATCH
7	1913	SURREY		Y							ALL	HILLSBORO DISPATCH
8	1913	SURREY		Y							ALL	HILLSBORO DISPATCH
8	1913	SURREY									ALL	HILLSBORO DISPATCH
8	1913				Y						ALL	HILLSBORO DISPATCH
8	1913				Y						ALL	HILLSBORO DISPATCH
9	1913	BUGGY REPAIR									FARMERS	HILLSBORO DISPATCH
9	1913	BUGGY REPAIR									ALL	HILLSBORO DISPATCH
9	1913	BUGGY TIRES		Y							ALL	HILLSBORO DISPATCH
9	1913	NO. 176 WINTER BUGGY		Y							ALL	HILLSBORO DISPATCH
10	1913	NO. 176 WINTER BUGGY		Y							ALL	HILLSBORO DISPATCH
10	1913	NO. 176 WINTER BUGGY	Y	Y							ALL	HILLSBORO DISPATCH
10	1913	NO. 176 WINTER BUGGY	Y	Y							ALL	HILLSBORO DISPATCH
11	1913	NO. 176 WINTER BUGGY	Y	Y							ALL	HILLSBORO DISPATCH
11	1913	NO. 176 WINTER BUGGY	Y	Y							ALL	HILLSBORO DISPATCH
11	1913	NO. 176 WINTER BUGGY	Y	Y							ALL	HILLSBORO DISPATCH
11	1913	NO. 176 WINTER BUGGY	Y	Y							ALL	HILLSBORO DISPATCH
12	1913	NO. 176 WINTER BUGGY		Y	Y						ALL	HILLSBORO DISPATCH
12	1913	NO. 176 WINTER BUGGY		Y	Y						ALL	HILLSBORO DISPATCH
12	1913	NO. 176 WINTER BUGGY		Y	Y						ALL	HILLSBORO DISPATCH

Month	Year	Vehicle Type	Specs	Prices	Sale	Other Services	Slogan Only	Promotions	TESTIMONIAL	Negro ID	Audience	Ad Source
4	1912										FARMERS	THE OHIO FARMER

Appendix B:

Sample of Client Locations Collected from Patterson Advertisements

Greenfield Republican and *Hillsboro Dispatch* ran the same advertisements simultaneously. To prevent duplication, the ads from only one or the other is documented here.

Ad Month	Ad Year	Vehicle Type	Client Location	Ad Source	Comments
1	1902	NO. 60 CUT-UNDER STANHOPE	GREENFIELD	GREENFIELD REPUBLICAN	FOR E. L. McCLAIN
1	1902	NO. 1 END SPRING PIANO BUGGY	GREENFIELD	GREENFIELD REPUBLICAN	
1	1902	NO. 2 END SPRING PIANO BUGGY	LEXINGTON, KY	GREENFIELD REPUBLICAN	HORSEMAN'S HDQTRS LEXINGTON
1	1902	NO. 1 END SPRING PIANO BUGGY	GREENFIELD	GREENFIELD REPUBLICAN	LEADING DAIRYMAN SOUTHERN OHIO
1	1902	NO. 1 END SPRING PIANO BUGGY	EAST MONROE	GREENFIELD REPUBLICAN	
1	1902	NO. 1 END SPRING PIANO BUGGY	NASHVILLE, TN	GREENFIELD REPUBLICAN	JEFF MARTIN LIVERY COMPANY
2	1902	NO. 1 END SPRING PIANO BUGGY	GREENFIELD	GREENFIELD REPUBLICAN	
2	1902	NO. 1 END SPRING PIANO BUGGY	GREENFIELD	GREENFIELD REPUBLICAN	
2	1902	NO. 1 END SPRING PIANO BUGGY	GREENFIELD	GREENFIELD REPUBLICAN	
2	1902	DOCTOR'S BUGGY	GREENFIELD	GREENFIELD REPUBLICAN	PHYSICIAN
2	1902	DOCTOR'S BUGGY	WINCHESTER, KY	GREENFIELD REPUBLICAN	PHYSICIAN
2	1902	DOCTOR'S BUGGY	NASHVILLE, TN	GREENFIELD REPUBLICAN	PHYSICIAN
3	1902	NO. 1 END SPRING PIANO BUGGY	GREENFIELD	GREENFIELD REPUBLICAN	
3	1902	NO. 1 END SPRING PIANO BUGGY	GREENFIELD	GREENFIELD REPUBLICAN	
3	1902	NO. 1 END SPRING PIANO BUGGY	LEESBURG	GREENFIELD REPUBLICAN	
3	1902	NO. 20 BOULEVARD	GREENFIELD	GREENFIELD REPUBLICAN	
3	1902	NO. 1 END SPRING PIANO BUGGY	GREENFIELD	GREENFIELD REPUBLICAN	
3	1902	NO. 1 END SPRING PIANO BUGGY	GREENFIELD	GREENFIELD REPUBLICAN	
3	1902	NO. 1 END SPRING PIANO BUGGY	GREENFIELD	GREENFIELD REPUBLICAN	
4	1902	NO. 119 ELLIPTIC SPRING BIKE WAGON	GREENFIELD	GREENFIELD REPUBLICAN	
5	1902	NO. 1 END SPRING PIANO BUGGY	NEW MARTINSBURG	GREENFIELD REPUBLICAN	
5	1902	NO. 1 END SPRING PIANO BUGGY	NEW MARTINSBURG	GREENFIELD REPUBLICAN	
5	1902	NO. 60 CUT-UNDER STANHOPE	GREENFIELD	GREENFIELD REPUBLICAN	
5	1902	NO. 61 CUT-UNDER STANHOPE	GREENFIELD	GREENFIELD REPUBLICAN	
5	1902	NO. 1 END SPRING PIANO BUGGY	GREENFIELD	GREENFIELD REPUBLICAN	
5	1902	NO. 1 END SPRING PIANO BUGGY	GREENFIELD	GREENFIELD REPUBLICAN	
5	1902	NO. 1 END SPRING PIANO BUGGY	GREENFIELD	GREENFIELD REPUBLICAN	
6	1902	NO. 1 END SPRING PIANO BUGGY	THORNTON	GREENFIELD REPUBLICAN	
6	1902	NO. 1 END SPRING PIANO BUGGY	GREENFIELD	GREENFIELD REPUBLICAN	
6	1902	NO. 1 END SPRING PIANO BUGGY	GREENFIELD	GREENFIELD REPUBLICAN	
6	1902	NO. 119 ELLIPTIC SPRING BIKE WAGON	GREENFIELD	GREENFIELD REPUBLICAN	

Ad Month	Ad Year	Vehicle Type	Client Location	Ad Source	Comments
6	1902	NO. 119 ELLIPTIC SPRING BIKE WAGON	GREENFIELD	GREENFIELD REPUBLICAN	
6	1902	NO. 1 END SPRING PIANO BUGGY	GREENFIELD	GREENFIELD REPUBLICAN	
6	1902	NO. 1 END SPRING PIANO BUGGY	GREENFIELD	GREENFIELD REPUBLICAN	
6	1902	NO. 1 END SPRING PIANO BUGGY	NEW MARTINSBURG	GREENFIELD REPUBLICAN	
7	1902	NO. 1 END SPRING PIANO BUGGY	LEESBURG	GREENFIELD REPUBLICAN	
7	1902	NO. 1 END SPRING PIANO BUGGY	NEW MARTINSBURG	GREENFIELD REPUBLICAN	
7	1902	NO. 1 END SPRING PIANO BUGGY	GREENFIELD	GREENFIELD REPUBLICAN	
3	1903	NO. 1 END SPRING PIANO BUGGY	EAST MONROE	GREENFIELD REPUBLICAN	
3	1903	NO. 1 END SPRING PIANO BUGGY	NEW MARTINSBURG	GREENFIELD REPUBLICAN	
4	1903	NO. 1 END SPRING PIANO BUGGY	GREENFIELD	GREENFIELD REPUBLICAN	
4	1903	NO. 1 END SPRING PIANO BUGGY	GREENFIELD	GREENFIELD REPUBLICAN	
4	1903	NO. 1 END SPRING PIANO BUGGY	GREENFIELD	GREENFIELD REPUBLICAN	
4	1903	NO. 1 END SPRING PIANO BUGGY	EAST MONROE	GREENFIELD REPUBLICAN	
4	1903	NO. 1 END SPRING PIANO BUGGY	EAST MONROE	GREENFIELD REPUBLICAN	
4	1903	NO. 1 END SPRING PIANO BUGGY	EAST MONROE	GREENFIELD REPUBLICAN	
5	1903	NO. 1 END SPRING PIANO BUGGY	GREENFIELD	GREENFIELD REPUBLICAN	
5	1903	NO. 1 END SPRING PIANO BUGGY	GREENFIELD	GREENFIELD REPUBLICAN	
5	1903	NO. 1 END SPRING PIANO BUGGY	GREENFIELD	GREENFIELD REPUBLICAN	
5	1903	NO. 1 END SPRING PIANO BUGGY	SOUTH SALEM	GREENFIELD REPUBLICAN	
5	1903	NO. 1 END SPRING PIANO BUGGY	LYNDON	GREENFIELD REPUBLICAN	
5	1903	NO. 1 END SPRING PIANO BUGGY	LYNDON	GREENFIELD REPUBLICAN	
5	1903	NO. 1 END SPRING PIANO BUGGY	GREENFIELD	GREENFIELD REPUBLICAN	
5	1903	NO. 1 END SPRING PIANO BUGGY	GREENFIELD	GREENFIELD REPUBLICAN	
5	1903	NO. 1 END SPRING PIANO BUGGY	GREENFIELD	GREENFIELD REPUBLICAN	
5	1903	NO. 1 END SPRING PIANO BUGGY	NEW MARTINSBURG	GREENFIELD REPUBLICAN	
5	1903	NO. 1 END SPRING PIANO BUGGY	JEFFERSONVILLE	GREENFIELD REPUBLICAN	
5	1903	NO. 1 END SPRING PIANO BUGGY	HIGHLAND	GREENFIELD REPUBLICAN	
3	1904	NO. 1 PHAETON SEATED BUGGY	EAST MONROE	GREENFIELD REPUBLICAN	
3	1904	NO. 1 PHAETON SEATED BUGGY	GREENFIELD	GREENFIELD REPUBLICAN	
3	1904	NO. 1 PHAETON SEATED BUGGY	SOUTH SALEM	GREENFIELD REPUBLICAN	
3	1904	NO. 1 PHAETON SEATED BUGGY	GREENFIELD	GREENFIELD REPUBLICAN	

Ad Month	Ad Year	Vehicle Type	Client Location	Ad Source	Comments
3	1904	NO. 1 PHAETON SEATED BUGGY	LYNDON	GREENFIELD REPUBLICAN	
4	1904	NO. 1 PHAETON SEATED BUGGY	LYNDON	GREENFIELD REPUBLICAN	
4	1904	NO. 1 PHAETON SEATED BUGGY	GREENFIELD	GREENFIELD REPUBLICAN	
4	1904	NO. 1 PHAETON SEATED BUGGY	GREENFIELD	GREENFIELD REPUBLICAN	
4	1904	NO. 1 PHAETON SEATED BUGGY	GREENFIELD	GREENFIELD REPUBLICAN	
5	1904	NO. 1 PHAETON SEATED BUGGY	GREENFIELD	GREENFIELD REPUBLICAN	
5	1904	NO. 1 PHAETON SEATED BUGGY	GREENFIELD	GREENFIELD REPUBLICAN	
5	1904	NO. 1 PHAETON SEATED BUGGY	GREENFIELD	GREENFIELD REPUBLICAN	
5	1904	NO. 1 PHAETON SEATED BUGGY	LYNDON	GREENFIELD REPUBLICAN	
5	1904	NO. 1 PHAETON SEATED BUGGY	GREENFIELD	GREENFIELD REPUBLICAN	
5	1904	NO. 1 PHAETON SEATED BUGGY	NEW MARTINSBURG	GREENFIELD REPUBLICAN	
5	1904	NO. 1 PHAETON SEATED BUGGY	GREENFIELD	GREENFIELD REPUBLICAN	
5	1904	NO. 1 PHAETON SEATED BUGGY	GREENFIELD	GREENFIELD REPUBLICAN	
5	1904	NO. 1 PHAETON SEATED BUGGY	NEW VIENNA	GREENFIELD REPUBLICAN	
6	1904	NO. 1 PHAETON SEATED BUGGY	BRIDGES	GREENFIELD REPUBLICAN	
6	1904	NO. 1 PHAETON SEATED BUGGY	BAINBRIDGE	GREENFIELD REPUBLICAN	
6	1904	NO. 1 PHAETON SEATED BUGGY	CAREYTOWN	GREENFIELD REPUBLICAN	
6	1904	NO. 1 PHAETON SEATED BUGGY	NEW VIENNA	GREENFIELD REPUBLICAN	
6	1904	NO. 1 PHAETON SEATED BUGGY	BLANCHESTER	GREENFIELD REPUBLICAN	
6	1904	NO. 1 PHAETON SEATED BUGGY	HILLSBORO	GREENFIELD REPUBLICAN	
6	1904	NO. 1 PHAETON SEATED BUGGY	BAINBRIDGE	GREENFIELD REPUBLICAN	
6	1904	NO. 1 PHAETON SEATED BUGGY	LYNDON	GREENFIELD REPUBLICAN	
6	1904	NO. 1 PHAETON SEATED BUGGY	SOUTH SALEM	GREENFIELD REPUBLICAN	
12	1905	NO. 1 PIANO BUGGY	GREENFIELD	GREENFIELD REPUBLICAN	
1	1906	NO. 1 PIANO BUGGY	LEESBURG	GREENFIELD REPUBLICAN	
1	1906	NO. 1 PIANO BUGGY	GREENFIELD	GREENFIELD REPUBLICAN	
1	1906	NO. 1 PIANO BUGGY	GREENFIELD	GREENFIELD REPUBLICAN	
1	1906	NO. 1 PIANO BUGGY	GREENFIELD	GREENFIELD REPUBLICAN	
1	1906	NO. 10	GREENFIELD	GREENFIELD REPUBLICAN	
2	1906	NO. 1 PIANO BUGGY	AUSTIN	GREENFIELD REPUBLICAN	
2	1906	NO. 1 PIANO BUGGY	CYNTHIANA	GREENFIELD REPUBLICAN	

Ad Month	Ad Year	Vehicle Type	Client Location	Ad Source	Comments
2	1906	NO. 1 PIANO BUGGY	GREENFIELD	GREENFIELD REPUBLICAN	
2	1906	NO. 1 PIANO BUGGY	GREENFIELD	GREENFIELD REPUBLICAN	
2	1906	NO. 1 PIANO BUGGY	GREENFIELD	GREENFIELD REPUBLICAN	
2	1906	NO. 1 PIANO BUGGY	GREENFIELD	GREENFIELD REPUBLICAN	
2	1906	NO. 1 PIANO BUGGY	CUBA	GREENFIELD REPUBLICAN	
3	1906	NO. 1 PIANO BUGGY	MARTINSVILLE	GREENFIELD REPUBLICAN	
3	1906	NO. 1 PIANO BUGGY	GREENFIELD	GREENFIELD REPUBLICAN	
3	1906	NO. 1 PIANO BUGGY	RUSSELLVILLE	GREENFIELD REPUBLICAN	
3	1906	NO. 1 PIANO BUGGY	GREENFIELD	GREENFIELD REPUBLICAN	
3	1906	NO. 1 PIANO BUGGY	RAINSBORO	GREENFIELD REPUBLICAN	
3	1906	SURREY	GREENFIELD	GREENFIELD REPUBLICAN	
3	1906	SURREY	GREENFIELD	GREENFIELD REPUBLICAN	
3	1906	NO. 1 PIANO BUGGY	GREENFIELD	GREENFIELD REPUBLICAN	
3	1906	NO. 1 PIANO BUGGY	GREENFIELD	GREENFIELD REPUBLICAN	
3	1906	NO. 1 PIANO BUGGY	GREENFIELD	GREENFIELD REPUBLICAN	J. A. PORTER'S NEW LIVERY
3	1906	NO. 1 PIANO BUGGY	GREENFIELD	GREENFIELD REPUBLICAN	J. A. PORTER'S NEW LIVERY
3	1906	NO. 1 PIANO BUGGY	GREENFIELD	GREENFIELD REPUBLICAN	J. A. PORTER'S NEW LIVERY
4	1906	NO. 1 PIANO BUGGY	BAINBRIDGE	GREENFIELD REPUBLICAN	
4	1906	NO. 1 PIANO BUGGY	BAINBRIDGE	GREENFIELD REPUBLICAN	
4	1906	NO. 1 PIANO BUGGY	BAINBRIDGE	GREENFIELD REPUBLICAN	
4	1906	NO. 1 PIANO BUGGY	JEFFERSONVILLE	GREENFIELD REPUBLICAN	
4	1906	NO. 1 PIANO BUGGY	JEFFERSONVILLE	GREENFIELD REPUBLICAN	
4	1906	NO. 1 PIANO BUGGY	JEFFERSONVILLE	GREENFIELD REPUBLICAN	
4	1906	NO. 1 PIANO BUGGY	JEFFERSONVILLE	GREENFIELD REPUBLICAN	
4	1906	NO. 1 PIANO BUGGY	JEFFERSONVILLE	GREENFIELD REPUBLICAN	
4	1906	NO. 1 PIANO BUGGY	JEFFERSONVILLE	GREENFIELD REPUBLICAN	
4	1906	NO. 1 PIANO BUGGY	GREENFIELD	GREENFIELD REPUBLICAN	
4	1906	NO. 1 PIANO BUGGY	GREENFIELD	GREENFIELD REPUBLICAN	
5	1906	NO. 1 PIANO BUGGY	WASHINGTON COURT HOUSE	GREENFIELD REPUBLICAN	
5	1906	NO. 1 PIANO BUGGY	LEESBURG	GREENFIELD REPUBLICAN	PHYSICIAN
5	1906	FINE STANHOPE	WASHINGTON COURT HOUSE	GREENFIELD REPUBLICAN	

Ad Month	Ad Year	Vehicle Type	Client Location	Ad Source	Comments
5	1906	FINE STANHOPE	GOOD HOPE	GREENFIELD REPUBLICAN	
5	1906	NO. 1 PIANO BUGGY	BAINBRIDGE	GREENFIELD REPUBLICAN	
5	1906	VERY FINE CARRIAGE	WASHINGTON COURT HOUSE	GREENFIELD REPUBLICAN	
5	1906	VERY FINE CARRIAGE	WASHINGTON COURT HOUSE	GREENFIELD REPUBLICAN	
5	1906	NO. 1 FINE STATION WAGON	GREENFIELD	GREENFIELD REPUBLICAN	
5	1906	NO. 41 BIKE RUNABOUT	BOWLING GREEN, KY	GREENFIELD REPUBLICAN	PHYSICIAN
5	1906	STANHOPE	GREENFIELD	GREENFIELD REPUBLICAN	
6	1906	NO. 1 PIANO BUGGY	SOUTH SALEM	GREENFIELD REPUBLICAN	
6	1906	NO. 1 PIANO BUGGY	GREENFIELD	GREENFIELD REPUBLICAN	
6	1906	NO. 1 PIANO BUGGY	GREENFIELD	GREENFIELD REPUBLICAN	
6	1906	NO. 1 PIANO BUGGY	GREENFIELD	GREENFIELD REPUBLICAN	
6	1906	NO. 1 PIANO BUGGY	GREENFIELD	GREENFIELD REPUBLICAN	
7	1906	FINE PHAETON	GREENFIELD	GREENFIELD REPUBLICAN	
7	1906	FINE STRAIGHT SILL SURREY	GOOD HOPE	GREENFIELD REPUBLICAN	
7	1906	NO. 1 PIANO BUGGY	GREENFIELD	GREENFIELD REPUBLICAN	
7	1906	NO. 1 PIANO BUGGY	GREENFIELD	GREENFIELD REPUBLICAN	
7	1906	FINE PHAETON	GREENFIELD	GREENFIELD REPUBLICAN	REVEREND J. F. COBAN
8	1906	NO. 1 PIANO BUGGY	NEW PETERSBURG	GREENFIELD REPUBLICAN	
8	1906	NO. 1 PIANO BUGGY	LYNDON	GREENFIELD REPUBLICAN	
8	1906	NO. 1 PIANO BUGGY	BUENA VISTA	GREENFIELD REPUBLICAN	
8	1906	NO. 1 PIANO BUGGY	GREENFIELD	GREENFIELD REPUBLICAN	
8	1906	NO. 1 PIANO BUGGY	LEESBURG	GREENFIELD REPUBLICAN	
8	1906	FINE LIGHT THREE QUARTER SURREY	GREENFIELD	GREENFIELD REPUBLICAN	
8	1906	NO. 1 PIANO BUGGY	GREENFIELD	GREENFIELD REPUBLICAN	
8	1906	NO. 1 PIANO BUGGY	GREENFIELD	GREENFIELD REPUBLICAN	
9	1906	NO. 1 PIANO BUGGY	GREENFIELD	GREENFIELD REPUBLICAN	
9	1906	NO. 1 PIANO BUGGY	WEST UNION	GREENFIELD REPUBLICAN	PHYSICIAN
9	1906	NO. 1 PIANO BUGGY	WEST UNION	GREENFIELD REPUBLICAN	
9	1906	NO. 1 PIANO BUGGY	FRUITDALE	GREENFIELD REPUBLICAN	
9	1906	NO. 1 PIANO BUGGY	GREENFIELD	GREENFIELD REPUBLICAN	
9	1906	NO. 1 PIANO BUGGY	DUNCANVILLE	GREENFIELD REPUBLICAN	

Ad Month	Ad Year	Vehicle Type	Client Location	Ad Source	Comments
9	1906	NO. 1 PIANO BUGGY	HAMILTON	GREENFIELD REPUBLICAN	
9	1906	NO. 1 PIANO BUGGY	HIGHLAND	GREENFIELD REPUBLICAN	
9	1906	NO. 1 PIANO BUGGY	AUSTIN	GREENFIELD REPUBLICAN	
9	1906	NO. 1 PIANO BUGGY	WASHINGTON COURT HOUSE	GREENFIELD REPUBLICAN	
10	1906	NO. 1 PIANO BUGGY	GREENFIELD	GREENFIELD REPUBLICAN	
10	1906	NO. 1 PIANO BUGGY	WASHINGTON COURT HOUSE	GREENFIELD REPUBLICAN	
10	1906	NO. 1 PIANO BUGGY	SOUTH SALEM	GREENFIELD REPUBLICAN	
10	1906	NO. 1 PIANO BUGGY	EAST MONROE	GREENFIELD REPUBLICAN	
10	1906	NO. 1 PIANO BUGGY	GREENFIELD	GREENFIELD REPUBLICAN	
10	1906	VERY FINE THREE QUARTER SURREY	EAST MONROE	GREENFIELD REPUBLICAN	
11	1906	NO. 1 PIANO BUGGY	GREENFIELD	GREENFIELD REPUBLICAN	
11	1906	NO. 1 PIANO BUGGY	AUSTIN	GREENFIELD REPUBLICAN	
11	1906	NO. 1 PIANO BUGGY	THORNTON	GREENFIELD REPUBLICAN	
11	1906	NO. 1 PIANO BUGGY	WASHINGTON COURT HOUSE	GREENFIELD REPUBLICAN	
11	1906	NO. 1 PIANO BUGGY	GREENFIELD	GREENFIELD REPUBLICAN	
11	1906	NO. 1 PIANO BUGGY	GOOD HOPE	GREENFIELD REPUBLICAN	
2	1907	NO. 1 PIANO BUGGY	GREENFIELD	GREENFIELD REPUBLICAN	
3	1907	NO. 1 PIANO BUGGY	GREENFIELD	GREENFIELD REPUBLICAN	
3	1907	NO. 1 PIANO BUGGY	NELSONVILLE	GREENFIELD REPUBLICAN	
4	1907	NO. 1 PIANO BUGGY	WILMINGTON	GREENFIELD REPUBLICAN	
5	1907	NO. 1 PIANO BUGGY	GREENFIELD	GREENFIELD REPUBLICAN	
5	1907	NO. 1 PIANO BUGGY	GREENFIELD	GREENFIELD REPUBLICAN	
5	1907	NO. 1 PIANO BUGGY	SOUTH SALEM	GREENFIELD REPUBLICAN	
5	1907	NO. 1 PIANO BUGGY	GREENFIELD	GREENFIELD REPUBLICAN	
5	1907	NO. 1 PIANO BUGGY	EAST MONROE	GREENFIELD REPUBLICAN	
5	1907	CURTAIN STATION WAGON	GREENFIELD	GREENFIELD REPUBLICAN	
5	1907	NO. 40 STRAIGHT SILL SURREY	GREENFIELD	GREENFIELD REPUBLICAN	
5	1907	NO. 40 STRAIGHT SILL SURREY	GREENFIELD	GREENFIELD REPUBLICAN	
6	1907	NO. 1 PIANO BUGGY	NEW MARTINSBURG	GREENFIELD REPUBLICAN	
6	1907	NO. 1 PIANO BUGGY	SINKING SPRING	GREENFIELD REPUBLICAN	
6	1907	NO. 1 PIANO BUGGY	GREENFIELD	GREENFIELD REPUBLICAN	

Ad Month	Ad Year	Vehicle Type	Client Location	Ad Source	Comments
6	1907	NO. 1 PIANO BUGGY	SOUTH SALEM	GREENFIELD REPUBLICAN	
6	1907	NO. 12 FINE AUTO SEAT RUNABOUT	GREENFIELD	GREENFIELD REPUBLICAN	
6	1907	NO. 12 FINE AUTO SEAT RUNABOUT	GREENFIELD	GREENFIELD REPUBLICAN	
6	1907	NO. 12 FINE AUTO SEAT RUNABOUT	GREENFIELD	GREENFIELD REPUBLICAN	
6	1907	NO. 12 FINE AUTO SEAT RUNABOUT	GREENFIELD	GREENFIELD REPUBLICAN	
6	1907	NO. 12 FINE AUTO SEAT RUNABOUT	GREENFIELD	GREENFIELD REPUBLICAN	
6	1907	NO. 12 FINE AUTO SEAT RUNABOUT	GREENFIELD	GREENFIELD REPUBLICAN	
6	1907	NO. 12 FINE AUTO SEAT RUNABOUT	GREENFIELD	GREENFIELD REPUBLICAN	
6	1907	NO. 12 FINE AUTO SEAT RUNABOUT	GREENFIELD	GREENFIELD REPUBLICAN	
6	1907	NO. 12 FINE AUTO SEAT RUNABOUT	UNKNOWN, GEORGIA	GREENFIELD REPUBLICAN	
6	1907	NO. 12 FINE AUTO SEAT RUNABOUT	UNKNOWN, RHODE ISLAND	GREENFIELD REPUBLICAN	
7	1907	NO. 1 PIANO BUGGY	GREENFIELD	GREENFIELD REPUBLICAN	
7	1907	NO. 1 PIANO BUGGY	GREENFIELD	GREENFIELD REPUBLICAN	
7	1907	NO. 1 PIANO BUGGY	GREENFIELD	GREENFIELD REPUBLICAN	
7	1907	NO. 1 PIANO BUGGY	GREENFIELD	GREENFIELD REPUBLICAN	
7	1907	NO. 1 PIANO BUGGY	LEESBURG	GREENFIELD REPUBLICAN	
7	1907	NO. 1 PIANO BUGGY	BAINBRIDGE	GREENFIELD REPUBLICAN	
7	1907	NO. 12 FINE AUTO SEAT RUNABOUT	LEESBURG	GREENFIELD REPUBLICAN	
7	1907	NO. 12 FINE AUTO SEAT RUNABOUT	EAST MONROE	GREENFIELD REPUBLICAN	
7	1907	NO. 12 FINE AUTO SEAT RUNABOUT	WILMINGTON	GREENFIELD REPUBLICAN	
7	1907	NO. 100 DOCTOR'S PIANO BUGGY	UNKNOWN	GREENFIELD REPUBLICAN	PHYSICIAN
7	1907	NO. 100 DOCTOR'S PIANO BUGGY	UNKNOWN	GREENFIELD REPUBLICAN	PHYSICIAN
7	1907	NO. 1 PIANO BUGGY	FRUITDALE	GREENFIELD REPUBLICAN	
7	1907	NO. 1 PIANO BUGGY	CYNTHIANA	GREENFIELD REPUBLICAN	
7	1907	NO. 1 PIANO BUGGY	RAINSBORO	GREENFIELD REPUBLICAN	
7	1907	NO. 1 PIANO BUGGY	BAINBRIDGE	GREENFIELD REPUBLICAN	
7	1907	NO. 1 PIANO BUGGY	BAINBRIDGE	GREENFIELD REPUBLICAN	PHYSICIAN
7	1907	NO. 1 PIANO BUGGY	NEW MARTINSBURG	GREENFIELD REPUBLICAN	
7	1907	NO. 1 PIANO BUGGY	WILMINGTON	GREENFIELD REPUBLICAN	
7	1907	NO. 1 PIANO BUGGY	BOURNEVILLE	GREENFIELD REPUBLICAN	
7	1907	NO. 1 PIANO BUGGY	GREENFIELD	GREENFIELD REPUBLICAN	

Ad Month	Ad Year	Vehicle Type	Client Location	Ad Source	Comments
7	1907	NO. 1 PIANO BUGGY	GREENFIELD	GREENFIELD REPUBLICAN	
3	1908	NO. 1 PIANO BUGGY	GREENFIELD	GREENFIELD REPUBLICAN	
3	1908	NO. 1 PIANO BUGGY	GREENFIELD	GREENFIELD REPUBLICAN	
3	1908	NO. 1 PIANO BUGGY	GREENFIELD	GREENFIELD REPUBLICAN	
3	1908	NO. 1 PIANO BUGGY	NEW VIENNA	GREENFIELD REPUBLICAN	
3	1908	NO. 1 PIANO BUGGY	LEESBURG	GREENFIELD REPUBLICAN	
3	1908	NO. 1 PIANO BUGGY	BAINBRIDGE	GREENFIELD REPUBLICAN	
3	1908	NO. 1 PIANO BUGGY	LYNDON	GREENFIELD REPUBLICAN	
3	1908	NO. 1 PIANO BUGGY	LYNDON	GREENFIELD REPUBLICAN	
3	1908	NO. 1 PIANO BUGGY	LYNDON	GREENFIELD REPUBLICAN	
3	1908	NO. 1 PIANO BUGGY	GREENFIELD	GREENFIELD REPUBLICAN	
3	1908	NO. 1 PIANO BUGGY	LEESBURG	GREENFIELD REPUBLICAN	
3	1908	NO. 1 PIANO BUGGY	GREENFIELD	GREENFIELD REPUBLICAN	PHYSICIAN
3	1908	NO. 1 PIANO BUGGY	LEESBURG	GREENFIELD REPUBLICAN	
3	1908	NO. 1 PIANO BUGGY	XENIA	GREENFIELD REPUBLICAN	
3	1908	NO. 1 PIANO BUGGY	GREENFIELD	GREENFIELD REPUBLICAN	
3	1908	NO. 1 PIANO BUGGY	WILBERFORCE	GREENFIELD REPUBLICAN	
3	1908	NO. 1 PIANO BUGGY	GREENFIELD	GREENFIELD REPUBLICAN	
4	1908	20 INCH BUGGIES	GREENFIELD	GREENFIELD REPUBLICAN	
4	1908	20 INCH BUGGIES	XENIA	GREENFIELD REPUBLICAN	
4	1908	20 INCH BUGGIES	MOWRYSTOWN	GREENFIELD REPUBLICAN	
4	1908	20 INCH BUGGIES	GREENFIELD	GREENFIELD REPUBLICAN	
4	1908	20 INCH BUGGIES	GREENFIELD	GREENFIELD REPUBLICAN	
4	1908	20 INCH BUGGIES	HILLSBORO	GREENFIELD REPUBLICAN	
4	1908	UNKNOWN	SOUTH SALEM	GREENFIELD REPUBLICAN	
4	1908	UNKNOWN	LEESBURG	GREENFIELD REPUBLICAN	
4	1908	UNKNOWN	WASHINGTON COURT HOUSE	GREENFIELD REPUBLICAN	
4	1908	UNKNOWN	GREENFIELD	GREENFIELD REPUBLICAN	
4	1908	UNKNOWN	BAINBRIDGE	GREENFIELD REPUBLICAN	
4	1908	UNKNOWN	GREENFIELD	GREENFIELD REPUBLICAN	
4	1908	UNKNOWN	MOUNT STERLING	GREENFIELD REPUBLICAN	

Ad Month	Ad Year	Vehicle Type	Client Location	Ad Source	Comments
4	1908	UNKNOWN	FRUITDALE	GREENFIELD REPUBLICAN	
4	1908	UNKNOWN	GREENFIELD	GREENFIELD REPUBLICAN	
4	1908	UNKNOWN	SPRINGFIELD	GREENFIELD REPUBLICAN	
4	1908	UNKNOWN	GREENFIELD	GREENFIELD REPUBLICAN	
4	1908	UNKNOWN	LYNDON	GREENFIELD REPUBLICAN	
4	1908	UNKNOWN	GREENFIELD	GREENFIELD REPUBLICAN	
4	1908	UNKNOWN	SOUTH SALEM	GREENFIELD REPUBLICAN	
4	1908	UNKNOWN	GREENFIELD	GREENFIELD REPUBLICAN	
4	1908	UNKNOWN	NELSONVILLE	GREENFIELD REPUBLICAN	
4	1908	UNKNOWN	GREENFIELD	GREENFIELD REPUBLICAN	BAKERY
4	1908	UNKNOWN	GREENFIELD	GREENFIELD REPUBLICAN	DAIRY
4	1908	UNKNOWN	GREENFIELD	GREENFIELD REPUBLICAN	BAKERY
4	1908	UNKNOWN	GREENFIELD	GREENFIELD REPUBLICAN	
4	1908	UNKNOWN	WEST UNION	GREENFIELD REPUBLICAN	
4	1908	UNKNOWN	WEST UNION	GREENFIELD REPUBLICAN	
4	1908	UNKNOWN	LEESBURG	GREENFIELD REPUBLICAN	
4	1908	UNKNOWN	LEESBURG	GREENFIELD REPUBLICAN	
4	1908	UNKNOWN	HILLSBORO	GREENFIELD REPUBLICAN	PHYSICIAN
4	1908	UNKNOWN	WASHINGTON COURT HOUSE	GREENFIELD REPUBLICAN	
4	1908	UNKNOWN	WELLSTON	GREENFIELD REPUBLICAN	LIVERY
4	1908	UNKNOWN	SOUTH SALEM	GREENFIELD REPUBLICAN	
4	1908	UNKNOWN	GREENFIELD	GREENFIELD REPUBLICAN	
4	1908	UNKNOWN	WASHINGTON COURT HOUSE	GREENFIELD REPUBLICAN	
4	1908	UNKNOWN	GREENFIELD	GREENFIELD REPUBLICAN	
4	1908	UNKNOWN	BAINBRIDGE	GREENFIELD REPUBLICAN	
4	1908	UNKNOWN	WASHINGTON COURT HOUSE	GREENFIELD REPUBLICAN	PHYSICIAN
5	1908	UNKNOWN	LEESBURG	GREENFIELD REPUBLICAN	
5	1908	UNKNOWN	GREENFIELD	GREENFIELD REPUBLICAN	
5	1908	UNKNOWN	BAINBRIDGE	GREENFIELD REPUBLICAN	
5	1908	UNKNOWN	EAST MONROE	GREENFIELD REPUBLICAN	
5	1908	UNKNOWN	NEW PETERSBURG	GREENFIELD REPUBLICAN	

Ad Month	Ad Year	Vehicle Type	Client Location	Ad Source	Comments
5	1908	UNKNOWN	LEESBURG	GREENFIELD REPUBLICAN	
5	1908	UNKNOWN	LYNDON	GREENFIELD REPUBLICAN	
5	1908	UNKNOWN	LEESBURG	GREENFIELD REPUBLICAN	
5	1908	UNKNOWN	CAREYTOWN	GREENFIELD REPUBLICAN	
5	1908	UNKNOWN	GOOD HOPE	GREENFIELD REPUBLICAN	
5	1908	UNKNOWN	CHILLICOTHE	GREENFIELD REPUBLICAN	
5	1908	UNKNOWN	WASHINGTON COURT HOUSE	GREENFIELD REPUBLICAN	
5	1908	UNKNOWN	GREENFIELD	GREENFIELD REPUBLICAN	
5	1908	UNKNOWN	GREENFIELD	GREENFIELD REPUBLICAN	
5	1908	UNKNOWN	WASHINGTON COURT HOUSE	GREENFIELD REPUBLICAN	
5	1908	UNKNOWN	GREENFIELD	GREENFIELD REPUBLICAN	
6	1908	UNKNOWN	FRANKFORT	GREENFIELD REPUBLICAN	
6	1908	UNKNOWN	GREENFIELD	GREENFIELD REPUBLICAN	
6	1908	UNKNOWN	LEESBURG	GREENFIELD REPUBLICAN	
6	1908	UNKNOWN	CEDARVILLE	GREENFIELD REPUBLICAN	
6	1908	UNKNOWN	GREENFIELD	GREENFIELD REPUBLICAN	
6	1908	UNKNOWN	GREENFIELD	GREENFIELD REPUBLICAN	
6	1908	UNKNOWN	EAST MONROE	GREENFIELD REPUBLICAN	
6	1908	UNKNOWN	WASHINGTON COURT HOUSE	GREENFIELD REPUBLICAN	
6	1908	UNKNOWN	WASHINGTON COURT HOUSE	GREENFIELD REPUBLICAN	
6	1908	UNKNOWN	CYNTHIANA	GREENFIELD REPUBLICAN	
6	1908	UNKNOWN	HILLSBORO	GREENFIELD REPUBLICAN	
6	1908	UNKNOWN	GREENFIELD	GREENFIELD REPUBLICAN	
6	1908	UNKNOWN	HIGHLAND	GREENFIELD REPUBLICAN	
6	1908	UNKNOWN	GREENFIELD	GREENFIELD REPUBLICAN	
6	1908	UNKNOWN	MOUNT STERLING	GREENFIELD REPUBLICAN	
6	1908	UNKNOWN	AUSTIN	GREENFIELD REPUBLICAN	
6	1908	UNKNOWN	MOUNT STERLING	GREENFIELD REPUBLICAN	
6	1908	UNKNOWN	GREENFIELD	GREENFIELD REPUBLICAN	
6	1908	UNKNOWN	LYNDON	GREENFIELD REPUBLICAN	
6	1908	UNKNOWN	GREENFIELD	GREENFIELD REPUBLICAN	

Ad Month	Ad Year	Vehicle Type	Client Location	Ad Source	Comments
6	1908	UNKNOWN	GREENFIELD	GREENFIELD REPUBLICAN	
6	1908	UNKNOWN	GREENFIELD	GREENFIELD REPUBLICAN	
6	1908	UNKNOWN	GREENFIELD	GREENFIELD REPUBLICAN	
6	1908	UNKNOWN	GREENFIELD	GREENFIELD REPUBLICAN	
6	1908	UNKNOWN	BAINBRIDGE	GREENFIELD REPUBLICAN	
6	1908	UNKNOWN	GREENFIELD	GREENFIELD REPUBLICAN	
6	1908	UNKNOWN	GREENFIELD	GREENFIELD REPUBLICAN	
6	1908	UNKNOWN	LEESBURG	GREENFIELD REPUBLICAN	
7	1908	LIGHT SURREY	GREENFIELD	GREENFIELD REPUBLICAN	
7	1908	LIGHT SURREY	GREENFIELD	GREENFIELD REPUBLICAN	
7	1908	LIGHT SURREY	GREENFIELD	GREENFIELD REPUBLICAN	
7	1908	LIGHT SURREY	GREENFIELD	GREENFIELD REPUBLICAN	
7	1908	LIGHT SURREY	WELLSTON	GREENFIELD REPUBLICAN	LEACH UNDERTAKING COMPANY
7	1908	LIGHT SURREY	NEW VIENNA	GREENFIELD REPUBLICAN	
7	1908	LIGHT SURREY	GREENFIELD	GREENFIELD REPUBLICAN	
7	1908	LIGHT SURREY	GREENFIELD	GREENFIELD REPUBLICAN	
7	1908	LIGHT SURREY	GREENFIELD	GREENFIELD REPUBLICAN	
7	1908	LIGHT SURREY	GREENFIELD	GREENFIELD REPUBLICAN	
7	1908	LIGHT SURREY	LYNDON	GREENFIELD REPUBLICAN	
7	1908	LIGHT SURREY	BOURNEVILLE	GREENFIELD REPUBLICAN	
7	1908	LIGHT SURREY	GREENFIELD	GREENFIELD REPUBLICAN	
7	1908	LIGHT SURREY	GREENFIELD	GREENFIELD REPUBLICAN	
7	1908	LIGHT SURREY	GREENFIELD	GREENFIELD REPUBLICAN	
7	1908	LIGHT SURREY	GREENFIELD	GREENFIELD REPUBLICAN	
7	1908	LIGHT SURREY	GREENFIELD	GREENFIELD REPUBLICAN	
7	1908	LIGHT SURREY	GREENFIELD	GREENFIELD REPUBLICAN	
7	1908	LIGHT SURREY	GREENFIELD	GREENFIELD REPUBLICAN	
8	1908	LIGHT SURREY	GREENFIELD	GREENFIELD REPUBLICAN	PHYSICIAN
8	1908	LIGHT SURREY	GREENFIELD	GREENFIELD REPUBLICAN	COUNTY COMMISSIONER
8	1908	LIGHT SURREY	GREENFIELD	GREENFIELD REPUBLICAN	
8	1908	LIGHT SURREY	GREENFIELD	GREENFIELD REPUBLICAN	

Ad Month	Ad Year	Vehicle Type	Client Location	Ad Source	Comments
8	1908	LIGHT SURREY	GREENFIELD	GREENFIELD REPUBLICAN	
8	1908	LIGHT SURREY	GREENFIELD	GREENFIELD REPUBLICAN	
8	1908	LIGHT SURREY	GREENFIELD	GREENFIELD REPUBLICAN	
8	1908	LIGHT SURREY	GREENFIELD	GREENFIELD REPUBLICAN	
8	1908	LIGHT SURREY	GREENFIELD	GREENFIELD REPUBLICAN	
8	1908	LIGHT SURREY	GREENFIELD	GREENFIELD REPUBLICAN	
8	1908	LIGHT SURREY	GREENFIELD	GREENFIELD REPUBLICAN	
8	1908	LIGHT SURREY	GREENFIELD	GREENFIELD REPUBLICAN	
8	1908	LIGHT SURREY	HILLSBORO	GREENFIELD REPUBLICAN	
8	1908	LIGHT SURREY	SOUTH SALEM	GREENFIELD REPUBLICAN	
8	1908	LIGHT SURREY	LEESBURG	GREENFIELD REPUBLICAN	
8	1908	LIGHT SURREY	GREENFIELD	GREENFIELD REPUBLICAN	
8	1908	LIGHT SURREY	AUSTIN	GREENFIELD REPUBLICAN	
8	1908	LIGHT SURREY	JACKSON	GREENFIELD REPUBLICAN	
8	1908	LIGHT SURREY	AUSTIN	GREENFIELD REPUBLICAN	
8	1908	LIGHT SURREY	LEESBURG	GREENFIELD REPUBLICAN	
8	1908	LIGHT SURREY	GREENFIELD	GREENFIELD REPUBLICAN	
8	1908	LIGHT SURREY	GREENFIELD	GREENFIELD REPUBLICAN	
8	1908	LIGHT SURREY	FRUITDALE	GREENFIELD REPUBLICAN	
8	1908	LIGHT SURREY	HILLSBORO	GREENFIELD REPUBLICAN	
8	1908	LIGHT SURREY	GREENFIELD	GREENFIELD REPUBLICAN	
9	1908	NO. 1 PIANO BUGGY	GREENFIELD	GREENFIELD REPUBLICAN	
9	1908	NO. 1 PIANO BUGGY	FRANKFORT	GREENFIELD REPUBLICAN	
9	1908	NO. 1 PIANO BUGGY	GREENFIELD	GREENFIELD REPUBLICAN	
9	1908	NO. 1 PIANO BUGGY	NEW MARTINSBURG	GREENFIELD REPUBLICAN	
9	1908	NO. 1 PIANO BUGGY	NEW PETERSBURG	GREENFIELD REPUBLICAN	
9	1908	NO. 1 PIANO BUGGY	GREENFIELD	GREENFIELD REPUBLICAN	
9	1908	NO. 1 PIANO BUGGY	NEW PETERSBURG	GREENFIELD REPUBLICAN	
9	1908	NO. 1 PIANO BUGGY	GREENFIELD	GREENFIELD REPUBLICAN	
9	1908	NO. 1 PIANO BUGGY	GREENFIELD	GREENFIELD REPUBLICAN	
9	1908	NO. 1 PIANO BUGGY	GREENFIELD	GREENFIELD REPUBLICAN	

Ad Month	Ad Year	Vehicle Type	Client Location	Ad Source	Comments
9	1908	NO. 1 PIANO BUGGY	GREENFIELD	GREENFIELD REPUBLICAN	
10	1908	NO. 1 PIANO BUGGY	GREENFIELD	GREENFIELD REPUBLICAN	
10	1908	NO. 1 PIANO BUGGY	SOUTH SALEM	GREENFIELD REPUBLICAN	
10	1908	NO. 1 PIANO BUGGY	LYNDON	GREENFIELD REPUBLICAN	
10	1908	NO. 1 PIANO BUGGY	GREENFIELD	GREENFIELD REPUBLICAN	
10	1908	NO. 1 PIANO BUGGY	HILLSBORO	GREENFIELD REPUBLICAN	
11	1908	NO. 1 PIANO BUGGY	GREENFIELD	GREENFIELD REPUBLICAN	
11	1908	NO. 1 PIANO BUGGY	GREENFIELD	GREENFIELD REPUBLICAN	
11	1908	FINE STANHOPE	GREENFIELD	GREENFIELD REPUBLICAN	
12	1908	LIGHT SURREY	GREENFIELD	GREENFIELD REPUBLICAN	
1	1909	STORM BUGGY	GREENFIELD	GREENFIELD REPUBLICAN	PHYSICIAN
1	1909	STORM BUGGY	GREENFIELD	GREENFIELD REPUBLICAN	PHYSICIAN
1	1909	STORM BUGGY	GREENFIELD	GREENFIELD REPUBLICAN	ATTORNEY
2	1909	UNKNOWN	GREENFIELD	GREENFIELD REPUBLICAN	
2	1909	UNKNOWN	GREENFIELD	GREENFIELD REPUBLICAN	
2	1909	UNKNOWN	GREENFIELD	GREENFIELD REPUBLICAN	
2	1909	UNKNOWN	GREENFIELD	GREENFIELD REPUBLICAN	
2	1909	UNKNOWN	GREENFIELD	GREENFIELD REPUBLICAN	
2	1909	UNKNOWN	GREENFIELD	GREENFIELD REPUBLICAN	
2	1909	UNKNOWN	GREENFIELD	GREENFIELD REPUBLICAN	
2	1909	UNKNOWN	RAINSBORO	GREENFIELD REPUBLICAN	
2	1909	UNKNOWN	GREENFIELD	GREENFIELD REPUBLICAN	
2	1909	UNKNOWN	ROXABELL	GREENFIELD REPUBLICAN	
2	1909	UNKNOWN	WELLSTON	GREENFIELD REPUBLICAN	JACKSON LIVERY & SALE COMPANY
2	1909	UNKNOWN	WELLSTON	GREENFIELD REPUBLICAN	JACKSON LIVERY & SALE COMPANY
2	1909	UNKNOWN	WELLSTON	GREENFIELD REPUBLICAN	JACKSON LIVERY & SALE COMPANY
2	1909	UNKNOWN	WELLSTON	GREENFIELD REPUBLICAN	JACKSON LIVERY & SALE COMPANY
2	1909	UNKNOWN	WELLSTON	GREENFIELD REPUBLICAN	JACKSON LIVERY & SALE COMPANY
2	1909	UNKNOWN	WELLSTON	GREENFIELD REPUBLICAN	JACKSON LIVERY & SALE COMPANY
2	1909	UNKNOWN	WELLSTON	GREENFIELD REPUBLICAN	JACKSON LIVERY & SALE COMPANY
2	1909	UNKNOWN	WELLSTON	GREENFIELD REPUBLICAN	JACKSON LIVERY & SALE COMPANY

Ad Month	Ad Year	Vehicle Type	Client Location	Ad Source	Comments
2	1909	UNKNOWN	WELLSTON	GREENFIELD REPUBLICAN	JACKSON LIVERY & SALE COMPANY
2	1909	UNKNOWN	WELLSTON	GREENFIELD REPUBLICAN	JACKSON LIVERY & SALE COMPANY
2	1909	RUNABOUT	PUERTO RICO	GREENFIELD REPUBLICAN	PUERTO RICO DEPT. OF EDUCATION
3	1909	UNKNOWN	GREENFIELD	GREENFIELD REPUBLICAN	
3	1909	UNKNOWN	GOOD HOPE	GREENFIELD REPUBLICAN	
3	1909	UNKNOWN	WASHINGTON COURT HOUSE	GREENFIELD REPUBLICAN	
3	1909	UNKNOWN	LEESBURG	GREENFIELD REPUBLICAN	
3	1909	UNKNOWN	GREENFIELD	GREENFIELD REPUBLICAN	
3	1909	UNKNOWN	HILLSBORO	GREENFIELD REPUBLICAN	
3	1909	UNKNOWN	PUERTO RICO	GREENFIELD REPUBLICAN	PUERTO RICO
3	1909	UNKNOWN	HILLSBORO	GREENFIELD REPUBLICAN	
3	1909	UNKNOWN	SOUTH SALEM	GREENFIELD REPUBLICAN	
3	1909	UNKNOWN	GREENFIELD	GREENFIELD REPUBLICAN	
3	1909	UNKNOWN	GREENFIELD	GREENFIELD REPUBLICAN	
3	1909	UNKNOWN	SOUTH SALEM	GREENFIELD REPUBLICAN	
3	1909	UNKNOWN	GREENFIELD	GREENFIELD REPUBLICAN	
3	1909	UNKNOWN	WEST UNION	GREENFIELD REPUBLICAN	
4	1909	UNKNOWN	GREENFIELD	GREENFIELD REPUBLICAN	
4	1909	UNKNOWN	GREENFIELD	GREENFIELD REPUBLICAN	
4	1909	UNKNOWN	GREENFIELD	GREENFIELD REPUBLICAN	
4	1909	UNKNOWN	GREENFIELD	GREENFIELD REPUBLICAN	
4	1909	UNKNOWN	GREENFIELD	GREENFIELD REPUBLICAN	
4	1909	UNKNOWN	GREENFIELD	GREENFIELD REPUBLICAN	PHYSICIAN
4	1909	UNKNOWN	GREENFIELD	GREENFIELD REPUBLICAN	
4	1909	UNKNOWN	GREENFIELD	GREENFIELD REPUBLICAN	
4	1909	UNKNOWN	GREENFIELD	GREENFIELD REPUBLICAN	
4	1909	UNKNOWN	GREENFIELD	GREENFIELD REPUBLICAN	
4	1909	UNKNOWN	GREENFIELD	GREENFIELD REPUBLICAN	
4	1909	UNKNOWN	GREENFIELD	GREENFIELD REPUBLICAN	
4	1909	UNKNOWN	GREENFIELD	GREENFIELD REPUBLICAN	
4	1909	UNKNOWN	GREENFIELD	GREENFIELD REPUBLICAN	

Ad Month	Ad Year	Vehicle Type	Client Location	Ad Source	Comments
4	1909	UNKNOWN	GREENFIELD	GREENFIELD REPUBLICAN	
4	1909	UNKNOWN	GREENFIELD	GREENFIELD REPUBLICAN	
4	1909	UNKNOWN	GREENFIELD	GREENFIELD REPUBLICAN	
4	1909	UNKNOWN	GREENFIELD	GREENFIELD REPUBLICAN	
4	1909	UNKNOWN	GREENFIELD	GREENFIELD REPUBLICAN	
4	1909	UNKNOWN	GREENFIELD	GREENFIELD REPUBLICAN	
4	1909	UNKNOWN	GREENFIELD	GREENFIELD REPUBLICAN	
4	1909	UNKNOWN	GREENFIELD	GREENFIELD REPUBLICAN	
4	1909	UNKNOWN	GREENFIELD	GREENFIELD REPUBLICAN	
4	1909	UNKNOWN	GREENFIELD	GREENFIELD REPUBLICAN	
4	1909	UNKNOWN	GREENFIELD	GREENFIELD REPUBLICAN	
4	1909	UNKNOWN	GREENFIELD	GREENFIELD REPUBLICAN	
4	1909	UNKNOWN	GREENFIELD	GREENFIELD REPUBLICAN	
4	1909	UNKNOWN	GREENFIELD	GREENFIELD REPUBLICAN	
4	1909	UNKNOWN	GREENFIELD	GREENFIELD REPUBLICAN	
4	1909	UNKNOWN	GREENFIELD	GREENFIELD REPUBLICAN	
4	1909	UNKNOWN	GREENFIELD	GREENFIELD REPUBLICAN	
4	1909	UNKNOWN	GREENFIELD	GREENFIELD REPUBLICAN	
4	1909	UNKNOWN	GREENFIELD	GREENFIELD REPUBLICAN	
4	1909	UNKNOWN	GREENFIELD	GREENFIELD REPUBLICAN	
4	1909	UNKNOWN	GREENFIELD	GREENFIELD REPUBLICAN	
4	1909	UNKNOWN	GREENFIELD	GREENFIELD REPUBLICAN	
4	1909	UNKNOWN	GREENFIELD	GREENFIELD REPUBLICAN	
4	1909	UNKNOWN	GREENFIELD	GREENFIELD REPUBLICAN	
4	1909	UNKNOWN	GREENFIELD	GREENFIELD REPUBLICAN	
4	1909	UNKNOWN	GREENFIELD	GREENFIELD REPUBLICAN	
4	1909	UNKNOWN	GREENFIELD	GREENFIELD REPUBLICAN	
4	1909	UNKNOWN	GREENFIELD	GREENFIELD REPUBLICAN	PHYSICIAN
4	1909	UNKNOWN	GREENFIELD	GREENFIELD REPUBLICAN	
4	1909	UNKNOWN	GREENFIELD	GREENFIELD REPUBLICAN	
4	1909	UNKNOWN	GREENFIELD	GREENFIELD REPUBLICAN	

Ad Month	Ad Year	Vehicle Type	Client Location	Ad Source	Comments
4	1909	UNKNOWN	GREENFIELD	GREENFIELD REPUBLICAN	
4	1909	UNKNOWN	GREENFIELD	GREENFIELD REPUBLICAN	
4	1909	UNKNOWN	GREENFIELD	GREENFIELD REPUBLICAN	
7	1914	RUNABOUT	GREENFIELD	GREENFIELD REPUBLICAN	
12	1909	STORM BUGGY	GREENFIELD	HILLSBORO DISPATCH	
12	1909	STORM BUGGY	GREENFIELD	HILLSBORO DISPATCH	
1	1910	STORM BUGGY	AUSTIN	HILLSBORO DISPATCH	
1	1910	STORM BUGGY	LEESBURG	HILLSBORO DISPATCH	
1	1910	STORM BUGGY	LEESBURG	HILLSBORO DISPATCH	
1	1910	STORM BUGGY	GREENFIELD	HILLSBORO DISPATCH	PHYSICIAN
3	1910	UNKNOWN	GREENFIELD	HILLSBORO DISPATCH	PHYSICIAN
3	1910	UNKNOWN	LEESBURG	HILLSBORO DISPATCH	
3	1910	UNKNOWN	HILLSBORO	HILLSBORO DISPATCH	
3	1910	UNKNOWN	AUSTIN	HILLSBORO DISPATCH	
3	1910	UNKNOWN	MARSHALL	HILLSBORO DISPATCH	PHYSICIAN
3	1910	UNKNOWN	GREENFIELD	HILLSBORO DISPATCH	PHYSICIAN
3	1910	UNKNOWN	UNKNOWN	HILLSBORO DISPATCH	
3	1910	UNKNOWN	UNKNOWN	HILLSBORO DISPATCH	
3	1910	UNKNOWN	UNKNOWN	HILLSBORO DISPATCH	
3	1910	UNKNOWN	UNKNOWN	HILLSBORO DISPATCH	
3	1910	UNKNOWN	UNKNOWN	HILLSBORO DISPATCH	
3	1910	UNKNOWN	UNKNOWN	HILLSBORO DISPATCH	
3	1910	UNKNOWN	UNKNOWN	HILLSBORO DISPATCH	
3	1910	UNKNOWN	UNKNOWN	HILLSBORO DISPATCH	
3	1910	UNKNOWN	UNKNOWN	HILLSBORO DISPATCH	
3	1910	UNKNOWN	UNKNOWN	HILLSBORO DISPATCH	
3	1910	UNKNOWN	UNKNOWN	HILLSBORO DISPATCH	
3	1910	UNKNOWN	UNKNOWN	HILLSBORO DISPATCH	
3	1910	UNKNOWN	UNKNOWN	HILLSBORO DISPATCH	
3	1910	UNKNOWN	UNKNOWN	HILLSBORO DISPATCH	
3	1910	UNKNOWN	UNKNOWN	HILLSBORO DISPATCH	

Ad Month	Ad Year	Vehicle Type	Client Location	Ad Source	Comments
3	1910	UNKNOWN	UNKNOWN	HILLSBORO DISPATCH	
3	1910	UNKNOWN	UNKNOWN	HILLSBORO DISPATCH	
3	1910	UNKNOWN	UNKNOWN	HILLSBORO DISPATCH	
3	1910	UNKNOWN	UNKNOWN	HILLSBORO DISPATCH	PHYSICIAN
3	1910	UNKNOWN	UNKNOWN	HILLSBORO DISPATCH	
3	1910	UNKNOWN	UNKNOWN	HILLSBORO DISPATCH	
3	1910	UNKNOWN	UNKNOWN	HILLSBORO DISPATCH	
3	1910	UNKNOWN	UNKNOWN	HILLSBORO DISPATCH	
3	1910	UNKNOWN	UNKNOWN	HILLSBORO DISPATCH	
3	1910	UNKNOWN	UNKNOWN	HILLSBORO DISPATCH	
3	1910	UNKNOWN	UNKNOWN	HILLSBORO DISPATCH	
3	1910	UNKNOWN	UNKNOWN	HILLSBORO DISPATCH	
3	1910	UNKNOWN	UNKNOWN	HILLSBORO DISPATCH	
3	1910	UNKNOWN	UNKNOWN	HILLSBORO DISPATCH	
3	1910	UNKNOWN	UNKNOWN	HILLSBORO DISPATCH	
4	1910	UNKNOWN	UNKNOWN	HILLSBORO DISPATCH	
4	1910	UNKNOWN	UNKNOWN	HILLSBORO DISPATCH	
4	1910	UNKNOWN	UNKNOWN	HILLSBORO DISPATCH	
4	1910	UNKNOWN	UNKNOWN	HILLSBORO DISPATCH	
4	1910	UNKNOWN	UNKNOWN	HILLSBORO DISPATCH	
4	1910	UNKNOWN	UNKNOWN	HILLSBORO DISPATCH	PHYSICIAN
4	1910	UNKNOWN	UNKNOWN	HILLSBORO DISPATCH	
4	1910	UNKNOWN	UNKNOWN	HILLSBORO DISPATCH	
4	1910	UNKNOWN	UNKNOWN	HILLSBORO DISPATCH	
4	1910	UNKNOWN	UNKNOWN	HILLSBORO DISPATCH	
4	1910	UNKNOWN	UNKNOWN	HILLSBORO DISPATCH	
4	1910	UNKNOWN	UNKNOWN	HILLSBORO DISPATCH	
4	1910	UNKNOWN	UNKNOWN	HILLSBORO DISPATCH	
4	1910	UNKNOWN	UNKNOWN	HILLSBORO DISPATCH	
4	1910	UNKNOWN	UNKNOWN	HILLSBORO DISPATCH	
4	1910	UNKNOWN	UNKNOWN	HILLSBORO DISPATCH	

Ad Month	Ad Year	Vehicle Type	Client Location	Ad Source	Comments
4	1910	UNKNOWN	UNKNOWN	HILLSBORO DISPATCH	
4	1910	UNKNOWN	UNKNOWN	HILLSBORO DISPATCH	
4	1910	UNKNOWN	UNKNOWN	HILLSBORO DISPATCH	
4	1910	UNKNOWN	UNKNOWN	HILLSBORO DISPATCH	
4	1910	UNKNOWN	UNKNOWN	HILLSBORO DISPATCH	
4	1910	UNKNOWN	UNKNOWN	HILLSBORO DISPATCH	
4	1910	UNKNOWN	UNKNOWN	HILLSBORO DISPATCH	PHYSICIAN
4	1910	UNKNOWN	UNKNOWN	HILLSBORO DISPATCH	
4	1910	UNKNOWN	UNKNOWN	HILLSBORO DISPATCH	
4	1910	UNKNOWN	UNKNOWN	HILLSBORO DISPATCH	
4	1910	UNKNOWN	UNKNOWN	HILLSBORO DISPATCH	
4	1910	UNKNOWN	UNKNOWN	HILLSBORO DISPATCH	
4	1910	UNKNOWN	UNKNOWN	HILLSBORO DISPATCH	
4	1910	UNKNOWN	UNKNOWN	HILLSBORO DISPATCH	
4	1910	UNKNOWN	UNKNOWN	HILLSBORO DISPATCH	PHYSICIAN
4	1910	UNKNOWN	UNKNOWN	HILLSBORO DISPATCH	
4	1910	UNKNOWN	UNKNOWN	HILLSBORO DISPATCH	
4	1910	UNKNOWN	UNKNOWN	HILLSBORO DISPATCH	
4	1910	UNKNOWN	UNKNOWN	HILLSBORO DISPATCH	
4	1910	UNKNOWN	UNKNOWN	HILLSBORO DISPATCH	
4	1910	UNKNOWN	UNKNOWN	HILLSBORO DISPATCH	
4	1910	UNKNOWN	UNKNOWN	HILLSBORO DISPATCH	
4	1910	UNKNOWN	UNKNOWN	HILLSBORO DISPATCH	
4	1910	UNKNOWN	UNKNOWN	HILLSBORO DISPATCH	
4	1910	UNKNOWN	UNKNOWN	HILLSBORO DISPATCH	
4	1910	UNKNOWN	UNKNOWN	HILLSBORO DISPATCH	
4	1910	UNKNOWN	UNKNOWN	HILLSBORO DISPATCH	
4	1910	UNKNOWN	UNKNOWN	HILLSBORO DISPATCH	
4	1910	UNKNOWN	UNKNOWN	HILLSBORO DISPATCH	
4	1910	UNKNOWN	UNKNOWN	HILLSBORO DISPATCH	
4	1910	UNKNOWN	UNKNOWN	HILLSBORO DISPATCH	

Ad Month	Ad Year	Vehicle Type	Client Location	Ad Source	Comments
4	1910	UNKNOWN	UNKNOWN	HILLSBORO DISPATCH	
4	1910	UNKNOWN	UNKNOWN	HILLSBORO DISPATCH	
5	1910	UNKNOWN	UNKNOWN	HILLSBORO DISPATCH	
5	1910	UNKNOWN	UNKNOWN	HILLSBORO DISPATCH	
5	1910	UNKNOWN	UNKNOWN	HILLSBORO DISPATCH	
5	1910	UNKNOWN	UNKNOWN	HILLSBORO DISPATCH	
5	1910	UNKNOWN	UNKNOWN	HILLSBORO DISPATCH	RURAL MAIL CARRIER
5	1910	UNKNOWN	UNKNOWN	HILLSBORO DISPATCH	
5	1910	UNKNOWN	UNKNOWN	HILLSBORO DISPATCH	
5	1910	UNKNOWN	UNKNOWN	HILLSBORO DISPATCH	
5	1910	UNKNOWN	UNKNOWN	HILLSBORO DISPATCH	
5	1910	UNKNOWN	UNKNOWN	HILLSBORO DISPATCH	
5	1910	UNKNOWN	UNKNOWN	HILLSBORO DISPATCH	
5	1910	UNKNOWN	UNKNOWN	HILLSBORO DISPATCH	
5	1910	UNKNOWN	UNKNOWN	HILLSBORO DISPATCH	
5	1910	UNKNOWN	UNKNOWN	HILLSBORO DISPATCH	
5	1910	UNKNOWN	UNKNOWN	HILLSBORO DISPATCH	PHYSICIAN
5	1910	UNKNOWN	UNKNOWN	HILLSBORO DISPATCH	
5	1910	UNKNOWN	UNKNOWN	HILLSBORO DISPATCH	
5	1910	UNKNOWN	UNKNOWN	HILLSBORO DISPATCH	
5	1910	UNKNOWN	UNKNOWN	HILLSBORO DISPATCH	
6	1910	UNKNOWN	UNKNOWN	HILLSBORO DISPATCH	
6	1910	UNKNOWN	UNKNOWN	HILLSBORO DISPATCH	
6	1910	UNKNOWN	UNKNOWN	HILLSBORO DISPATCH	
6	1910	UNKNOWN	UNKNOWN	HILLSBORO DISPATCH	
6	1910	UNKNOWN	UNKNOWN	HILLSBORO DISPATCH	
6	1910	UNKNOWN	UNKNOWN	HILLSBORO DISPATCH	
6	1910	UNKNOWN	UNKNOWN	HILLSBORO DISPATCH	
6	1910	UNKNOWN	UNKNOWN	HILLSBORO DISPATCH	
6	1910	UNKNOWN	UNKNOWN	HILLSBORO DISPATCH	
6	1910	UNKNOWN	UNKNOWN	HILLSBORO DISPATCH	

Ad Month	Ad Year	Vehicle Type	Client Location	Ad Source	Comments
6	1910	UNKNOWN	UNKNOWN	HILLSBORO DISPATCH	
6	1910	UNKNOWN	UNKNOWN	HILLSBORO DISPATCH	
6	1910	UNKNOWN	UNKNOWN	HILLSBORO DISPATCH	
6	1910	UNKNOWN	UNKNOWN	HILLSBORO DISPATCH	
6	1910	UNKNOWN	UNKNOWN	HILLSBORO DISPATCH	
6	1910	UNKNOWN	UNKNOWN	HILLSBORO DISPATCH	
6	1910	UNKNOWN	UNKNOWN	HILLSBORO DISPATCH	
6	1910	UNKNOWN	UNKNOWN	HILLSBORO DISPATCH	
6	1910	UNKNOWN	UNKNOWN	HILLSBORO DISPATCH	
6	1910	UNKNOWN	UNKNOWN	HILLSBORO DISPATCH	
6	1910	UNKNOWN	UNKNOWN	HILLSBORO DISPATCH	
6	1910	UNKNOWN	UNKNOWN	HILLSBORO DISPATCH	
6	1910	UNKNOWN	UNKNOWN	HILLSBORO DISPATCH	
6	1910	UNKNOWN	UNKNOWN	HILLSBORO DISPATCH	
6	1910	UNKNOWN	UNKNOWN	HILLSBORO DISPATCH	
6	1910	UNKNOWN	UNKNOWN	HILLSBORO DISPATCH	
6	1910	UNKNOWN	UNKNOWN	HILLSBORO DISPATCH	
6	1910	UNKNOWN	UNKNOWN	HILLSBORO DISPATCH	
6	1910	UNKNOWN	UNKNOWN	HILLSBORO DISPATCH	
6	1910	UNKNOWN	UNKNOWN	HILLSBORO DISPATCH	
6	1910	UNKNOWN	UNKNOWN	HILLSBORO DISPATCH	
6	1910	UNKNOWN	UNKNOWN	HILLSBORO DISPATCH	
6	1910	UNKNOWN	UNKNOWN	HILLSBORO DISPATCH	
6	1910	UNKNOWN	UNKNOWN	HILLSBORO DISPATCH	
12	1910	STORM BUGGY	UNKNOWN	HILLSBORO DISPATCH	
12	1910	STORM BUGGY	UNKNOWN	HILLSBORO DISPATCH	
12	1910	STORM BUGGY	UNKNOWN	HILLSBORO DISPATCH	
12	1910	STORM BUGGY	UNKNOWN	HILLSBORO DISPATCH	
12	1910	STORM BUGGY	UNKNOWN	HILLSBORO DISPATCH	
12	1910	STORM BUGGY	UNKNOWN	HILLSBORO DISPATCH	
12	1910	STORM BUGGY	UNKNOWN	HILLSBORO DISPATCH	

Ad Month	Ad Year	Vehicle Type	Client Location	Ad Source	Comments
12	1910	STORM BUGGY	UNKNOWN	HILLSBORO DISPATCH	
12	1910	STORM BUGGY	UNKNOWN	HILLSBORO DISPATCH	
12	1910	STORM BUGGY	UNKNOWN	HILLSBORO DISPATCH	
12	1910	STORM BUGGY	UNKNOWN	HILLSBORO DISPATCH	
12	1910	STORM BUGGY	UNKNOWN	HILLSBORO DISPATCH	
12	1910	STORM BUGGY	UNKNOWN	HILLSBORO DISPATCH	
12	1910	STORM BUGGY	UNKNOWN	HILLSBORO DISPATCH	
12	1910	STORM BUGGY	UNKNOWN	HILLSBORO DISPATCH	
12	1910	STORM BUGGY	UNKNOWN	HILLSBORO DISPATCH	
12	1910	STORM BUGGY	UNKNOWN	HILLSBORO DISPATCH	
12	1910	STORM BUGGY	UNKNOWN	HILLSBORO DISPATCH	
12	1910	STORM BUGGY	UNKNOWN	HILLSBORO DISPATCH	
12	1910	STORM BUGGY	UNKNOWN	HILLSBORO DISPATCH	
12	1910	STORM BUGGY	UNKNOWN	HILLSBORO DISPATCH	
10	1911	STORM BUGGY	UNKNOWN	HILLSBORO DISPATCH	
10	1911	STORM BUGGY	UNKNOWN	HILLSBORO DISPATCH	
10	1911	STORM BUGGY	UNKNOWN	HILLSBORO DISPATCH	
10	1911	STORM BUGGY	UNKNOWN	HILLSBORO DISPATCH	
10	1911	STORM BUGGY	UNKNOWN	HILLSBORO DISPATCH	
10	1911	STORM BUGGY	UNKNOWN	HILLSBORO DISPATCH	
10	1911	STORM BUGGY	UNKNOWN	HILLSBORO DISPATCH	
10	1911	STORM BUGGY	UNKNOWN	HILLSBORO DISPATCH	
10	1911	STORM BUGGY	UNKNOWN	HILLSBORO DISPATCH	
10	1911	STORM BUGGY	UNKNOWN	HILLSBORO DISPATCH	
10	1911	STORM BUGGY	UNKNOWN	HILLSBORO DISPATCH	
10	1911	STORM BUGGY	UNKNOWN	HILLSBORO DISPATCH	
10	1911	STORM BUGGY	UNKNOWN	HILLSBORO DISPATCH	
10	1911	STORM BUGGY	UNKNOWN	HILLSBORO DISPATCH	
7	1912	BUGGY RUG	HILLSBORO	HILLSBORO DISPATCH	
7	1912	UNKNOWN	MARENGO	HILLSBORO DISPATCH	
8	1912	RUNABOUT	NASHVILLE, TN	HILLSBORO DISPATCH	PHYSICIAN

Ad Month	Ad Year	Vehicle Type	Client Location	Ad Source	Comments
4	1913	UNKNOWN	NORTH HAMPTON	HILLSBORO DISPATCH	
4	1909	NO. 1 PIANO BUGGY	NEWELLTON, LA	JOURNAL NAT. MEDICAL ASSN.	PHYSICIAN
1	1910	UNKNOWN	ATLANTA, GA	JOURNAL NAT. MEDICAL ASSN.	PHYSICIAN
1	1910	UNKNOWN	FERNANDINA, FL	JOURNAL NAT. MEDICAL ASSN.	PHYSICIAN
1	1910	UNKNOWN	VALDOSTA, GA	JOURNAL NAT. MEDICAL ASSN.	PHYSICIAN
1	1910	UNKNOWN	JACKSONVILLE, FL	JOURNAL NAT. MEDICAL ASSN.	PHYSICIAN
1	1910	UNKNOWN	CHATTANOOGA, TN	JOURNAL NAT. MEDICAL ASSN.	PHYSICIAN
1	1910	UNKNOWN	CLARKSVILLE, TN	JOURNAL NAT. MEDICAL ASSN.	PHYSICIAN
1	1910	UNKNOWN	SEDALIA, MO	JOURNAL NAT. MEDICAL ASSN.	PHYSICIAN
1	1910	UNKNOWN	LITTLE ROCK, AR	JOURNAL NAT. MEDICAL ASSN.	PHYSICIAN
1	1910	UNKNOWN	MEXIA, TX	JOURNAL NAT. MEDICAL ASSN.	PHYSICIAN
1	1910	UNKNOWN	ATLANTA, GA	JOURNAL NAT. MEDICAL ASSN.	PHYSICIAN
1	1910	UNKNOWN	PURCELL, OK	JOURNAL NAT. MEDICAL ASSN.	PHYSICIAN
1	1910	UNKNOWN	JACKSONVILLE, FL	JOURNAL NAT. MEDICAL ASSN.	PHYSICIAN
1	1910	UNKNOWN	UNIONTOWN, AL	JOURNAL NAT. MEDICAL ASSN.	PHYSICIAN
1	1910	UNKNOWN	BALDWIN, LA	JOURNAL NAT. MEDICAL ASSN.	PHYSICIAN
4	1910	UNKNOWN	LAFAYETTE, LA	JOURNAL NAT. MEDICAL ASSN.	PHYSICIAN
4	1910	UNKNOWN	LITTLE ROCK, AR	JOURNAL NAT. MEDICAL ASSN.	PHYSICIAN
4	1910	UNKNOWN	WILMINGTON, NC	JOURNAL NAT. MEDICAL ASSN.	PHYSICIAN
4	1910	UNKNOWN	GREENFIELD	JOURNAL NAT. MEDICAL ASSN.	PHYSICIAN
4	1910	UNKNOWN	MT. STERLING, KY	JOURNAL NAT. MEDICAL ASSN.	PHYSICIAN
4	1910	UNKNOWN	UNION CITY, TN	JOURNAL NAT. MEDICAL ASSN.	PHYSICIAN
4	1910	UNKNOWN	NATCHEZ, MS	JOURNAL NAT. MEDICAL ASSN.	PHYSICIAN
4	1910	UNKNOWN	CYNTHIANA	JOURNAL NAT. MEDICAL ASSN.	PHYSICIAN
4	1910	UNKNOWN	OWENSBORO, KY	JOURNAL NAT. MEDICAL ASSN.	PHYSICIAN
4	1910	UNKNOWN	SALISBURY, NC	JOURNAL NAT. MEDICAL ASSN.	PHYSICIAN
4	1910	UNKNOWN	MARSHALL	JOURNAL NAT. MEDICAL ASSN.	PHYSICIAN
4	1910	UNKNOWN	NEWPORT, AR	JOURNAL NAT. MEDICAL ASSN.	PHYSICIAN
4	1910	UNKNOWN	UNION, SC	JOURNAL NAT. MEDICAL ASSN.	PHYSICIAN
4	1910	UNKNOWN	OWENSBORO, KY	JOURNAL NAT. MEDICAL ASSN.	PHYSICIAN
4	1910	UNKNOWN	JACKSONVILLE, FL	JOURNAL NAT. MEDICAL ASSN.	PHYSICIAN

Ad Month	Ad Year	Vehicle Type	Client Location	Ad Source	Comments
4	1910	UNKNOWN	GREENFIELD	JOURNAL NAT. MEDICAL ASSN.	PHYSICIAN
4	1910	UNKNOWN	NAPOLEONVILLE, LA	JOURNAL NAT. MEDICAL ASSN.	PHYSICIAN
4	1910	UNKNOWN	WAXAHACHIE, TX	JOURNAL NAT. MEDICAL ASSN.	PHYSICIAN
4	1910	UNKNOWN	NASHVILLE, TN	JOURNAL NAT. MEDICAL ASSN.	PHYSICIAN
4	1910	UNKNOWN	PADUCAH, KY	JOURNAL NAT. MEDICAL ASSN.	PHYSICIAN
4	1910	UNKNOWN	LEXINGTON, KY	JOURNAL NAT. MEDICAL ASSN.	PHYSICIAN
7	1910	UNKNOWN	NORFOLK, VA	JOURNAL NAT. MEDICAL ASSN.	PHYSICIAN
10	1910	UNKNOWN	NEWPORT, AR	JOURNAL NAT. MEDICAL ASSN.	PHYSICIAN
1	1911	STANHOPE	TALLAPOOSA, GA	JOURNAL NAT. MEDICAL ASSN.	PHYSICIAN
4	1911	NO. 70 DOCTOR'S WAGON	ATLANTA, GA	JOURNAL NAT. MEDICAL ASSN.	PHYSICIAN
7	1911	UNKNOWN	NOTTOWAY, VA	JOURNAL NAT. MEDICAL ASSN.	PHYSICIAN
10	1911	STORM BUGGY	LEXINGTON, KY	JOURNAL NAT. MEDICAL ASSN.	PHYSICIAN
1	1912	NO. 65 STANHOPE	TEMPLE, TX	JOURNAL NAT. MEDICAL ASSN.	PHYSICIAN
10	1913	STORM BUGGY	TUSCALOOSA, AL	JOURNAL NAT. MEDICAL ASSN.	PHYSICIAN
1	1916	PATTERSON GREENFIELD	LEXINGTON, KY	JOURNAL NAT. MEDICAL ASSN.	PHYSICIAN
4	1916	PATTERSON GREENFIELD	TALLADEGA, AL	JOURNAL NAT. MEDICAL ASSN.	MERCHANT GROCER

Appendix C:

Greenfield Properties Purchased by the Patterson Family and Company

VOL	PAGE	MONTH	YEAR	LOT (S)	GRANTEE	GRANTOR	NOTES
30	393	1	1862	174	CHARLES R. PATTERSON	PETER L. FINDLEY	BOUGHT FOR $35. FIRST PROPERTY ON RECORD TO BE PURCHASED BY C. R. PATTERSON. NOTE YEAR IS 1862.
33	448	6	1865	51	CHARLES R. PATTERSON	JOHN WRIGHT	BOUGHT FOR $420. CORNER OF LAFAYETTE AND 2ND STREETS. SECOND PROPERTY OF C. R.
42	2	2	1869	10 232	CHARLES R. PATTERSON	JOSEPH WILSON	BOUGHT FOR $325.
40	638	2	1869	127 142	CHARLES R. PATTERSON CALVIN P. HACKERT	WILHELMINA WEIDENOUR	BOUGHT FOR $300. NE CORNER OF NORTH AND 4TH STREETS.
50	557	3	1876	28	CHARLES R. PATTERSON	W. H. DYKE	BOUGHT FOR $875. FRONTS ON JEFFERSON ST.
82	232	8	1890	99 108	CHARLES R. PATTERSON	WILLIAM R. TEMPLETON	CARRIAGE FACTORY. 18,562 SQ. FEET
90	86	4	1895	28	JOSEPHINE PATTERSON	CHARLES R. PATTERSON	FRONTS ON JEFFERSON ST. DOWNTOWN. PAYMENT TO W. H. WILSON. FILE NO. 3196.
88	425	8	1895	99 108	FREDERICK D. PATTERSON JOSEPHINE PATTERSON SAMUEL C. PATTERSON	CHARLES R. PATTERSON	CARRIAGE FACTORY. $1400 TO HOME BUILDING AND LOAN CO. AND $200 TO WILLIAM TEMPLETON PAYABLE ON AUG. 21, 1895. FILE NO. 2448.
88	33	9	1897		JOSEPHINE PATTERSON	CEDAR LODGE NO. 17 F & AM	AFRICAN MASONIC LODGE SOLD THEIR SHARE OF THE 3RD FLOOR OF THE HIGHLAND COUNTY BANK DUE TO INDEBTEDNESS. JOSEPHINE AND 6 OTHERS PURCHASED THE SHARES FOR $625.

VOL	PAGE	MONTH	YEAR	LOT (S)	GRANTEE	GRANTOR	NOTES
90	106	6	1899	760	FREDERICK D. PATTERSON	MARGARET BABB ET AL.	FILE NO. 3216
91	524	5	1900	991	FREDERICK D. PATTERSON	M. IRWIN DUNLAP	BOUGHT FOR $185. FILE NO. 4135.
92	319	9	1900	18	FREDERICK D. PATTERSON	SUSAN STEELE	FILE NO. 4461
92	483	12	1900	94 83	FREDERICK D. PATTERSON JOSEPHINE PATTERSON	JULIA A. WARE	SE CORNER OF LAFAYETTE & WASHINGTON FILE NO. 4598
93	32	12	1900	9	JOSEPHINE PATTERSON WILLIAM GAY OUTZ	MARY WEST	CORNER OF SOUTH AND 2ND ST. FILE NO. 4702.
93	317	3	1901	99 108	FREDERICK D. PATTERSON	DOLLIE PATTERSON KATIE PATTERSON	FREDERICK BOUGHT HIS SISTER'S INTEREST IN THE FACTORY.
94	377	9	1901	1144	FREDERICK D. PATTERSON	M. IRWIN DUNLAP	BOUGHT FOR $750. FILE NO. 5505.
96	351	7	1902	1005	JOSEPHINE PATTERSON	O. F. BEATTY	FILE NO. 6496.
96	456	9	1902	94 83	FREDERICK D. PATTERSON	W. P. SHERBLE	FACTORY ACROSS STREET. $2,000. FILE NO. 6590
99	101	11	1903	94	JOSEPHINE PATTERSON	FREDERICK D. PATTERSON	1/2 INTEREST IN LOT 94 FOR $2,084. FILE NO. 7908. 2,296 SQ. FEET.
103	286	12	1905	457	FREDERICK D. PATTERSON	M. IRWIN DUNLAP	BOUGHT FOR $850. FILE NO. 10190.
104	483	11	1906		JOSEPHINE PATTERSON	JOHN MANN ET AL.	JOSEPHINE BOUGHT OUT ONE OF THE PARTNERS OF THEIR SHARE OF THE CEDAR GROVE LODGE.

VOL	PAGE	MONTH	YEAR	LOT (S)	GRANTEE	GRANTOR	NOTES
104	485	11	1906		JOSEPHINE PATTERSON	W. D. PARKER	JOSEPHINE BOUGHT OUT ONE OF THE PARTNERS OF THEIR SHARE OF THE CEDAR GROVE LODGE.
106	212	3	1907	546	FREDERICK D. PATTERSON	M. IRWIN DUNLAP	FILE NO. 12658
111	520	11	1910	22	FREDERICK D. PATTERSON	L. C. NICHOLSON	CORNER OF 4TH AND SOUTH STREETS. $2,535 PAID TO HOME BLDG. AND LOAN CO.
111	267	3	1911	99 108	JOSEPHINE PATTERSON	JOSIAH RENICK	CARRIAGE FACTORY. PLUS EASEMENT 10' WIDE AND 83' LONG FROM LAFAYETTE ST. AND ROOM FOR STAIRS TO 2ND FLOOR OF BUILDING. SEVERAL PROPERTIES WERE SOLD TO RAISE THE MONEY TO PAY OFF THE FACTORY. FILES NO. 16341-16344 SHOW PROPERTIES SOLD-LOTS 1144, 9, 1005, AND 18 BETWEEN FEB. 25TH AND FEB. 28TH BY FREDERICK D., ESTALLINE, JOSEPHINE, WILLIAM AND CARRIE OUTZ.
114	144	3	1913	24	CHARLES W. NAPPER	ELIZABETH GRASSLEY	FILE NO. 18383
117	416	8	1914	368	JOSEPHINE PATTERSON	ARTHUR JENKINS ELLA JENKINS	PROPERTY FRONTS ON NORTH ST. AND IS JUST WEST OF 7TH ST.
122	92	8	1916	149	ESTALLINE P. PATTERSON	CHARLES F. HEAD NELLIE HEAD	FILE NO. 196. PROPERTY ON CORNER OF 4TH AND MIRABEAU STREETS
125	105	12	1919	78	FREDERICK D. PATTERSON JOHN SIMMS JOHN R. RUDD	E. B. WATTS FRANCES DUNLAP	WATTS AND DUNLAP EACH OWNED PORTION OF LOT 78. 8,231 SQ. FEET. FILES NO. 1431 & 1433
131	360	12	1921	78	FREDERICK D. PATTERSON	JOHN MANN	8,231 SQ. FT., FILE NO. 8226

VOL	PAGE	MONTH	YEAR	LOT (S)	GRANTEE	GRANTOR	NOTES
131	364	7	1923	593	FREDERICK D. PATTERSON	J. D. VARNEY	FAMILY HOME ON JEFFERSON ST. FILE NO. 8231
131	365	7	1923	593	ESTALLINE P. PATTERSON	FREDERICK D. PATTERSON	FAMILY HOME ON JEFFERSON ST. FILE NO. 8232
131	361	9	1923	78	JOHN SIMMS	FREDERICK D. PATTERSON	FACTORY TRANSFER TO SIMMS. FILE NO. 8227
137	501	1	1928	663 664 665 666 694 695 696 697	FREDERICK D. PATTERSON	GREENFIELD BD. OF EDUCATION	"SOUTH SIDE SCHOOL BUILDING PROPERTY", BOUGHT AT AUCTION ON JAN. 7, 1928 FROM GREENFIELD BOARD OF EDUCATION WITH INTENT TO MODIFY AS A BUS FACTORY, SEVERAL LOTS FRONTED BETWEEN LINDEN AVE. AND DICKEY AVE. AND CENTERED BETWEEN 5TH AND 6TH STREETS. FILE NO. 3529
138	479	7	1928	663 664 665 666 694 695 696 697	FREDERICK P. PATTERSON POSTELL PATTERSON	FREDERICK D. PATTERSON	TRANSFER OF "SOUTH SIDE SCHOOL BUILDING" TO SONS. $3,000 MORTGAGE PAYABLE TO THE FIDELITY BUILDING AND LOAN COMPANY OF GREENFIELD. FILE NO. 4808

Appendix D:

Greenfield Properties Mortgages by the Patterson Family and Company

VOL	PAGE	MONTH	YEAR	GRANTEE	GRANTEE	COLLATERAL	AMOUNT	RELEASED	NOTES
5	356	6	1865	CHARLES R. PATTERSON JOSEPHINE PATTERSON	JEROME B. DANIELS	LOT 51	$357.48	Nov-66	DUE NOV 1866
9	441	1	1872	CHARLES R. PATTERSON JOSEPHINE PATTERSON	CHARLES CROTHERS	LOT 10	$171.38	Jan-73	DUE JAN 1873. NO INTEREST.
9	501	2	1872	CHARLES PATTERSON	CHARLES R. PATTERSON	LOT 143 LOT 158	$1,000.00	Nov-73	DUE NOV 1873. PLUS 8%. C.R.'S FATHER.
13	590	7	1876	CHARLES R. PATTERSON JOSEPHINE PATTERSON	JOHN O'DELL	LOT 28	$792.35	Jul-77	DUE JULY 1877. PLUS 8%.
24	92	12	1885	CHARLES R. PATTERSON JOSEPHINE PATTERSON	T. M. ELLIOTT	LOT 28	$369.77	Mar-87	DUE AFTER 15 MONTHS. PLUS 6%.
26	360	2	1888	CHARLES R. PATTERSON JOSEPHINE PATTERSON	W. H. WILSON	LOT 28	$875.00	May-95	
32	435	12	1895	CHARLES R. PATTERSON JOSEPHINE PATTERSON	FREDERICK D. PATTERSON DOLLIE PATTERSON SAMUEL C. PATTERSON	LOT 28	$1,200.00	Dec-97	DUE IN 2 YEARS
36	205	4	1895	CHARLES R. PATTERSON JOSEPHINE PATTERSON	W. H. WILSON	LOT 28	$1,239.51	Aug-01	DUE IN 3 YEARS. PLUS 8%.
36	405	8	1895	CHARLES R. PATTERSON JOSEPHINE PATTERSON	HOME BLDG. & LOAN	LOT 99 LOT 108	$1,400.00	Jan-99	$1.61 WEEK INTEREST
36	408	8	1895	CHARLES R. PATTERSON JOSEPHINE PATTERSON	WILLIAM R. TEMPLETON	LOT 99 LOT 108	$200.00	Mar-96	DUE IN 1 YEAR. SHARE OF LOTS 99 & 108.

VOL	PAGE	MONTH	YEAR	GRANTEE	GRANTEE	COLLATERAL	AMOUNT	RELEASED	NOTES
37	354	2	1896	CHARLES R. PATTERSON JOSEPHINE PATTERSON	HOME BLDG. & LOAN	LOT 99 LOT 108	$2,000.00	?-99	$2.30 WEEK INTEREST
38	21	2	1896	CHARLES R. PATTERSON JOSEPHINE PATTERSON	JAMES P. LOWE	LOT 99 LOT 108	$500.00	Feb-97	DUE IN 1 YEAR. PLUS 8%.
42	38	1	1899	CHARLES R. PATTERSON JOSEPHINE PATTERSON FREDERICK D. PATTERSON SAMUEL C. PATTERSON	HOME BLDG. & LOAN	LOT 99 LOT 108	$3,200.00	Mar-01	$3.69 WEEK INTEREST.
43	24	9	1899	CHARLES R. PATTERSON JOSEPHINE PATTERSON	HOME BLDG. & LOAN	LOT 28	$400.00	YES/DATE?	$0.46 WEEK INTEREST
43	76	10	1899	FREDERICK D. PATTERSON	HOME BLDG. & LOAN	LOT 760	$400.00	YES/DATE?	$0.46 WEEK INTEREST
45	46	10	1900	FREDERICK D. PATTERSON	HOME BLDG. & LOAN	LOT 18	$400.00	YES/DATE?	$0.46 WEEK INTEREST
45	172	12	1900	FREDERICK D. PATTERSON CHARLES R. PATTERSON JOSEPHINE PATTERSON	JULIA A. WARE	LOT 94 LOT 83	$1,600.00	YES/DATE?	PLUS 6%. $600 DUE 1 YEAR, $1,000 DUE 2 YEARS.
45	467	3	1901	FREDERICK D. PATTERSON CHARLES R. PATTERSON JOSEPHINE PATTERSON	HOME BLDG. & LOAN	LOT 99 LOT 108	$3,200.00	Jan-02	$3.68 WEEK INTEREST
46	405	9	1901	FREDERICK D. PATTERSON ESTALLINE PATTERSON	HOME BLDG. & LOAN	LOT 1144	$400.00	May-05	

VOL	PAGE	MONTH	YEAR	GRANTEE	GRANTEE	COLLATERAL	AMOUNT	RELEASED	NOTES
47	141	1	1902	FREDERICK D. PATTERSON	JOHN FULLERTON	LOT 99	$3,000.00	Jan-03	DUE IN 3 YEARS. PLUS 6.5%.
				CHARLES R. PATTERSON	J. L. FULLERTON	LOT 108			
				JOSEPHINE PATTERSON					
48	437	12	1902	CHARLES R. PATTERSON	HOME BLDG. & LOAN	LOT 99	$3,000.00	May-05	$3.45 WEEK INTEREST
				JOSEPHINE PATTERSON		LOT 108			
				FREDERICK D. PATTERSON					
				ESTALLINE PATTERSON					
50	64	11	1903	CHARLES R. PATTERSON	FAY BALDWIN	LOT 94	$4,084.40	Aug-04	PLUS 8%
				JOSEPHINE PATTERSON					
52	262	5	1905	CHARLES R. PATTERSON	HOME BLDG. & LOAN	LOT 99	$5,000.00	Feb-08	$5.75 WEEK INTEREST
				JOSEPHINE PATTERSON		LOT 108			
				FREDERICK D. PATTERSON					
				ESTALLINE PATTERSON					
53	58	10	1905	FREDERICK D. PATTERSON	HOME BLDG. & LOAN	LOT 1144	$600.00	YES/DATE?	$0.69 WEEK INTEREST
				ESTALLINE PATTERSON					
53	233	1	1906	CHARLES R. PATTERSON	HIGHLAND COUNTY BANK	LOT 94	$1,000.00	Aug-06	DUE AUG. 1, 1906
				JOSEPHINE PATTERSON		LOT 83			
				FREDERICK D. PATTERSON					
				ESTALLINE PATTERSON					
53	355	2	1906	FREDERICK D. PATTERSON	HOME BLDG. & LOAN	LOT 457	$600.00	YES/DATE?	$0.69 WEEK INTEREST
				ESTALLINE PATTERSON					
55	88	3	1907	FREDERICK D. PATTERSON	HOME BLDG. & LOAN	LOT 18	$400.00	YES/DATE?	$0.46 WEEK INTEREST
				ESTALLINE PATTERSON					

VOL	PAGE	MONTH	YEAR	GRANTEE	GRANTEE	COLLATERAL	AMOUNT	RELEASED	NOTES
55	466	8	1907	FREDERICK D. PATTERSON	HOME BLDG. & LOAN	LOT 546	$800.00	YES/DATE?	$0.92 WEEK INTEREST
				ESTALLINE PATTERSON					
56	153	1	1908	CHARLES R. PATTERSON	FIDELITY BLDG. & LOAN	LOT 28	$1,000.00	Jan-13	$1.15 WEEK INTEREST
				JOSEPHINE PATTERSON					
56	205	1	1907	CHARLES R. PATTERSON	HIGHLAND COUNTY BANK	LOT 94	$1,000.00	Apr-12	PLUS 6%. DUE 210 DAYS AFTER DATE.
				JOSEPHINE PATTERSON		LOT 83			
				FREDERICK D. PATTERSON					
				ESTALLINE PATTERSON					
56	206	12	1906	CHARLES R. PATTERSON	HIGHLAND COUNTY BANK	LOT 94	$1,000.00	Apr-12	PLUS 6%. DUE 8 MONTHS AFTER DATE.
				JOSEPHINE PATTERSON		LOT 83			
				FREDERICK D. PATTERSON					
				ESTALLINE PATTERSON					
56	229	2	1908	CHARLES R. PATTERSON	HOME BLDG. & LOAN	LOT 99	$5,000.00	May-11	$5.75 WEEK INTEREST
				JOSEPHINE PATTERSON		LOT 108			
				FREDERICK D. PATTERSON					
				ESTALLINE PATTERSON					
60	10	1	1911	FREDERICK D. PATTERSON	HOME BLDG. & LOAN	LOT 457	$600.00	Oct-15	$0.69 WEEK INTEREST
				ESTALLINE PATTERSON					
60	233	5	1911	JOSEPHINE PATTERSON	HOME BLDG. & LOAN	LOT 99	$5,000.00	Jan-13	$5.75 WEEK INTEREST
				FREDERICK D. PATTERSON		LOT 108			
				ESTALLINE PATTERSON					
61	231	4	1912	JOSEPHINE PATTERSON	HOME BLDG. & LOAN	LOT 99	$3,000.00	YES/DATE?	PLUS 6%. DUE IN 1 YEAR.
						LOT 108			

VOL	PAGE	MONTH	YEAR	GRANTEE	GRANTEE	COLLATERAL	AMOUNT	RELEASED	NOTES
61	232	4	1912	JOSEPHINE PATTERSON FREDERICK D. PATTERSON ESTALLINE PATTERSON	HOME BLDG. & LOAN	LOT 94 LOT 83	$3,000.00	YES/DATE?	PLUS 6%. DUE IN 1 YEAR.
61	373	7	1912	JOSEPHINE PATTERSON	HOME BLDG. & LOAN	LOT 99 LOT 108	$5,200.00	Jan-14	$5.98 WEEK INTEREST.
62	73	1	1913	JOSEPHINE PATTERSON	FIDELITY BLDG. & LOAN	LOT 28	$1,000.00	Jan-17	$1.15 WEEK INTEREST
62	100	1	1913	JOSEPHINE PATTERSON FREDERICK D. PATTERSON ESTALLINE PATTERSON	HOME BLDG. & LOAN	LOT 99 LOT 108	$5,000.00	May-16	$5.75 WEEK INTEREST
62	101	1	1913	FREDERICK D. PATTERSON ESTALLINE PATTERSON	HOME BLDG. & LOAN	LOT 22	$2,600.00	Jan-18	$2.99 WEEK INTEREST
63	104	1	1914	JOSEPHINE PATTERSON	HOME BLDG. & LOAN	LOT 99 LOT 108	$5,200.00	YES/DATE?	$5.98 WEEK INTEREST
64	248	1	1915	FREDERICK D. PATTERSON ESTALLINE PATTERSON	HOME BLDG. & LOAN	LOT 22	$2,600.00	Mar-16	$2.99 WEEK INTEREST
64	249	1	1915	JOSEPHINE PATTERSON	HOME BLDG. & LOAN	LOT 99 LOT 108	$5,200.00	?-1916	$5.98 WEEK INTEREST
65	503	3	1916	FREDERICK D. PATTERSON ESTALLINE PATTERSON	FIDELITY BLDG. & LOAN	LOT 22	$3,800.00	Aug-16	$4.37 WEEK INTEREST
66	60	5	1916	JOSEPHINE PATTERSON FREDERICK D. PATTERSON ESTALLINE PATTERSON	HOME BLDG. & LOAN	LOT 99 LOT 108	$5,200.00	Jul-17	$5.98 WEEK INTEREST

VOL	PAGE	MONTH	YEAR	GRANTEE	GRANTEE	COLLATERAL	AMOUNT	RELEASED	NOTES
66	217	8	1916	FREDERICK D. PATTERSON ESTALLINE PATTERSON	HIGHLAND COUNTY BANK	LOT 149	$1,500.00	Aug-17	PLUS 6%. DUE IN 1 YEAR.
66	494	1	1917	JOSEPHINE PATTERSON	FIDELITY BLDG. & LOAN	LOT 28	$1,200.00	Jan-21	$1.38 WEEK INTEREST
67	308	7	1917	JOSEPHINE PATTERSON FREDERICK D. PATTERSON ESTALLINE PATTERSON	HOME BLDG. & LOAN	LOT 99 LOT 108	$5,200.00	YES/DATE?	$5.98 WEEK INTEREST
67	448	9	1917	JOSEPHINE PATTERSON FREDERICK D. PATTERSON	HIGHLAND COUNTY BANK	LOT 99 LOT 108	$3,954.46	YES/DATE?	PLUS 6%
70	189	1	1920	FREDERICK D. PATTERSON JOHN SIMMS KATE SIMMS JOHN R. RUDD DOLLIE RUDD	FRANCES DUNLAP	LOT 78	$1,000.00	Dec-21	PLUS 7%
70	190	1	1920	FREDERICK D. PATTERSON JOHN SIMMS KATE SIMMS JOHN R. RUDD DOLLIE RUDD	E. B. WATTS	LOT 78	$1,000.00	Dec-21	PLUS 7%
71	260	6	1920	JOSEPHINE PATTERSON	HIGHLAND COUNTY BANK	LOT 28	$500.00	YES/DATE?	PLUS 8%

VOL	PAGE	MONTH	YEAR	GRANTEE	GRANTEE	COLLATERAL	AMOUNT	RELEASED	NOTES
72	392	12	1921	FREDERICK D. PATTERSON JOHN SIMMS KATE SIMMS JOHN R. RUDD DOLLIE RUDD JOHN MANN ARABELLA MANN	FIDELITY BLDG. & LOAN	LOT 78	$1,000.00	Aug-24	$1.15 WEEK INTEREST
74	41	10	1922	JOSEPHINE PATTERSON FREDERICK D. PATTERSON	HOME BLDG. & LOAN	LOT 99 LOT 108	$8,000.00	?	$9.28 WEEK INTEREST
74	42	10	1922	JOSEPHINE PATTERSON FREDERICK D. PATTERSON	HOME BLDG. & LOAN	LOT 94 LOT 83	$5,000.00	Jul-24	$5.75 WEEK INTEREST
75	219	3	1923	JOSEPHINE PATTERSON FREDERICK D. PATTERSON	HIGHLAND COUNTY BANK	LOT 99 LOT 108	$2,500.00	?-24	PLUS 7%
75	410	8	1923	FREDERICK D. PATTERSON ESTALLINE PATTERSON	FIDELITY BLDG. & LOAN	LOT 593	$3,000.00	YES/DATE?	$3.45 WEEK INTEREST
76	113	12	1923	FREDERICK D. PATTERSON ESTALLINE PATTERSON	FIDELITY BLDG. & LOAN	LOT 593	$1,000.00	Jun-27	$1.15 WEEK INTEREST
77	61	7	1924	JOSEPHINE PATTERSON FREDERICK D. PATTERSON ESTALLINE PATTERSON	FIDELITY BLDG. & LOAN	LOT 94 LOT 83	$6,000.00	COURT	$6.90 WEEK INTEREST. COURT CASE NO. 12253. RELEASED BY COURT ON 26 MAY 1936. SOLD AT SHERIFF'S SALE. OVERLAPS OTHER MORTGAGES.

VOL	PAGE	MONTH	YEAR	GRANTEE	GRANTEE	COLLATERAL	AMOUNT	RELEASED	NOTES
77	455	12	1924	JOSEPHINE PATTERSON FREDERICK D. PATTERSON	JOHN O'DELL	LOT 99 LOT 108 LOT 94 LOT 83	$6,000.00	?	PLUS 6%. OVERLAPS OTHER MORTGAGES.
79	473	4	1926	JOSEPHINE PATTERSON FREDERICK D. PATTERSON	HIGHLAND COUNTY BANK	LOT 94 LOT 83 LOT 99 LOT 108	$19,000.00	COURT	PLUS 7%. COURT CASE NO. 12253. RELEASED BY COURT ON 26 MAY 1936. SOLD AT SHERIFF'S SALE. OVERLAPS OTHER MORTGAGES.
81	380	?	1926	FREDERICK D. PATTERSON ESTALLINE PATTERSON	FIDELITY BLDG. & LOAN	LOT 593	$4,000.00	?	$4.60 WEEK INTEREST
82	316	1	1928	FREDERICK D. PATTERSON	HIGHLAND COUNTY BANK	LOT 663 LOT 664 LOT 665 LOT 666 LOT 694 LOT 695 LOT 696 LOT 697	$3,000.00	May-28	PLUS 7%. "SOUTH SCHOOL BUILDING PROPERTY"
83	99	5	1928	FREDERICK D. PATTERSON ESTALLINE PATTERSON	FIDELITY BLDG. & LOAN	LOT 663 LOT 664 LOT 665 LOT 666 LOT 694 LOT 695 LOT 696 LOT 697	$3,000.00	Oct-28	PLUS 8%. "SOUTH SCHOOL BUILDING PROPERTY"

VOL	PAGE	MONTH	YEAR	GRANTEE	GRANTEE	COLLATERAL	AMOUNT	RELEASED	NOTES
83	484	?	1928	FREDERICK P. PATTERSON BERNICE PATTERSON POSTELL PATTERSON MARGUERITE PATTERSON	THE SUPREME LIFE INS. CO.	LOT 663 LOT 664 LOT 665 LOT 666 LOT 694 LOT 695 LOT 696 LOT 697	$6,000.00	Oct-33	PLUS 6.5%. PAYMENTS EACH YEAR ON OCT. 19TH $600 FROM 1929-1932 AND $3,600 IN 1933. COMPANY FROM CHICAGO. "SOUTH SCHOOL BUILDING"
86	97	5	1931	FREDERICK P. PATTERSON BERNICE PATTERSON POSTELL PATTERSON MARGUERITE PATTERSON	THE SUPREME LIFE INS. CO.	LOT 663 LOT 664 LOT 665 LOT 666 LOT 694 LOT 695 LOT 696 LOT 697	$6,150.00	COURT	PLUS 6.5%. COMPANY FROM CHICAGO. PAYMENTS EACH OCTOBER 1931 = $2000, 1932 = $1,000, 1933 = $1,000, 1934 = $1,000, 1935 = $1,150. SOLD AT SHERIFF'S SALE IN 1933. SUPREME LIFE INSURANCE COMPANY BROUGHT A SUIT FOR $7013.31. "SOUTH SCHOOL BUILDING"
B&L2	295	10	1932	ESTALLINE PATTERSON	FIDELITY BLDG. & LOAN	LOT 593	$3,200.00	COURT	$3.68 WEEK INTEREST. COMBINED WITH B&L2 - 297. COURT CASE NO. 13467 IN JAN 1938. SOLD AT SHERIFF'S SALE IN MARCH 1938. SUIT BROUGHT FOR $5,102.13.
B&L2	297	10	1932	ESTALLINE PATTERSON	FIDELITY BLDG. & LOAN	LOT 593	$110.00	COURT	PLUS 6%. DUE IN 60 DAYS. THIS IS COMBINED WITH B&L2-295 AND WAS SETTLED BY THE COURT CASE.

Made in the USA
Columbia, SC
30 July 2024

39283419R00221